Half a million spectators gather and crowd along the long stretches of the river banks to watch

Boats of different sizes, shapes and historical significance participate in the biggest regatta in the world

A truly unique folks-festival, carrying rich history, traditions, and scintillating competitions

A delightful sight to behold and cherish !

Arpo, hurray !

Readers's comments :

- A book with many stories within, proves your point "the regatta embodies and symbolizes many things at once!"

- I read your very interesting commentary on Desi and Paradesi (Des - North Indian and Desam - South Indian). It's very interesting that North Indians and Pakistanis have many things in common. I liked your comment on Diaspora and "desi, paradesi diaspora". I agree that it's important to promote ideas of "Humanities" such as knowledge, philosophy, literature and art .

- The speech made by the Sister principal is uplifting to all women gender in the whole wide world. To my young daughters and me, it is the best part of your very interesting book. Are you a professor? Meaningful captions too.

- The launching ceremony for the chundan, the cuisine, sadhya, and music mela is fascinating. The showman drummer and the flashy dancing boyz were hilarious, dramatic and beautifully explained. Is it a real story? We learned many things about Indian and Western styles of music, very nice.

- The Futuristic master plan is absolutely super and detailed. Please share the link to the company website. I like to learn more. The technology is absolutely necessary and urgently needed for the world. It is difficult to understand everything the first time. Everyone should read it more times.

- I saw the many things on a sightseeing excursion in a houseboat. Watching the sceneries alone cannot give us a good understanding. You talked like a tour guide, even about the ducks, frogs, and turtles in keen observation and humor. The place we saw was really beautiful. If I come another time, I rent your boat and like to talk to you. Good job!

- Exciting running commentary. I'm a rower too. I shall come to the next regatta. Be there! I'm sure I'll enjoy it, though I am a snobbish Englishman, you might say. I like to taste your best toddy on the planet, better be as good as Martini. I didn't like what you said about us. We like both crown jewels, the one on the crown and the jewel in our old empire, I like India, you see? My great grandpa was stationed in Shimla. I have BW pictures. I want to visit there too. Cheers!

- I must tell you that in Book 1 "The Biggest Regatta" is what I liked the very best. This is something dear to my heart. The "Vallam Kali" of Kerela. You see, when I was in New Zealand, I was very lucky and privileged to be asked by the native Maori's to help paddle with them in a traditional Waka (Maori canoe) which can be traced back to the year 1400. We rowed to a Maori Wharenui, which is a communal house for ceremonial Maori traditions

- The fairytale stories from mythology, Maharajah, and the poems are very nice to read. Your take on genesis is queerly interesting. I am a Science enthusiast myself, my best!

- The soliloquy of the dejected fan (You?) and the boat's journey home is a classic, man.

- The Voyage of the coconut and its story tied to your land is very imaginative and interesting

- The wedding story of Ravi and Ramani itself is a fabulous love story. Plaudits!

Overview

This book narrates a unique and exotic paddling regatta held in folks festival style, by over 900 farming communities in coastal Kerala, India, in great passion, fervor, and intense competition of more than 100 boats, and 5000 paddlers, attracting half a million zealous on-site spectators, as well as potentially over 2 Billion TV Viewers, half the globe around. In this context, it reveals many of the unexplained senses and sensibilities of an ancient culture, and its traditions. There is a deeper meaning and surprising sagacity beneath its appearance. The culture, traditions, rituals, festivities, myths, and mythology are fathomed with plausible scientific explanations, and/or logical deductions. Implicitly and explicitly, they tend to compare oriental, and occidental philosophies, and attitudes. The race, festivities, the marine warfare legacy of each category of the race boats, the live video production & transmission with behind the scene activities, and the thrilling running commentary are a few of the many highlights. The traditional way to build the boats using natural resources and the resourceful architecture codes dating back to the Vedic times are elaborately explained through its entire process "from a tree in the rainforest to a floating masterpiece!". Along the way, readers are taken on a journey of many events, reclamation projects, myths, mythology, history, fallacy, and themes of great significance, and consequences. The nuances, rituals, the communal banquet 'Sadhya;' the euphoric, and hilariously eventful 'music mela' are narrated in authentic and fascinating accounts. It compares the traditions and performance styles of Indian & Western music.

The human ecology, the biosphere consisting of the flora, fauna, and fountains, as well as the sustainable lifestyle of over thousand village communities, should provide an acceptable model for world living, especially when the world is ominously staring at an imminent existential threat, and predicament. The plight of farming communities, the degradation of the environment, and the consequent global trend of 'migration in desperation' that are now widely seen, are passionately discussed offering the real causes that are often ignored in media discourses, and government policies. In this context, many scientific, technological, economic mitigation measures, and practical solutions are discussed in an 'interactive presentation' event at the race's exhibition venue. The book highlights the disparities of various social classes, genders, and the underprivileged. Faith & colonialism; exploitation & desecration; falsification of historical events for self-promotion and glorification, etc are candidly pointed out, perhaps drawing some ire, and consternation. They are truly eye-openers and deeply thought-provoking.

Several anecdotal stories, inspirational speeches, and eloquent poems spice up the reading pleasures. The many picture collages build reality and imagination to the descriptions, and narrations. Among the mixes of many subjects, each chapter provides the readers with a piece of information, an inference; a wit, wisdom; a taste of rustic humor, a serious thought; a surprise, a riddle, an expectation, or anticipation of an unraveling mystery, a potential solution; and certainly, a moral to cherish, and share with friends & families. The biggest regatta depicts a real world, a population, and a traditional lifestyle, a culture unknown to many, but under the circumstances may very well come to appreciate, and emulate. The theme embodies and symbolizes many subject matters all at once!

Everything that was said, was said in pursuit of truth

That is usually concealed, or mostly corrupted

Just in good jest, but with a lot of zest for earnest scrutiny

No offense intended!

Meant neither hurt, harm, nor malice!

The book is dedicated to the people of my native state of Kerala, India
For their sufferings in the wake of recurring floods in the lowlands
And landslides in the highlands.

The 33 countries currently living below the Mean Sea Level, as well as the island nations scattered around the oceans and seas; low-lying landmasses of Bangladesh, Sri Lanka, Thailand, the Philippines, Indonesia, Taiwan; the islands of the Maldives, Andaman & Nicobar; the Polynesian, Caribbean, Canary, the Channel Islands that are constantly threatened and bear the brunt, fury, and consequences of the ominous climate disasters horribly leading humanity to the precipitous cliff of impending extinction

Author: *Aurora Bobbyalis*

I've been working on some pieces of writing for quite a long time now and managed to arrive at a comfortable point to share them with others. At first, they started out as some scribbled manuscripts of recollections of certain fond but fading memories of childhood, adulthood, and other life events, experiences, and encounters. More incidents and thoughts started to flow in; many incoherent, some obscure, others silly but funny and worth telling. Gradually and unwittingly, it developed into an intriguing hobby and pastime; told a few people about it. Those who heard the gist of it said: "as far as I heard, it's interesting; very much like to see it when finished!". It dispelled some self-doubts; at the same time, boosted inner confidence. Lo, and behold! it soon evolved into a fascinating and engaging undertaking that I couldn't stop thinking about. Constantly, thoughts and ideas began rushing into the mind with one triggering yet another in powerful synergistic chain reactions and streams. Now, I am put into a challenging predicament to find a suitable way to express them by connecting ancient and antiquated concepts & styles with understandable lingo in the present-day situations and conversations.

The book you are about to read is simply the first of its kind resulting out of my earnest attempts at achieving this ambitious feat. In the process, it revealed some amazing phenomena, faculties, and under-utilized capabilities of mind, as well as the acute consciousness when attention is focused on a particular task. They must all be thoughtful, informative, engaging, and intriguing as well, to capture interest. Eventually, it transforms me into an author! Although, until this time, only as a self-publishing author. In fact, to date, I was able to write a total of 4 books and several other literary and artistic pieces of work including poetry, essays, song lyrics, music scores, video tutorials, and music videos.

But, this book, "The biggest regatta, Kerala Vallam Kali" is the first one to be published in print mainly for certain reasons. This was the first book I started soon after my retirement and the theme was well in tune with my sentimental and nostalgic psyche. This one has been the one that garnered the most interest and enthusiastic feedback. I was fortunate to live and work in several countries distributed in varied geographical regions; also in different responsible roles. In the process, I was exposed to various cultures, languages, experiences, challenges, and contrasting points of views which I thought were all invaluable knowledge in a constantly changing world moving at frenetic pace. I would love to share them with as many people as possible. Currently, the demography of multicultural people have become a dynamic, and burgeoning group because of migration due to wide-ranging reasons, events, and changes taking place in the world. I take great pleasure to relate the themes of the books and writings to this particular and resourceful group. Many of the issues that are causing these changes are discussed in this book. Most of this community whom I would like to identify as 'Desi-paradesi-diaspora-migrant' population speak English well, albeit not their mother-tongue. They are resourceful people who want to 'make a difference' in their own lives as well as others, and thereby to the world at large. The positive effects are already manifesting in the fields of Science, technology, healthcare, manufacturing, farming, sports, games, and several other disciplines. Sympathy, empathy, humanity, sensitivity, and traditions are the hallmarks of their character. Remarkably, many countries have transformed into multicultural nations embracing people of other cultures and ethnicities enriching their own civilization.

In summary, this documentary nonfiction story is mainly about a unique regatta of many superlatives and the one that is steadily gaining in worldwide attention, interest, and participation. People from over 900 villages in the state of Kerala: the tropical land of 44 rivers, 2 vast lakes; lagoons & coconut groves-lined backwaters with vibrant ecosystem; intricate networks of canals supporting great biodiversity; and 64 unique cultural artforms, and most importantly a state boasting 'Hundred percent literacy' lying along the Arabian Sea coast in the South Western strip of India, flock together in great anticipation and joy to attend the annual regatta staged in magnificent, scenic, and natural settings. The main theme of the book is the regatta festival & boat races, but it's also about humanity, communities, lives & life-styles, sociology dealing with village societies who habitually live in harmony with one another and the environment. Other important matters are also discussed such as, Education, Science & technology, plight of farmers, history, philosophy, music, fiction, mundane & esoteric themes, governance, ecology, our existence & survival in this world of 7 Billion people with steadily shrinking resources, and frightful future.

The book narrates absorbing anecdotal stories and episodes from ancient mythologies related to the Genesis & creation; Science, Astrology; kings & kingdoms; mundane & divine, as well as age-old traditions & rituals narrated and explained in plausible, and sometimes facetious and entertaining accounts. Furthermore, the book describes in great depth and details many wide-ranging subjects and facts which are previewed and highlighted in the 'preface'. Therefore, I would like to guide your kind attention and reference towards it.

This book recounts a contest unlike any other before which is riddled with drama and suspense bringing out a number of interesting facts never discussed before in sufficient details.

Caution: Please maintain a relaxed and congenial disposition; keep your good sensibility and sense of humor sharp, as such events, figure of speech, similes, and metaphors pop up periodically, end to end, throughout the reading!

The fabulous festivities, regatta, and competitions of the Kerala 'Vallam Kali'
Embody and symbolize several subject matters all at once!

"Truly 'enigmatic', captivating, thought-provoking, and interesting till the end!"

The Preface

This is a documentary, non-fiction story, and an illustrated book with actual photo images helping to build authentic imagination during reading. It's mainly and truly about the biggest paddling regatta as well as the biggest team sport in the world. Lately, it has become one of the biggest tourist attraction; some arriving several weeks early to actively participate in the races. In the boat races, there are no substitutes or bench players. Those who are in the boat are all active participants during the entire race! The event is described in many contexts of the race, competition, accompanying festivities, age-old rituals, the sociology & anthropology of the close-knit communities living in the secluded hamlets of the region, the diverse biosphere & ecology, natural beauty of the land, the prolific number of rivers, canals & reclaimed farmlands; the flood and miseries caused by them; traditional & superstitious beliefs; the myriad of customs; myths & Mythologies. The book provides a parochial as well as global perspective on many themes affecting, people, geopolitical, 'socioeconomic & cultural' systems prevailing. It is hilariously entertaining, at the same time, seriously thought-provoking; praising the good; exposing, vilifying the bad, and the antisocial elements & entities. The book highlights many of the unique systems and beliefs that elevate the status of this exceptionally pristine ecosystem, the inhabitants who strive to preserve in the face of many setbacks and adversities. Ultimately, it raises awareness and consciousness among the readers to diligently explore and find better ways to maintain a sustainable world that we could be proud to hand over to the succeeding generations exactly identical to the goal and intentions practiced by the many global indigenous tribes.

The theme of the book and contents are formed mostly from the actual experiences, interactions, observations, stories told by elders, eyewitness accounts heard in the past, present, as well as media documentaries on a variety of subjects. The facts are complemented with analogies, anecdotal stories, and comparisons providing plausible explanations. The author attempts to conjure up an immersive, 'virtual reality' experience of the events, describing lifestyles of the past in interesting, informative, nostalgic, passionate, and authentic style & details. It is the genuine hope that the readers will enjoy and appreciate the natural beauty of the land, Vallam-Kali, festivals, and the events in several more ways, when they actually get a chance to see this superlative event of joy, happiness, spirit, participation, passion, and history.

A courtesy reminder: In addition to the general reading public, the author is also mindful and cognizant of the readers in various professional fields. Therefore, readers may occasionally come across several specialized terms and terminologies, besides the vernacular words, and phrases. In such instances, the contextual meaning is given in parentheses next to the words itself. Furthermore, a look-up glossary of words is also available in the very last pages for reference and explanation!

Ebook readers may already have their E-READER devices set up for their favorite dictionary! (Highlight the word and Right click to access online dictionary websites!)

The book contains references and description of:

VallmKali as a folk festival, the passion, and exuberance surrounding the regatta. The regatta involves several boat races & categories of boats. Queer as it may seem, it takes place in a carnival and cultural atmosphere becoming a huge travel & tourist attraction

The Genesis & creation, 'Kerala, God's own Country' how did it all came about according to mythology, explained in periodic, scientific, and captivating chain of cataclysmic events

The many scenes and backdrop of the Kuttanad region in the Vembanad lake basin; the ecosystem with its Paddy fields & Palm groves; aquatic, terrestrial, and evolutionary ecology; flora & fauna; fish & fowl; cows & cattle; birds & animals; sights & scenes

The cosmopolitan port city of Kochi, "Queen of the Arabian Sea"; the many schools, colleges, and institutes situated in each and every town throughout the state of Kerala, leading the nation in education, and 'full literacy' in the tiny, densely populated state

Explanation of the various centuries old customs & traditions; ceremonies & rituals and myths & Mythology associated with the annual regatta, explaining its origin, implied meaning, and great significance

Typical idyllic hamlet by the water; harmonious communal relationship of its inhabitants, the agrarian lifestyle, traits, social structure, the congenial community spirit; their affinity to water; tide & waves; enormous sufferings from flood; ensuing destruction & misery they face during the annual Monsoon season

Interesting events chosen from the lives in the hamlet in a realistic, facetious manner. A fascinating tale of a village life- giving credence to the saying "the world is a village!"

Historical significance, nuances, and subtleties associated with VallamKali; its collateral influence on building strong communities

Some musings from the naive mind of a child who was seeking truth, meaning and righteousness

A truly amazing human endeavor!

The reclamation of land 'Pata-Sekha-rams' for cultivation from the depths of the turbulent lake basin, also the longest in India; the resultant transformation of the region & the enormous economic benefits; tasks, travails, setbacks, and struggles encountered in the generations-long, and epic undertaking

A glowing tribute to the ancestors and history makers for their thoughtful vision, adventurous venture, heritage, entrepreneurial spirit, and the enduring legacy

A passionate discussion of the alarming plight of farmers & farming, in both local and global perspectives; climate change in the world and mass displacement of people seen as 'migration in desperation'

The betrayal by the government and its long-lasting ramifications; reckless policies that extinguish the glowing spirit of entrepreneurship, leading to economic stagnation

A candid discussion of law & order situation and rampant dishonesty among politicians, bureaucrats and business people; the curse and scourge of frequent "Harthals", a total shutdown of everything making life miserable

"From the forest to a floating masterpiece". The evolutionary chain in the building of a ChundanVallam, the preeminent raceboat; the art, Science, passion and craftsmanship. The boats are built according to the book on Architecture & Engineering written by the Lord and the master Viswakarma, in the *Vedic times of the ancient past.

Homage to the carpenter who painstakingly designed & built the boat

* The period 1500-500 BCE (Before Common Era) when the 4 main Sanskrit Vedas were authored and revealed, that themselves took a few centuries of work in many fields to finish in its entirety.

An alternate method of learning trade and skills by traditional practices rather than by education in trade schools & laboratories. A system of 'templates' for everything!

The ceremonious and ritualistic launching of the ChundanVallam with community participation in great passion and fanfare

An absorbing & eventful musical performance of 'Chenda melam & Pancha Vadhyam' explicated in humorous, musical, and 'show-business' perspectives; an interesting comparison of Western & Indian music in the styles and character of performance

The "The incredible women, their eminent role in society" as expounded by the principal of a reputed convent run women's college

The esteemed place of women in the community; their tremendous contributions benefiting the whole humanity but seldom recognized

The significant contributions the Nursing, NRI, and diaspora communities continuously making to the economic stability, & well-being of the state of Keralam

"Sadhya", the communal banquet before the launching of the Chundan, the cuisine, nuances, manner and style, explained in eyewitness account

Science and physics of rowing, sculling, Paddling & canoeing

Vanchipatt, the 'Kerala Oarsman's song', its significant role in the races, its themes, rhymes & rhythm, and articulation. A remarkable, fact-based story about its history

The cultural heritage festivals taking place at the venue in the spirit of 'Renaissance'

The arrival of the great many fans in fanfare and expectation of victory and glory

The description of the categories of traditional race boats that are participating

Analytical discussions, juxtaposing occidental & oriental philosophies of the 'great divide', among all the stunning similarities

Keralam boasts having 64 unique art-forms; many are ancient. They are enacted at the venue in the 'cultural heritage' festival

An insightful, descriptive documentary of the 'end to end' Television broadcast operations helping readers understand the behind the scene activities and applied technologies used in bringing 'the picture home'

An enthralling lyrical poetry by eminent 'Nimisha Kavi', 'poet laureate-extempore' about the scenes, sceneries, sights, settings, and the ambiance at the venue

Scintillating play by play, boat by boat, spot by spot, scene by scene Running commentary'. Color & analytical commentaries by Science and rowing experts and specialists; Psychologist highlights the anxiety & ecstasy of the adoring fans as well as the frenetic activities of the betting syndicates

Some remarkable childhood memories, stories, experiences, thoughts, interwoven with anecdotes & incidents, narrated with nostalgic feelings

The drama & delay as well as the gripping suspense encountered in the announcement of the race results in several dramatic twists & turns; incidents & procedures in entrancing episodes

The glory of winning and miseries of loosing

The return journey of the victorious Chundan Vallam back home

Gulf Malayalees & diaspora watching the race, the economic impact and international cordiality created by the NRI (Indian expatriates) communities

A Futuristic proposal for the development, revitalization, and preservation of the Vembanad heritage region; elucidating various schemes, new technologies available today for renewable energy production & power generation, permanently ending the debilitating power outages, mitigation of urban crowding; saving planet earth

A 'call to action' in research, innovation, good governance & leadership

Picture collages from the scenes of VallamKali, other interesting visuals

The staging & veneration of the triumphant ChundanVallam. Victory celebrations fit for the epic win and the rustic village community with Music mela, fireworks and community events, described in fascinating eyewitness account.

Chapters & Index

Chapter 1 17

The Scene & Background

Onam, Onam procession, celebration, Maveli

Chapter 2 21

A splendid spectacle in the waters!

Kera, Keralam, tree of Life, Toddy shops, police constables

Chapter 3 36

A typical Evening in the village

Drunkards in ditches; the lives of alcohol addicts,
The miseries perpetuated, cases of domestic violence

Chapter 4 39

Water,Water everywhere!!

The scenes, water logged villages, life, affinity to water,
the indispensable Vallams for transport

Chapter 5 42

The Vast Kuttanad Basin

The Vembanad lake, canals & waterways,
Ekkal nutrients, paddy fields, farmers

Chapter 6 44

The Monsoons & floods

The bleak and miserable days of Monsoons,
Marooned & isolated villagers

Chapter 7 55

The Epic Saga of great human Endeavor!
Reclamation of land from the Vast lake
For Rice cultivation
A glowing tribute to the forefathers for their fabulous legacy

The great land reclamation project; the indomitable spirit, adventure;
Transformation of the region;
The social systems & community spirit & cooperation;
The setbacks & struggles faced; the monumental legacy that was passed on;
The unsung heroes, the resourceful indigenous communities;
The advent of mechanization, its efficiencies, resulting displacement of workers

Chapter 8 72

Betrayal by the Government
Plight of Farmers

The government, the menace; destructive policies & actions,
Ramifications, plight of entrepreneurs
The revolutionary government, wrecking institutions & systems;
Sowing seeds of chaos & anarchy perpetuating poverty,
Destitution, stagnation & backwardness of the great state

Chapter 9 79

The Magnificent Vembanadu Kayal
The picturesque port city, Kochi
The 'Queen of the Arabian Sea', seat of the highest justice, education,
Business, industry, shipbuilding, transportation, naval base;
The delta, palm groves, causeways & bridges
Bustling, residential, international, cosmopolitan, progressive, and beautiful!

Chapter 10 93

Those days of the forgotten past
Vallams, custom-made for different applications,
Effective & important for transportation.
Life stories of the Vallam punters, cargo transporters,
The lonely oarsman's songs

Chapter 11 100

The *Ulbhavam*, Genesis of
"God's own Country"
The 'beginning', evolution, cosmic activities, 'Yugams'
The birth of Keralam according to the fascinating mythology

Chapter 12 102

A PRIMAL CHAOS BEFORE CREATION!
The Sun, solar flares, "big bang", cosmic cycles,
Origin of DNA, 'Life begins on earth!'
Valiant acts of Lord Parasuraman.
Keralam, how was it reclaimed from the depths?

Chapter 13 107

From the mind of a child who was seeking
Truth, meaning & righteousness
Folklore, the Paathalam, the doomed abysmal dungeon,
Anxiety & fear created in the minds sins & atonements; The holy & the Saints, imps & devils;
Gods & demigods, causing a confused state of mind in children; Life in an extended family;
A defining moment of faith, belief, and realities

Chapter 14 120

The joyful Festival Season
The most joyful time in the Kuttanaadan villages follows the rice harvest
And cessation of all agricultural activities

Chapter 15 121

Early preparation
Before the Vallam Kali
Every village comes together for the communal festival season of
Onam & VallamKali
Getting the Chundan ready for the races

Chapter 16 *123*

Making of a Chundan Vallam
Tribute to the Ashaari, the carpenter
from a tree in the forest to a proud treasure of the entire village,
the art, precision skill & dedication
with which the master carpenter go about his task

Chapter 17 *135*

The ceremonious launching of the ChundanVallam
The entire village gathers to attend rituals, ceremonies,
The excitingly eventful ChendaMelam performance at and the launch;
Comparing the music & instruments with those of the Western
The women, the pride of humanity, a tributary poem of great significance "oh, Eves!"
Commencement address by the inspirational principal of the convent college

Chapter 18 *172*

The training for the races
The traditional communal Sadhya,
Training & practice for the races
The versatile and facetious "Kavi" Kunjan Nambiar,
Also, an inimitable 'Mizhavu' player

Chapter 19 *183*

The many Categories of participating Vallams,
The race boats
Chundan, Vaippu, Oaadi, Churulan, IruttuKuthi, Kochu Vallam,
Women's races; Different shapes, styles, color, character, speed, traditions;
The paddlers, prestige, pride, and the stakes

Chapter 20 *195*

The grand arrival of the enthusiastic spectators
Festive & boisterous crowd arriving in flotilla, by many other means
All comme determined to enjoy the races and other cultural events

Chapter 21 *199*

The sights & sounds
In and around the racecourse
The magnificent ambiance, exuberance, & expectations
At the biggest regatta in the world
The panorama of sceneries, backgrounds,
The blending of people, & nature in the natural settings

Chapter 22 *203*

The organization & preparation
The staged and orchestrated activities
The elaborate & extensive arrangements and efforts needed to stage the event
It involves social, government, business, media, travel & tourism entities;
Cultural societies, organizations, committees, race event judges;

Volunteers, performers & participants

Chapter 23 *206*

A 'Cultural Heritage' Show!

Enactment & staging of various cultural art forms in traditional setting and meaning
The destructive Harthals, chaotic governance, appalling law & order situation;
Gradually improving transport infrastructure, amenities

Chapter 24 *219*

The significant role played by the Malayalee Diaspora

The spirit, sacrifices & contribution made by the expatriate community the so-called NRI
The exceptional role of the nursing community in expanding new horizon and avenues

Chapter 25 *224*

The psyche of Keralites

The psyche, temperament, and mindset of the village communities

Chapter 26 *228*

A Mecca of Opportunities

New opportunities of many kinds open up;
need to make certain changes to capitalize on it
Understanding the dynamic world & the changing world order

Chapter 27 233

Around The Venue

Many things to see, participate, enjoy and experience, don't be just a spectator!

Chapter 28 *236*

Topics, conjectures, science & culture
Discussions & interaction between people

The distinctions between paddling, rowing, sculling
Major distinctions in perception, thought processing, philosophies in different cultures

Chapter 29 *244*

The Pavilion & Reception center

The decorated pavilion, the traditional greeting rituals,
Beautiful receptionists dressed in gilded Sarees, hair-tops, ornaments, bangles & adornments
Offering garlands & courteous "Namasthe" greetings

Chapter 30 250

THE TELEVISION BROADCAST OPERATIONS
An informative & in-depth technical Analysis

The elaborate set up for the satellite broadcast operations
The video, audio, sound, graphics & technology, commentary, automation, coordination

Chapter 31 *258*

The Preview & Hype

The preview and grand introduction of boats with details & descriptions,
No. of paddlers, track record, the village representing, the location,
The fans; introduction of boat captains, odds of winning, the sponsors, etc.

Chapter 32 *268*

The Malayalee diaspora gather to watch the race On satellite Television

Gathering of the international expatriate communities
From different countries, cultures & religions; enjoying the race and potluck dinner
In relaxed, casual, and cordial camaraderie

Chapter 33 *272*

The Race Starts!

All about the races, the course, distance, time needed
A contest between boats of many categories, sizes, shapes and unique historical significances
A race for glory, prize & prestige

Chapter 34 *274*

The running commentary

The scintillating commentary by specialists, color,
Analytical, science commentators,
Play by play, boat by boat, spot by spot, scene by scene, statistics, actions, projections
In between, an extempore poet laureate describes the scenes & sceneries,
Beauty of the nature; the expectations & ecstasy of the boisterous crowd, and
The many things he sees and observes at the venue
All in flowering language and fluent oratory

Chapter 35 *306*

Trip to the VallamKali A nostalgic reminiscence

Reminiscing on the trips to VallamKali as a child

Chapter 36 *319*

THE SOMBRE SCENE

The close, thrilling finish, agonizing uncertainties, the nerve racking wait
The many twists, turns, & tantrums

Chapter 37 *322*

A Futuristic Master plan

An elaborate plan & proposal for the revitalization, beautification
Of a sustainable, vibrant Vembanad region
With Renewable Energy, marine research & technology complexes, marina, fishing,
Residential /industrial/commercial/agricultural centers, parks, recreation & amenities centers
The end of the detested 'current cut'! Bringing prosperity to people & farmers with it
Importantly, sustainable living,; introduction of wide-ranging green technology initiatives

Chapter 38 *356*

The S.u.s.p.e.n.s.e!

The inordinate, unbearable delay in announcing the official results,
The nervous and edgy crowd

Chapter 39 363

Euphoria of winning and the Agony of defeat!

The pictures worth thousand words tell the whole story!
The victorious and the vanquished feel entirely contrasting emotions

Chapter 40 366

Flotilla escorts the Chundan back home
To its usual resting place

The triumphant return journey home escorted by flotilla of exultant fans
Celebrations of the victory
With sumptuous feasting all along the way;
The folks living along the banks of the waterways join in the procession

The Final chapter 369

The veneration & coronation
The staging of the triumphant Chundan

The Chundan floats in majestic state, the proud village celebrates its historic victory
With ChendaMelam, PanchaVadhyam & fireworks, in great jubilation and veneration
The Chundan, the pride & joy of the villagers, earns a well deserved rest
Until the time for the race in the coming year!

In a few days, it goes back to its modest hanger
And remains there out of view, and unattended!

A BOOK, TRULY INFORMATIVE, CAPTIVATING, AND INTERESTING THROUGHOUT
CLIMAXING IN SUSPENSE, DRAMA, A HASTILY-ARRANGED WEDDING
IMPORTANT DISCOVERIES & FOLKS-FESTIVITIES!

You are cordially invited for an absorbing reading!

Chapter 1

The Scene & Background

The rice harvest season has ended ahead of the onset of Monsoon rains. Now, there is a long spell of inactivity, lack of income, the inclement and dreadful monsoon season, consequential floods that bring enormous gloom and destructions that's to endure, before normalcy can return in time for the preparation for the next crop cycle. This year, the harvest was not particularly good. Years running, the debt has been steadily accumulating in alarming proportions. Coveted possessions are either sold off, mortgaged; gold possessions of the women households, and heirlooms were pawned in desperate efforts to forestall merciless and dire foreclosure actions from the banks and creditors. Grown-up daughters are to be married off soon. Dowry is illegal, but the tradition and practice continue since its beginning in history. The cost of education and healthcare is steeply soaring. Inflation runs rampant. The fertility of the soil is deteriorating fast. Farmers find it difficult to sustain even a modest lifestyle with the means they possess. Nevertheless, right now, people want to forget the maladies afflicting them. They are looking for ways to extricate themselves from such woes and worries. They are entitled to a little bit of cheer in their lives too. The folks-festival of 'Onam' is obviously something everyone can look forward to. Traditionally, it's a 10-day-long celebration [1st Day – Attam; 2nd Day – Chithira; 3rd Day – Chodhi; 4th Day – Visakam; 5th Day – Anizham; 6th Day – Thriketa; 7th Day – Moolam; 8th Day – Pooradam; 9th Day – Uthradam; 10th Day – **Thiru-Onam**] As a prelude to it, are the water festivals and regatta taking place at various locations. The "Moolam" Vallam Kali on the Pampa river at Champa-Kkulam flags off the season in its traditional grandeur and historic importance. The Aranmula Vallam Kali, which is actually not a competition but more of pageantry, takes place during the Onam festivals. The ceremonious regatta and procession have religious, historical, and devotional meanings and significance. That is also held on the placid waters of Pampa, at an upstream location. The boats for this particular event are known as 'Palliyodams' in the historical, or reverential meaning, though they look very much like the race boat, Chundan Vallam. Usually, about 50 or more of them made of different wood, color, tint, carvings, and other distinguishing characteristics participate in the festivities. The actual color variations of the wood are affected by its natural sap, the soil, sunlight, and climatic conditions where the trees grow. Each boat showcases and projects its own distinct heritage, status, and personality. The Palliyodams are shaped slightly differently from the type of ChundanVallams built for racing. Their bow as well as the stern are 'cantilever' shaped and arch higher over the surface of the water, or have higher 'waterline' clearance- the two ends never touch the water even with the full load. Therefore, the total contact area of the boat with water, or in other words, the 'draft' is minimal. Paddlers have specially carved extra-long paddles to do the paddling. The chundan has a "round-bottomed", 'keel-less hull', designed to move smoothly through the water with relative ease and faster at the expense of a little loss of stability and control. The chundan is built more like a canoe than a typical sailboat or speed boat. But the design has seldom caused problems with slip, skim, or rollover with 110+ paddling crew on board and therefore of little concern. Since ancient times, the marine engineers and artisans belonging to the 'ViswaKarma' community have mastered and designed many innovative structures for different applications. They inherited the traditional wisdom and blessings from the lord deity, Viswakarma who postulated the theory, practiced, and laid out precise measurements and specifications for each component, joint, and assembly of the pieces. He 'wrote the book' on this subject! The enduring template is still in vogue without any significant alterations.

Each class of boats has its own fans and affiliation, similar to the Football clubs in countries around the world. The other major attraction in connection with the boat festival is the famous 'Valla Sadhya' (The public community- luncheon, or the grand feast-quite different from a banquet) serving all the major 50+ traditional Keralite food items serving more than 200,000 people, the biggest of its kind in the whole world! It rivals the daily lunch offered to the public at the Sikh "Golden temple" at Amritsar, Punjab. People say: "Yum...yummy...oohm...! you must taste it at least once in life. It's the "mother of all Sadhyas!"

"Onam" is the folks' festival to celebrate the annual visit of their 'legendary and benevolent King' "MahaBali", reverentially addressed by the citizens as "Maveli", in his spiritual form. The king loved his subjects and cared so much for their welfare and happiness, that he asked for his 'ultimate wish' for 'a unique permission to visit his subjects once every year. Seeing his warmth, humility, and affection for his subjects, his wish was granted by the gods. His period of reign was the golden era in the history of Keralam; marked by egalitarian principles, peace, prosperity, contentment, and happiness for all. His legacy was freedom, fairness, equality, charity, kindness, and the practice of virtues in all its manifestations and forms. People didn't face any undue sufferings, struggles, or discrimination of any kind. During Onam, he visits every family and every house. (This is in contrast to the visit by Santa Claus who visit only houses in the West during Christmas Eve!).

The Onam celebrations, a folks' festival, uniquely of Kerala style; A truly happy and joyous occasion for all, No worries whatsoever!

The schools are given a week-long holiday. It's fun and playtime for children. They perform entertainment and traditional artforms, undertake social and community work. It is also a public holiday for four consecutive days. So everyone could prepare for a fitting welcome to the visiting king according to their means. They can also actively participate in the accompanying celebrations and festivals. People make very elaborate preparations to give the king a pleasant welcome. They try to show and impress upon him that they are presently living happily, and well-fed; life continues the same way as it used to be during his reign! They go to extreme lengths and efforts to prove this point. People make beautiful flower decorations of many kinds and put them up on their verandha, at the entrance to the house to receive their beloved king. Households start cooking nonperishable dry foods, fries, fritters, and sweets well in advance, joyfully sharing them with friends, neighbors, and co-workers. People celebrate, participate, enact plays, and play games in a radiantly cordial communal spirit, as it was prevalent during the golden era. Such was also the theme of the ancient “Olympic Games" in Greece, which now is corrupted by commercialism, doping, and judging scandals. Everyone is dressed in traditional white costumes with golden borderlines. Women folks wear freshly plucked jasmine garlands and gold & yellow ‘Jamayanthi’ flower tops; look more beautiful than ever. Everyone enjoys the most sumptuous meal they could afford for the 'OnaSadhya', on “Thiru Onam”, the principal and the most auspicious day of the week-long celebrations to show and assure the king that they have plenty indeed, not lacking in any of the necessities. The citizens do everything possible to make the king's visit pleasant, memorable, reassuring, and comforting. They pray that their former king may return to his 'other world' fully convinced and greatly pleased with the living conditions of the descendants of his erstwhile subjects! He shall have good news to report to the gods and will have his ‘visiting rights’ readily renewed!

The king begins his processional visit from his own erstwhile capital “Trikka-Kara” in the suburbs of the modern, magnificent port city of Kochi. In reality, today, the sublime principles of his time have become sinfully depraved and blatantly violated. The precocious organizers deliberately re-route the procession from entering the present capital “Thiru-Vanantha-Puram” and the premises of the highest seat of jurisprudence, the High court of Keralam, to avoid any embarrassment and grief to the visiting king. During his reign, they were the sublime citadels of truth, ethics, and correctness. Thankfully, the king is so magnanimous and fully amenable to the wishes of his escort; place no undue burden on the hosts, nor ask any prying questions. People fear, once he finds out the truth of what is really happening at these places, he will be heart-broken and may not want to come back ever again! That'll be a huge disappointment to the entire population of Keralam, a great pity, deplorable setback, and a big humiliation for them!

Chapter 2

A spendid spectacle in the waters!

Typically, the state of Keralam has a tropical climate. In fact, there are only two seasons- the 'hot & humid' summer and the drenching, irrepressible Monsoon season. The highly anticipated S/W Monsoon makes its landfall right here in the middle of June, usually without fail, in case it does, people become anxious and alarmed. Then. it methodically moves and spreads towards the North, shedding rain and heavy precipitation all along the way. The second one, the S/E monsoon is relatively shorter, but similarly, heavily packed with torrential rains, incessant and deafening thunder with intense bolts of lightning. Both are extremely beneficial and at the same destructive as one might imagine. On the average, annually, lightning strike kill more than 2000 people in India.

Because of the long spell of rain, plenty of rivers and canals are naturally formed across the entire land, the largest number in the world per land area. Water is fluid, wants to flow, and flows freely over and around any hindrance. It packs two tremendous forces of energy, potential & kinetic. Like the wild elephants, they knock off all obstacles in their way and make a path of their own with fast and furious force. In all, there are 44 rivers crisscrossing the stretch of this relatively small state. A river, canal, creek, or body of water is never too far away from any location. People are nature-loving; fiercely protective of the biosphere and ecology, the same way as Native Americans, Inuits of Alaska, the Maoris of New Zealand, the Aborigines of Australia, Bushmen in Africa, various Amazon tribes in Brazil, the indigenous tribes in Peru, Bolivia and Central America. The land is sacred for all of them; will not scar or pillage; meticulously preserve it for posterity. They believe that mankind has no right to own land on their own, but to pass it on for the benefit of future generations to live and prosper. But alas, the greedy colonialists cum marauding invaders destroyed most of it for selfish ends, occupied the conquered lands for good, and treated the natives with supercilious disdain and utterer disregard for the vanquished. In only a few hundred years of domination, their reckless and wanton behavior bring the world close to extinction. It is the same world the indigenous tribes lived and kept pristine for tens of thousands of years!

Naturally, all the people make their living, derive their culture, and pursue normal pursuits of life adjusting to the environments they live in. Because of regular rainfall, shining of inexorable sunlight, and the resulting humid conditions prevalent in Kerala, water-intensive plants, crops, and vegetation such as coconut palms, Paddy, and tropical fruits grow widely and in abundance. The name Keralam is derived from “Kera”, the coconut palm, and aptly means: '*The magnificent land of the coconut groves*'.

Interestingly, the coconut has a Polynesian origin in the pacific islands. The fallen coconuts from the trees used to simply float and drift; lived their lives and perished in the water. However, one time in a twist, following a cataclysmic event of the sort, a bunch of them were inextricably entangled and were swept away by sustained, monstrous storms and accompanying current in a certain direction. It turned out to be an epic journey. The tangled bunches were further pushed and prodded by strong ocean currents, tides, and waves. Unnoticed and unceremoniously, the bunch entered the Indian Ocean from the Pacific through the narrow passage between Australia and East Timor island. They hugged the coastlines of Indonesian islands. Then the wind changed its course and blew it past the islands of Andaman & Nicobar. The bunches were temporarily brought to the coast of the Bay of Bengal. They stalled there for a while. Then again, resumed its ‘voyage’ protected by the Eastern shoreline of the Indian coast, traveled around Sri Lanka right along the present shipping route, bypassing the perennially turbulent ‘Palk strait’ and splashed onto, and deflected around the cape of ‘Kanyakumari’ (the cape Comorin) entering the Arabian Sea.

The nightly Kerala fishermen trapped some floating ones in their nets. In the dim visibility of the fading moonlight, they noticed some whiskers on the underside that was immersed in the water. It looked very similar to the whiskers/barbels seen on catfish (koori), carp, sturgeon, zebrafish, and certain kinds of sharks. In great glee; brought them to the home ashore, thinking that it is a specie of new-found fish. They thought they could surprise everyone. Later on, when they found that it was not any fish at all, and fearing that these things might have torn their nets, in anger, they threw them out as useless things. Soon the shore was littered along and around with the unwanted coconuts. Fishermen like only fish; care less for anything else! Few years passed as uneventful. In the great scheme of things, every living thing has its ‘natural intelligence’ and the instinct to grow and multiply ingrained in them. The 'whiskers' were not the barbels of the fish, but were actually the sprouting young roots. When they were thrown around the shore, they came in contact with the fertile soil, its natural habitat. They settled in the sand, sank, and gradually got 'buried'. The young and smart roots soon sensed the presence of the necessary elements for its growth,- water and nutrients in the soil. Vigorously, they extended their reach to the deeper ground; hastily carried the ingredients to its seed where it was collected, processed, and made into nutrients that caused small leaves to sprout. Though tiny, the green leaves, in turn, manufactured food, the carbohydrates (sugar & other organic compounds) by the process of 'Photosynthesis' using light energy from the sun and synthesizing water collected by the roots along with the carbon dioxide captured from the surrounding atmosphere. The byproduct of photosynthesis is the most essential element needed for all the living beings on earth- oxygen! Besides food for the plant to grow, the other dual, concurrent, and extremely beneficial outcomes are the purification of the atmospheric air by supplying pure oxygen while greatly reducing the harmful carbon dioxide. The functions the plants perform, are quite astonishing as well as sustaining. It' should be one of the greatest wonders of nature, after the creation of the universe and life on earth!

The roots ran deep and grabbed the surrounding soil, establishing a strong footing to stand up and withstand the wind, weight, and resisting the uprooting forces of storms. Serendipity! Right there, they sprouted vigorously into green plants in the soil in a warm and growth-inducing environment. Each day, they grew steadily thick and tall. Unlike other trees and plants, it started to grow branches, but only at the apex; not anywhere else. It must be a queer and special kind of tree! On them, green leaves began to grow long in rows at precise intervals with gaps in between. They looked vcry prctty. When they swung and danced in the wind, they also fanned the trapped air. Surprisingly, it had a cooling effect underneath due to the moisture it was dispersing that came from the chemical reaction between oxygen, the byproduct of photosynthesis, and the Hydrogen in the surrounding atmosphere. In effect, the coconut trees served as a natural open-air 'air conditioner'. Totally amazing!

This must have given us the idea of ceiling fans in our homes!

The line of the newly grown trees presented a new sight, highlight, and identity to the shoreline. The fishermen could easily spot their sweet home shore from far away as they returned in the dim light of the very early mornings after the nightly fishing expeditions. Besides, they provided shade; made the air

cooler underneath- a large umbrella of sorts that need not be held! The trees provided comfort and shelter from the scorching sun, as well as keeping the daily catch of fish fresh for a while longer. The fish survived longer out of the water; they flounder and flop around inside the container basket, trying to jump out and rejoin the water. They looked healthy and fresh; buyers salivated upon seeing them; they were sold fast, fetched premium price to the fishermen. They began to love the coconuts; gathered all the fallen ones and enthusiastically replanted them with the help of neighbors, and family members.
Every time they went out to sea, they soured the seas for more of them, as much as they did for fish!

The tree garnered a lot of attention and curiosity from everyone. It fascinated people very much. They found out that there are many things they cannot see from simply standing on the ground and looking up. It blossoms fully, only after growing mature and tall! One has to climb to the very top in order to see its apex, crown, and flowers which develop into young buds as well as all the other wonderful but otherwise invisible parts. So, the curious folks climbed the trees; looked; examined; cut, and fell item by item. People on the ground split; cracked open, chewed, tasted every part of it. There have been wonderful and exciting discoveries all along and each and every day. Some parts are edible, some sweet and tasty. The nectar was refreshing. It quenched their thirst; refreshed the body; gave them energy and taste. There were certain parts that satisfied hunger too, so on and so forth in ever more astonishing ways and fascinating manner! In their laboratory of kitchens, the housewives extracted juice/milk from it; grated and shredded part by part; methodically tested each part in cooking; and tasted them like the fastidious and discriminating food 'connoisseur'. The households liked the new taste. The results were simply astonishing!

The children ate the white flesh (Endosperm), after which they skipped lunch indicating that it can satisfy hunger. Parents liked the fact. People started experimenting with it. They broke it open; cannibalized and casually left it there. Under the sunlight, it changed color and became brownish; chewed it; found out that it has oily taste. They ran to the miller with it and had it squeezed; took it back home; the housewives fried banana slices in it and found it very delicious. Then did the same with fish. They also found it was also a preservative; kept it for the rainy days of the Monsoons. People were fiercely racing to discover new things about the coconut. No discovery is trivial at this time! Soon, the tree and its products assumed legendary and mythical status. It became the major subject matter of any conversation. Religious people claimed it was the "Manna" they were talking about all along, that fall from heaven. For common people it was an invaluable gift and bounty in multiple ways, all begotten by the providence and care of mother nature, ultimately, coining the saying: "Nature protects and maintains what it creates!" & "Man shall not live by bread alone!"

The news spread like wildfire. Learned-men came to know about the wonder tree. It spiked their curiosity and interest. Unfortunately, they were not capable to climb the trees to see and examine everything for themselves and to their full satisfaction. They had the apex chopped and gently brought down to the ground without any harm or injury. What they could see at their first sight itself, marveled them greatly. They made a powerful pronouncement to the people that we must always protect everything that we find in nature, take care of them; everything has a purpose and use; they are provided to us by the bountiful nature. Nature, the mightiest of all, seems to know what the living beings need. This is to maintain and sustain all lives on earth. Live, prosper and procreate to the time of eternity. This is the all important purpose and intend of life. This is how nature communicates with us, showing, providing in strange and mysterious ways. Its language is mute, but deeply meaningful! The obedient followers took the knowledge and message to their hearts and homes: in its letter and spirit. They began to spread this wisdom by 'word-of-mouth. This is the first message and principle of "Conservation", which is "the careful and concerted action to preserve and protect something; especially: planned management of natural resources to prevent exploitation, destruction, degradation, or neglect". The idea was passed on successfully and successively through the entire generations, unbroken and unaltered. Now, as their ancestors did, the people of Keralam believe faithfully, and religiously in the practice of the same principle since the time forward!

Well, people began to plant them wherever they see a little ground to dig a hole. The coconut seeds grew vigorously into tall trees; produced more coconuts and proliferated into hundreds of thousands and millions. The religious leaders sensed an opportunity here. One by one, they claimed that the coconut seeds were directly and secretly given to them. In identical messaging stories, each narrated what had happened during an intense "Thapass/meditation' in a dreadful night of floods, torrential thunderstorm accompanied by ferocious thunder and lightning. They told the people in frightening manner as to how they were suddenly shaken by a loud command followed by the appearance of a bright figure that streaked across the skies by the North-star. It descended fast in lightning speed with a flare; landed smoothly right in front of them; gave this gift from a huge pair of shining hands with the divine utterance: "plant it!". Instantly thereafter, in a flash of puff of wind, flash, flame, and plume of trailing smoke, the figure vanished in thin air. The gullible folks believed every bit of the story heard repeatedly. Some were terrified; some startled; others rued the fact that perhaps they were tired and fell asleep earlier on that eventful night. Many others felt that they were terrible sinners for missing such an astonishing sight and the act of receiving a divine gift directly from the heavens.

The actual account from the local fishermen dissipated fast, as they were not good communicators. Common people would rather believe in heavenly occurrences and utterances than their own eyes and experiences. Furthermore, people began to ponder why Keralam was chosen by the gods to deliver this 'tree of life'. Finding no clear answers, they began to attach a lot of mystical stories to Keralam, simply calling it "God's own country!"

[The above is an excerpt from my hastily arranged bedtime story narrated to my children. They wanted to hear a different one each time and something other than the cartoon stories- preferably an actual story that happened in ancient past in a faraway place- like their parent's mystical homeland. It is worth mentioning that both my son and daughter are zealous conservationists!]

The coconut palm is appropriately called the "tree of life" as every part of it is used for daily life and consumption. Grated/shredded coconut and milk are essential ingredients in every Keralite curry recipe: its milk helps build HDL, good cholesterol, and is lactose tolerant. The Exo, endo, and pericarps of the coconuts are extensively found in dietary use. Cooking oil from the desiccated pericarp called 'Copra'- the sun-dried white husk is prolifically used in frying food items & ingredients; also in Ayurvedic oil as hair tonic & body massages balms; as mouthwash, reducing gingivitis- which causes plaque build-up. Regularly, the local population apply a specially prepared oil to the hair and body before taking daily warm bath. It keeps the hair dark and shining, and the skin smooth like the feathers of the passerine tropical bird, Drongo. While ceremoniously applying the treated coconut oil to hair, grandmothers usually tell little girls the oil will groom their hair thick and long just like the tail of the bird now perching on the banana stem. Readily, the naughty and the innocent comply! The copra, being parched and processed without actual delay, never require any application of Sulfur dioxide (sulfate) preservatives that are widely used in dried fruits in grocery stores to prevent molds and to prolong shelf-life. The preservative chemicals are known to exacerbate asthmatic conditions. That's bad!

Coconut, the "Tree of Life" provides many products for living

All edible parts of the coconut tree possess numerous medicinal properties of fighting the harmful 'free radicals'. [Free radicals are highly reactive and unstable molecules that are produced in the body naturally as a byproduct of metabolism, or by exposure to toxins in the living/working environment. Though, free radicals have an extremely short lifespan of only a fraction of a second. Within that time, they can damage DNA, sometimes resulting in the mutations that can lead to the dreaded disease- cancer. Thankfully, antioxidants in the foods we eat can neutralize the unstable molecules by reacting with them forming different compounds. Thus, drastically reducing the deleterious effects!]. Coconut water is a rich source of natural electrolyte and used as energy booster rivaling the chemically formulated, Coke, PepsiCo's 'Gatorade' & similar drinks that are aggressively mass-marketed through sports professionals; commercially branded, and sold as 'energy drinks', that are actually known to cause obesity. Natural coconut water is antiviral, antibacterial, anti-inflammatory as well as antioxidant. It is an instant remedy for diarrhea and dehydration, so much so, coconut water had been used as an IV solution during scarcity and in emergency situations, especially in wartimes! (ask the islanders!)

The mineral-rich coconut sap containing amino acids is useful for the neural functions of the brain. Other parts of the coconut tree are also very useful. The seasoned leaves are woven for thatching roofs. Tree trunks (palm wood) are used as timber, beams, posts, and lumber for structural work. The raw, barked logs are used to bridge small irrigation canals in the coconut groves. Well-polished wood pieces are used in household furniture as well. Its unique crimson color, tinge, and thin longitudinal fiber strands stand out among others.

Coconut products, all of its parts are useful items in daily life

The coconut shells, stems & roots find mundane uses as natural, cost-free household cooking and heating fuels, reducing consumption of costlier and polluting greenhouse gas and electricity. The outermost greenish, fibrous 'exocarp' is spun into strong natural fiber for ropes, yarns, mats, and coir products; find everyday uses. The “Vatam", the thickly inter-woven, braided rope, is so strong that it is used for tugging barges, logging, wrapping, lifting, and fastening heavy loads. The roots are used in mouthwash and dye-making.

The tender, ivory-yellow "Pookkula" (palm cabbage, the clusters of apical flowers) are symbols of purity. These items are in high demand in ceremonies, wedding decorations, as well as daily Pooja in Hindu religious rituals.

The naturally fermented nectar, the sweet 'n sour-tasting "Toddy", the iconic & favorite Keralite alcoholic beverage, though an anathema (severely detested thing) to housewives, is another sought-after product derived from the apex of the coconut trees. Charcoal from coconut shells, and burned rice chaff have lately become favorite beauty product ingredients among Hollywood & Bollywood celebrities, in a way, emulating the age-old practice of Keralites. They use them for skin purification, natural teeth whitening, and similar cosmetic uses. Charcoal is widely used in water purification filters in the 'whole-house' water treatment systems. Supposedly, they neutralize minerals and other substances from the well-water pumped out of the deep stream aquifer, which is commonly contaminated by debris, pathogens, etc. The ubiquitous ironing device in Kerala households called "Theppu petty" (the non-electric ironing box) is heated by the slow-burning shell-charcoal. It retains heat for a long while, almost as much as the polluting coal does. There is nothing better to press the laundered pure cotton clothes that are pre-treated in starch for retaining form, and "Neelam" (an effective brightening blue colored 'off-the-shelf' laundry concoction) for brightening up the white-colored clothes as most commonly seen worn in Keralam (as the white LED lights at 5000^{0} K). The charcoal serves as a slow-burning fuel in tea, & "Samovar" kettles, a favorite of tea-drinkers in Arabic & East European countries, as it keeps the tea warm for hours longer.

On our planet earth, there is really no tree more versatile, productive and useful to life than the bountiful **coconut palm**!

It lives long, serving humanity in its entire life.

Simply stupendous!

THOUGH MUTE, IT SHOWS THE WAY FORWARD FOR HUMANITY;

HELPS US TO BE SELF-SUFFICIENT, AND EARTH-FRIENDLY!

The drunkards have funny 'multipolar' character, existence, and bizarre behavior patterns

The dusk to late-night time is somewhat eventful and comical in villages and the surrounding neighborhoods that have Toddy shops. Alcohol affects the Cerebellum that controls body movements and balance. The drunkards stagger and swagger. They experience delusions and hallucinations. Their Cerebral cortex suffers impairment; loses inhibitions; neuron activity slows down; brain functions and sensory powers retard. They talk, sing, and argue, all in loud voices; thoughts and speech become slurred, vision blurred. They talk about a lot of things, but all complete nonsense. A host of functions controlled by the Hypothalamus, pituitary glands, and Medulla are adversely affected. Their eyes roll and see things around them spinning constantly; feel the earth has steeply fallen off its axis by an additional 23.5° and spinning faster. They think it is the work of their jealous neighbors; hurl profanity and hatred at them, violently rattle their front gates made of heavy cast iron, and chained at night. They attempt to enter the compound, confront them, and settle this injustice then and there! The neighbors become annoyed at the nuisance; summon the police. The police respond to the scene; approach the shop owner directly, where they pin the 'root of all the evils' to; make serious-looking dramatic gestures, shout out commands and noises; scold him in front of the customers. Sensing the severity of the situation, the patrons gulp the leftovers. In the act, some fluid goes into the wrong orifice- the windpipe (trachea); 'aspiration' happens; cough & burp; they suppress the discomfort; quickly head for the exits. Actually, there is only one opening, it is used as an entrance as well as an exit. Most secretive and wretched places are designed to operate that way! The patrons hide their faces with their handkerchiefs and/or umbrella; furtively stagger to the exit and suddenly vanish in the darkness of the mature night. Meekly, the shop owner escorts the police chief to an inner room, a 'sanctorum' of sort; opens the cash drawer, and makes several bundles of currency notes, secure them with tight rubber bands. Ironically, the portrait of Mahatma Gandhi, the 'father of the nation' who practiced strict 'abstinence' and 'teetotalism' is on the face of the high denomination currency notes, as can be seen through the cracks in the dilapidated door. The owner thrusts the bundles deep into the pockets of the commander's uniform that hold many other official papers too.

Here, the chief doesn't handle the cash! Technically, he doesn't commit any wrongdoing. All well and good!

The police party comes out of the shop; circles around as if searching for any other miscreants; makes angry scenes, makes themselves well heard in the neighborhoods shouting orders; lingers around for a while longer to see the lights in the neighborhood homes go off. They pretend that they want to put an immediate end to this kind of 'nuisance & nonsense' once and for all; quickly return to the station somewhat richer and adequately 'rewarded' for their nightly mission.

Instead of doing this exercise of travesty sporadically, the chief strikes a long-term deal with the Toddy-shop owner for a certain sum of money to be periodically paid as discreet 'gratuity' payment. Actually, some officers have been asking for transfers and/or promotions. Understandably, they wanted to be closer to their family and home. To get that done, they know they have to climb the proverbial '18 step-ladder' (a parochial term describing 'taken to the ultimate task'); each higher 'rung' of the ladder with increasing 'pay-off' rates, all the way up to His Highnesses- the greedy and the unscrupulous ministers. The officers have been on the prowl to find some free money from anywhere to offset the 'incidental cost' burden in connection with the anticipated "transfer order".

Having returned to the station, the constables go straight to the lounge, indulge in some casual chat and loose talk, where a framed picture of the Police minister gravely and demandingly 'looks down' upon them.

"God knows everything; providence never fails! Oh.. ah.. ah.. haaa"!

"Everything happens for a reason!"

"One has to take, to pay another; the evil spreads; spreads fast and furious as a virus. This is the fact of life and is a common knowledge, my dear mates!"

"Look at this! This is just a fistful of money, a pittance compared to the sacs-full our ministers are eliciting. This is only a venial offense, a petty theft! Our sin shall be overlooked and easily forgiven by gods as trivial, a desperate act of survival: He may even feel pity on us; throw some more towards us...like what happened tonight!"

"You say,... no 'hell time' for us! ahhhhh... haaa ha.." One of the constables interjected. Hum... a deep sigh!, "PadBhanaBha! (Lord!) Have mercy on us!".

Another has a 'prick of conscience' "Yeh,...In re-incarnation, you'll come back as a dog or donkey for this offense,... How fitting that will be!"

"Oh, No! I won't be in the same kennel with you; barking all the time; like now; shut up, you silly.I want peace and dignity in the next life at least!"

"Bribe-takers never get any... lose their own self-respect too...you fool!"

In the colloquial dialect and casual mannerism, the constables discuss their 'pragmatic philosophy' in cynical manner. Nobody feels any shame nor harbors any qualms about it afterward. The practice is institutionalized, ingrained, immortalized; prevalent throughout the national culture of the officialdom!

At this time, the sound of slow, steady, perfectly timed footsteps of heavy boots is heard approaching. This is none other than the Chief inspector himself, the top police and law enforcement official at the station, and who successfully led the team this evening. The slouching constables straighten up, stand up erect in line in solemn 'official salute'.

"Stand at ease!" he commands.

"Listen!, I happened to overhear your conversations. Your feeling of guilt and regret is understandable. I feel the same too. He takes up his cellphone from his chest pocket and says "everything was recorded. As we speak, it is being uploaded to the 'cloud' in Switzerland. He shows the phone display, the ".ch" as the country code Top Level Domain (ccTLD). The 'upload' progress bar can be seen still halfway to completion.

"Well, now, none of you have to be ashamed of anything. We are not invited to go to the "Dalal Street" or Rotary clubs and give policy speeches; charge 'speaking fees' which would make everything look noble, decent, and acceptable. Nor, can we issue liquor or other licenses, earning some money on the side. I cannot get my wife and children to listen to my own simple speeches on ethics & morality, let alone other people. We are damned to do this menial job and survive somehow under the current conditions of galloping inflation and living expenses. When we take money, it is called bribes while in the former cases they are euphemistically termed speaking/ consultation/ facilitation fees, etc., etc... deliberately avoiding any mention of the word 'bribery'. I call this bundle of money 'law & order' fees because we responded fast and put an end to the nuisance in the neighborhood. People can now live and sleep peacefully. We render an essential public service to the community- promptly and sincerely. These speech makers have three versions of the same topic, one for the media press release which is for public consumption, the second for the paid audience. Most importantly the third one, also the real one for the backstage, confidentially given to the CEO of the sponsoring organization to help him profit from the real facts. On the strength of the privileged information, he/she eventually earns a hefty bonus from the generous corporate board. This is the "world order" of today's reality, folks!

You have felt that every time you go grocery shopping, the prices increase by the day. This is because of the corrupt politicians and the business people are spending their ill-gotten free money lavishly, having parties in posh hotels where they carry out more of the unholy 'wheelings and dealings'. This causes "inflation", making the lives of common people even more miserable. They don't care a hang! Well, this is the OS (Operating System) of the entire world today. We must play by it, or shall we perish. We are entitled to make a bare minimum standard of living, and for that, we are forced into doing this. I'm not proud of it, nor are you from what I heard from your conversations. Now, first thing first, most importantly, expect the Toddy man to visit the station every 4th of the month for 'being interrogated'; send him in without any fuss. But before that, send away all the visitors waiting in line. For him, no security search is needed; make sure that he is well treated as soon as he is inside my office. Stay a solemn guard at the gate. If citizens arrive, politely send them away by saying the 'Aemaan' (the boss) is interrogating a serious offender and is in an angry mood; if you want to get anything done, 'it's just not a good time!' try another time, sir/ma'am! Come another day, please! I delegate the head constable to make sure that the Toddy man gets enough refreshments. His favorite is 'Parippu Vadai', banana fritter (snacks), and hot coffee. I shall distribute your 'bonus' immediately after he leaves, thereafter, it's 'all clear!'. We shall be open for routine business, serving the public. Understand that clearly?... Yeh,..All of you?".

The chief continues "this one contains dossier and shenanigans of ministers, judges, and officials; kept similarly on the cloud. It is safer in Switzerland. It is beyond the jurisdiction of foreign governments. Besides, they have a penchant for secrecy. It is no secret that it has become the haven for money launderers and scoundrels; despots & tyrants; corrupt heads of states; drug and human traffickers, evil operators of all kinds. All the looted money from war, stolen goods, stolen masterpieces of art, gold nuggets, etc. find their way into the big vaults and tunnels built underneath the Alps mountain in that country, safely kept beside the internationally funded giant atom Collider/smasher to find the 'god particle'. Got that?...

Well, If anything happens to me, it is so programmed to automatically disseminate information to the selected media outlets, Wiki-freaks, 'Murky Leaks', etc. All the focus will be on the big shots. You guys will be spared as the least of all priorities; being at the bottom of the totem pole, have no value or media sensation. I care for your well-being! Your loyalty and allegiance must always rest with me, not to the ministers or the other petty people, called MLAs (Member of Legislative Assembly)! Life may improve for everybody afterward!

The officers nodded in tacit consent and obedient obeisance.

The constables trust the intelligence and the scheme the boss has devised, as they haven't heard anybody saying these things so candidly and clearly. Now, they know that everyone is deeply aware of what's going on in the land. It confirms many of their own doubts and suppositions; clears their conscience off feeling guilt; provides some peace of mind; ultimately, makes everyone feel that there 'on the same track'.

Chapter 3

A typical Evening in the village

All the toddy shops must close at the stroke of midnight. Some of the more intoxicated already experience blackouts; fall into ditches or culverts on their way home; spend the night right there, mired in mud, and stinking. They are frequently awakened by rain, thunder, and lightning; but cannot '*find their feet*' nor bearings; have no option but to remain there, fully soaked and shivering. Early in the morning, worshipers, farmers, and fishermen hear groans and murmurs; find familiar people in unfamiliar places. They identify their own relatives and prominent faces in the community; inform immediate family members just for the reward of gratitude, friendship, or simply being a good neighbor. The unrecognized and unresponsive ones are left alone; walk away from them in disgust, cursing profusely. The previous night, little children had asked mothers about fathers when they were put to bed. Lovingly and tenderly, they were told that their father will return, soon after they had fallen asleep. So, as innocent and gullible they are, they try to fall asleep as quickly as possible!

Now, as they wake up, the mystery of the previous night slowly unravels. They hear a totally different story; feel the hurt and enormous embarrassment; sob and weep in mothers' laps. The families had attempted several remedies. They tried Detoxification, Rehabilitation, and Counseling therapy; spiritual pursuits of charity, social work; offered Novena; "*VazhiPadu & BaliPadu*"; "*Namaskara", "Nishkara*" & "Samaskara" rituals. Out of sheer desperation, some had even sought help from exorcists for emancipation from the wicked witches, demons, and the haunted spells. Some were forcibly administered witches' brew to ingest; had "Koodo-thram" or occult/Voodoo witchcraft sorceries, and incantations performed on them. Nothing worked! All were proven futile. The drunkards continue their habits without any compunction while their families live in shame, guilt, and financial ruin.

Instinctively, the housewives check their jewelry boxes, notice a few items missing. Horrible thoughts haunt their minds; fear that it may have been already sold off, pawned, or perhaps gifted to those notorious persons of '*ill repute*' in the neighborhood. Suspicion of certain individuals afflicts their minds. This is the gold jewelry their parents had lovingly given to them as dowry, exceedingly beyond their financial capabilities. The mothers, in turn, have been keeping them as treasures for the weddings of their own daughters in the future. They rage; impatiently wait for the obvious culprits to come home, confront them right at sight, at the front door itself. The women explode in arguments, quarrels, altercations; fights ensue; hell breaks loose at home. Mayhem, bewilderment, and horrible bedlam pervades; children cry their lungs out, not knowing the exact cause or reason of it all. The pent-up anger ultimately resulting in serious cases of domestic violence, causing visible bodily injuries and bruises. They have to remain out of sight of neighbors and relatives for some while!

Rightfully or wrongfully, in the strange scheme of things and blame game, the mute coconut tree is blamed for producing its sap-nectar that happens to be alcoholic when fermented, as well as for all the indiscretions the toddy addicts are causing to the families!

The family treasure, woman's best friend, the jewelry collection, a major part of the dowry

The coconut is a fruit, a nut, and a seed. Even remotely, there is none like it. The cashew, Almond, Walnut, Fig trees, and date palms must all get an honorable mention for the nutritive value of the nuts they produce. There is no tree as useful and beneficial to mankind as coconut palms in many more ways. There is not a shred of item of the tree that may be discarded or left unused! They teach humanity a great lesson. Keralites take this lesson deep into their souls, lifestyle, and attitude. They (also known as, 'Malayalees' named after their language and mother tongue) are by nature and philosophically minimalistic. According to their ancient teachings, mankind must possess and consume only the essentials, whether they are belongings, materials, or pleasures. This is in great contrast to the modern world where unbridled consumption and consumerism prevail, and pervade; even lauded; devour everything inexorably, totally disregarding the welfare, and necessities of fellow beings and future generations. In this context, the avaricious consumption of food items, trashing excess cooked food after parties, and from restaurants, as well as the voracious consumption of fossil fuels, unbridled activities in mining, oil drilling, hydraulic fracking, deforestation, etc. should rush to our consciousness. 'Being wasteful is sinful'!

It's a lesson taught from the childhood, cherished and practiced thereafter in the course of the entire life!

Chapter 4

Water, Water everywhere !!

......not a drop to drink!

Water is in plenty, not only in the rivers and canals, but also in the land. If one digs a ditch a meter or so deep, water comes oozing out, similar to what metaphorically said about Gulf states of finding oil while digging a hole anywhere, as a reference to its abundance. Here, people have their drinking water from open ponds or comparatively shallow wells. The same water source is conveniently used for drinking, irrigation, laundry, fish farming, and other daily uses. Water quality is seldom tested, therefore suffers enormous consequences. Nowadays, public health officials make casual visits and conduct cursory inspections. Rainwater flows and seeps into the ponds carrying ground foliage, dirt, decay, fertilizers, and minerals with it. Heavy toxic metals occur naturally in soil and found in potentially dangerous concentrations due to human activities. The small body of water serves as breeding grounds for mosquitoes, insects, bacteria, and microbes, causing frequent waterborne diseases in epidemic proportions.

People living along the river banks often construct their cabanas or bathhouses by building granite steps extending into the river. A frame is immersed onto the river bed and heavy gravel is deposited, creating a floor and flat surface. This provides a great method to enjoy an 'immersed', stretched out, full-body bath in discreet privacy, convenience, and safety. The temperature of the water is always comfortable. There's a vernacular saying "the flowing water carries no filth!". The 'Karimeen' or the "silver chromide" fish, find the shade and flock to this shelter giving the folks yet another opportunity for a comfortable catch. Under the sweaty climatic conditions, people take frequent dips in the water. It opens up the pores in the skin, making it absorbent of fresh air; feel quite refreshing. In the process, they get a natural, gentle massage from the waves too. But, now being educated about pollution and its ill effects, people have, quite reluctantly, moved this facility indoors for the comfort of 'running water' under the 'rain shower' faucets.

Typical Kuttanadan scenes: the ubiquitous water, affinity to water, and it's use in every way & walk of life

Every household generously offers its share of easements for walkways and waterways for public use and other conveniences. Most of the essential infrastructure is built by "Srama Danam" or volunteer community work. They maintain a very informal and casual but cordial, caring neighborly contacts. Neighbors are invited for social and family functions. When in need, they borrow and share items and resources. Objections and disputes are settled by consensus or advice from respected 'elders' in the village. The relative physical isolation of communities into small villages make villagers mutually dependent for many mundane things. Good education makes people cultured, tolerant and congenial. This arrangement, of course, lends itself to intrusion on privacy, prying into other people's affairs, rumor- mongering, and jealousy of prosperous relatives, and neighbors.

The peaceful ambiance & natural rhythms of simplistic lifestyle in Kuttanaad

The lifestyle is mostly leisurely and casual. Everyone is known to each other. There is only one of each facility or institution available in a village, whether it is a type of school, dispensary, place of worship, shop, bank, or playground for people to interact, make acquaintance, or do business with. People of different walks of life live in the village make it an inclusive community of cooperation and caring.

People are not pressured to do anything beyond their normal needs and capacities. Not everyone has to file taxes and be concerned with tedious rules, clauses, exclusions, and limits for tax deductions. They don't have to deal with numerous insurance forms and documents - Life, Home, Auto, Medical, Drug, Fire, Liability, Hospitalization, etc. that need legal and expert help to file in other places. Voting eligibility and registration are universal rights; just simple and routine procedures. Police and people need not fear each other as neither carry gun or taser; armour or ammunition.

Lives lived in simplicity and with minimum fuss!

Chapter 5

The Vast Kuttanad Basin

Kuttanad finds its place in the lowlands in the Vembanad lake basin that lies 1–2 m below mean sea level

The low region of Kutta-naad lies in the middle of the gigantic Vembanad lake basin, which is contributed by as many as 10 rivers and/or tributaries. It is the longest lake in India covering just over 2000 Sq. Km with average depth of 12 m, dwarfing most others. The river streams, bring primary soil nutrients; phosphate, nitrate, potash, secondary nutrients of calcium, magnesium, sulfur, and the micro nutrients (collectively called "Ekkal" in Malayalam) from the highlands, and make the lowlands fertile, though a great bulk of it washes away into the Arabian Sea. Some of it becomes settled on the bed of the basin, making it shallower as time goes by. This is conducive to new species of algae (Payal) which clogs the water-ways, blankets the surface cutting off vital sunlight & oxygen supply to the fish population killing many. There's thirst in the water too! Besides, it contaminates the water. Little action is undertaken by the state to remedy this effect. Many practical and technology based solutions are extensively discussed in this book in a subsequent chapter, "A futuristic Master plan". The inhabitants conserve fresh water and natural resources that in turn preserve the extraordinary diversity of species that rivers, lakes, and wetlands can sustain.

Quaint, pristine Kuttanad in its natural rustic splendor, and sublime simplicity

Chapter 6

The Monsoons & floods

During the two distinct Monsoon seasons, the waters become rough; the waves lash the banks; the monstrous ones can rise by several meters. It is also the annual flood season, inundating vast contiguous territories, marooning communities, causing disruptions in life, heaping miseries one after another, and a lot of damages in its cruel reign. The originally built causeways across some of the fields which were since upgraded to roads lie submerged in water, disrupting road traffic. Some vehicles are seen abandoned here and there on roadsides. The region is transformed into a wide, turbulent lake with water continuously rising with the rain. The torrents of on-rushing drain from the slopes of the highlands also contribute to the rise in water level. The Kuttanad region is transformed into a basin, collecting water. Arguably, the lake becomes the largest one in the world for this period of time.

It bares the need for dredging, which has never been carried out, except during the period of large-scale reclamation needs and occasionally for some manual work for personal use. Despite the damage, everyone forgets the urgency for water/flood management, soon after the water level recedes after a couple of miserable months. During the flood season, the food stores become depleted fast. If anything is available at all, it cannot be cooked or prepared for lack of fuel or other essentials. Drinking water becomes contaminated. Poverty rages rampant; becomes excruciating; people struggle to conserve strength by avoiding non-essential chores. Usually, some horrible epidemics or diseases break out in the aftermath.

The tall coconut trees are the only visible landmarks that identify strips of land from water. During the flood season, when we were heading home from somewhere, all we could see was water as far as the eyes could see; the horizon appeared in the far, against the palm-lined strips of lands, cluttered together; could hardly distinguish one from the other. Thankfully, during the daytime, we could spot our own tiny hamlet recognized by a lone, age-less 'date palm' tree, that looks like a California fan palm standing tall. It is the only date palm in the whole region. According to elders, it was there since antiquity. It serves as a unique, unmistakable landmark! From the distance, it appeared like a tree standing erect, unyielding to the storms, on the surface of the water. We learned from the Science classes that the water takes the shape of its container and that it cannot be 'bent' around. It may appear to be bending when thrown out of a glass or forced through a hose/pump due to the pull of gravity. At this juncture, it was difficult for us to visualize that the earth is round and spherical, although this fact was known from the time of Pythagoras in the 6th century BC. We also could not believe in Nicolas Copernicus's theory and its confirmation by Galileo Galilei that the earth and other planets in the solar system revolve around the sun. We, in our simple minds or as taught by the church, argued that, if that were true, we won't be able to stand; will be constantly falling down with the rotation of the earth!

Our childhood suppositions didn't go far enough into the realms of Sciences!

Oftentimes, I think I haven't yet outgrown my powerful childhood naiveté, notions and myths.

We were led to believe that these astronomers deservedly earned the punishment and imprisonment for derision, heresy, and hoax until we saw satellite photos of earth and outer space. There was immediate consternation, an unpardonable grievance, a crisis of faith, and a question on the claim of '*divine infallibility*'. We asked among ourselves how many other myths and falsehoods might have been perpetrated on gullible believers over the years gone by. How many more still remain rooted in our consciousness as indelible?. Naturally, the report of the Russian 'sputnik' spiked my interest in space. It seems that it traveled through my DNA, judging from the fact that my son became an Aerospace engineer-without any kind of persuasion or guidance from me whatsoever!

It's a terrible thing to hijack people's minds, brainwash, and stuff them with tall tales & lies as divine revelations. Such minds that are wrongly inculcated, drilled, and riddled will be hard to be ever fully liberated and become capable of 'true thinking'. Religions, royalties, monarchs, and corporations do not ever apologize for their mistakes. It's beneath them as they suppose it will erode their influence, status, invincibility, infallibility, and credibility. They don't want to be disproved by themselves! Periodically, they tell the faithful to make the 'sacrament of confession' showing perfect contrition and true penitence for the salvation of souls.

It's perplexing "Why do they refuse to do it at least once in human history, for the atonement of the grave 'crimes they committed against humanity'?".
'Truth & reconciliation' is necessary to maintain credible justness, and rectify colossal faults!

Gathering dark clouds and lashing winds herald impending doom and the onset of the menacing Monsoons

The fishing birds have all long gone. They don't possess the skills to fish in severely agitated waters as the seagulls and cormorants can. An abandonment, isolation, and eerie apprehension prevails in the region. The horizon, which looked distant, seems much closer now. It is difficult to imagine that normal life is going on beyond it. The only transportation possible is by Vallams. The water is now so deep and the current becoming stronger than ever that even an extra-long pole is of little use in punting/navigation. Paddling is the only option; consequently, the Kochu-Vallam (single-seater/paddle-boat) becomes the queen of the waters and becomes an essential means of transport. The tiny thing, the manned

water scooter, is recognized for its place in the water festivals of today. This also exemplifies the affinity of '*Kuttanadiens*' to the Vallam.
Caution! Before climbing into one of them, one has to be circumspect; must first look skywards for any sign of imminent storm or rain; check the waters for large waves before leaving home! Because it can easily rock the boat and topple you into the water. The treacherous current can curtail your life, my friend!

The giant shutters or gates of the Thotta-Palli '*spillway*' to the Arabian seas cannot be open, as the sea is raging and the sea tide is at higher level. For the same reason, also the ones in "*Thannir- Mukkam Causeway*" bridge across the embouchure of the Vembanad Kayal and the Arabian Sea remain shut. Many poisonous and stinging underwater creatures migrate to the area. The snakes and other reptiles desperately search for scarce food, prey, and shelter. The nocturnes pursue them, because their territory is violated. The disparate living beings dare to enter dwelling places, scaring people, causing frequent commotion and screams at homes. [Frightfully, my mother often recounted that a poisonous snake crawled over me and between me and my brother while we were her only children and sleeping side by side as infants.] The schools are closed. The children cannot get to schools; most schools in the region are submerged anyway.

In most countries, floods are relatively a short-term event. Water cannot stay still. Its mass carries both potential and kinetic energy with it. It has to flow and therefore washes away fast. But not here! It's a season, a season that's long, unkind, and stubborn. Over most countries the lightning strike on the average above a km high; stretches 15 km long, and lasts a second or so; but here it seems to occur at lower levels, and last longer than a second. [South American countries of Brazil, Uruguay, and Argentina record higher numbers] Satellite tracking technology alone can ascertain the actual measurements. People find themselves helpless against the ferocity and brutality of these natural forces. It commands respect; instills fear; makes people revere nature and God, not knowing who's doing this to them!

The rainbow appears over the horizon in the shape of a giant arc, over everything visible, both ends appear touching the waters, its reflection makes it look twice in size; offers a rare sign of imminent respite. When the sun gleans faintly through the thick dark clouds, people become sanguine, only to be dashed moments later. The '*heavens open up*' spilling an abundance of unwanted precipitation. People send desperate prayers and petitions, many complaints too in the form of hushed curses in the opposite direction of the rainfall, hoping they could interdict the rains.

Nature has the ultimate sway; in an autocratic way!

The inevitable floods rage mercilessly, following the irrepressible Monsoons

The principal occupation of the people is rice cultivation. According to the phyto-genesis of rice, its cultivation dates back to more than 5000 years, continuously providing food for humanity. In Keralam, most families maintain coconut palm groves, banana plantations, vegetable, and fruit orchards too with an abundance of assortments. Then, there are fruit trees of rich varieties of Mango, jack fruit, and several shrubs bearing different berries and fruits. Some people go fishing. A rich assortment of fish species live and thrive in the natural and pristine waters and in the constant currents. There is plenty of food for them to feed on, grow, and spawn. One may casually throw a net and retrieve it a little later and come up with enough fish for a curry that goes well with rice and vegetable side-dishes. There is no meal in Kuttandu without fish. "If you can catch a fish, you won't starve!" No wonder, all households have one or more fishing nets to suit different fishing methods.

Some rich sweet water fish varieties commonly found in the Vembanad lake basin

A 'catch' is not very difficult and frequently attempted using various techniques. "Kari-meen", (green chromide) the official state fish of Kerala, possibly belonging to the Tilapia family; 'Varal' a type of Carp, 'Kanambu', a Mullet, 'Koori' from Catfish, "Valah" from Swordfish families are all indigenous, well-known fish varieties; command good demand in the fish market. As the

islanders around the world, Filipinos, Japanese, Indonesians, Norwegians, the people here also rely predominantly on a fish diet. That is their main protein source; contains many vitamins and minerals as well. Like meat is to most Western populations, fish is to these populations and the islanders. The ongoing and controversial 'Whaling war' must be viewed in this context. A day's meal in Kuttanad will have invariably included fish. It may be apt to mention here that the Malayalee expatriate community in the Gulf countries begins to show a higher incidence of cardiac problems when, conveniently, they change their eating habits to the 'meat-based diet'. Their diet has been altered because of reasons of the relative cost-savings and convenience in cooking. According to a Keralite medical researcher, this is due to the fact that their genetic and metabolic constitution fail to produce, adequately, the required enzymes to break down the meat particles. In the Gulf countries, beef is imported from Australia, New Zealand, Ireland, and from anywhere; while mutton and camel meat are indigenous.
He cautions: "meat must only be consumed only sparingly!"

A traditional village household in Kuttanad strives to stay self-sufficient and self-sustaining. They try hard to avoid having to pay recurring costs of monthly bills of rent, utility fees and the like. They cultivate their own parcel of land for the essentials: rice, produces, vegetables, banana, mango, vegetable and other fruit orchards; also a variety of edible roots, Rhizomes (rootstock) such as tapioca several food & grocery items. Tapioca, belonging to cassava/yucca specie, probably brought by the Portuguese, is usually planted where ever the soil is dry and poor. It is used as supplemental food to rice. Sometimes, it's nicknamed 'poor man's food' as it can provide only Carbohydrate and no significant amount of other nutrients. Nevertheless, it is a staple food for many in tropical countries, avoiding famines. This tropical land is blessed with biodiversity of many species and subspecies of trees, plants, animals, and crops. The cows and cattle graze safely on their pastures, provide milk and several dairy products for the family and manure for the orchards & fields. 'The cows give 'more than it takes in' (ignoring the green-house gas, methane!). They are gentle and considered domestic animals. People take passionate care of the cows. They are consciously aware of its favored place in the Hindu religion, but are not fanatics like many in North India. In Keralam, currently, there is no vexation towards 'eating beef'.

It must also get a due mention in this context, that some farmers in the highlands raise pigs as a secondary occupation and income. The pigs are fed with table scraps and surplus garden vegetables. The farmers consume enough; prepare brined and smoked hams & bacon for later use; sell costlier parts such as sirloin, tenderloin, loin roast, loin chops, and the Boston Butt or shoulder, from "*High-on-the-Hog*". They make "*Vindaloo*", also a Goan delicacy, marinated in vinegar with ground onion, ginger, garlic, mustard seeds, cumin seeds, cinnamon and cloves, also plenty of curry leaves for a leafy flavor. They make sausage, naturally smoked hot dogs, and other delicacies out of it. Pork is a rich source of 'thiamine' which is essential to break down carbohydrates into energy. The body and brain need the same for their stamina and vitality. (The vegetarians may dispute it vehemently. Generally speaking, they claim to exhibit sharper mental faculties!). Besides, pork meat is good for the muscles, and nervous system functions. (keep in mind that the heart is also a muscle besides being a pump!). An interesting factoid is that 'Schweinefleisch' is to Germans, as beef is to their counterparts in Western countries. (*In Detchland, das Rindfleisch ist auch teurer; und die Brätwurst schmeckt gut!*). Now, pig products find uses in many households, industrial, pharmaceutical commodities. For a certain religious group, the consumption of pork meat is taboo. In the colonial days, the use of pork products in the British army started the infamous "Sepoy revolt". The British were notorious for fomenting religious conflicts in secular, colonized countries. It's' among the cornerstones of their sinister policies of "divide and rule" doctrine.

Meat products contain calories, protein, iron, B_{12}, also some of the most undesirables- fat and cholesterol. Bacon gives much higher calories than most other meat. Cricket fans in India lament thc 'religious ban' on meat products, as they become aware of the strong relationships of physical strength, stamina, and the 'animal spirit of competitiveness' associated with the meat diet. They want their pace bowlers bowl faster and the batsmen hit more 'Sixes'! Their cricket nemesis Australians have proved this point. "You're what you eat!'. It's a well-known fact! Some ambitious Indian players secretly indulge, despite the severe repercussions they could face, if caught in the 'act of eating'!

People usually demand others to take risks and do the things, for the things they want to leisurely watch, and casually enjoy!

Poultry is leaner than most cuts of beef and pork; Turkey provides the highest amount of protein among poultry; ducks and chicken take succeeding ranks. Fish is a lean protein alternative to red meat; besides being rich in omega-3 fatty acids, which support brain and neural development.

A traditional self-sustaining model of a household practiced for many generations

Most of the households maintain an animal farm of hatcheries of chicken, ducks, fowl, rabbits, and fish, thus needing a minimum of things from the grocery shelves or external markets. It's a routine chore for housewives. They can stay at home, take care of children, mind other homely business, above all, earn or save money right from home, sparing the hassle and time in the shopping and carrying goods home.

An effective and clever '*hedge*' against galloping grocery price inflation!

Occasionally, health hazards involving infections from animal to human do occur. Public health professionals assert that the pathogens (any disease-producing agent, esp. a virus, bacterium, or other microorganisms) are transmitted by the nocturnes such as bats and reptiles those prey on such sheltered domestic animals; and also due to poor hygiene in, and around the cages. They make house visits only when epidemics erupt; recommend basic precautionary measures. Their job is finished!
Inflation is a severe health hazard too!

In essence, a parcel of land is transformed into an abode, a farm, a barn, a pasture, a field, groves, hatcheries, a pond, or well, a yard with a flower garden and orchard. Most sheds and structures needed for this purpose are constructed out of the readily available, renewable resources from the yard or plantations, such as wooden posts and other coconut materials. The people use traditional skills as well as the trusted & tested methods developed over generations. Almost everything is recycled, or find good use somewhere.
'*Renewable resources*' is not a new catchword in this society.
Though not quite glamorous, it's certainly an ingenious and amazingly sustainable life style!

The highlanders are relatively more affluent and prosperous than the lowlanders. Somehow, that seems to be the case the whole world over! A German folk song "Drunten im Unterland, Da ist's halt fein!" comes to mind. The soil is much more fertile and stable. The graded slopes are fortified with "Kayyala", (retaining walls) of cobbled stone barriers to protect soil erosion and preserves soil nutrients. Thus, the hillside is turned into a series of graded *terraces* for farming, where high-yielding Rubber trees are planted that provide a rich yield for about 25 continuous years with moderate cost for 'tapping' expenses of rubber milk. Consistently, the rubber products command good international and industrial demand; fetch a premium price. Unlike low-lying Kuttanad, they need not have to pump water out of their fields, as they are never submerged. Also, they don't have to deal with the destructive floods. The rainwater quickly glides down the slopes, leaving only a few puddles here and there. Sometimes the rivers in the area swell and cause temporary problems. The soil is also very absorbent. But, periodically, they have to face severe dangers and damages from mudslides, uprooted trees, piles of falling scree (broken pieces of rock). This is caused by the bursting of earthen dams in the hidden hilltops at very high altitudes. The dams may not be anything nearly comparable to the largest earth-filled *Tarbela Dam* in Pakistan. Here, there are a number of scattered,

naturally formed ones of sufficient volume enough to cause disaster. Occasionally, heavy thunderstorms or some natural phenomena denude (make bare) the buried rock formations and push them down the slopes without warning in great speed and force causing severe damage. In fact, no one knows the exact number, or sizes of even the visible ones, as there is no government survey conducted at these places and heights. Most of the explorations are carried out by the homesteaders and enterprising farmers with only rudimentary tools, measuring instruments, or meager individual resources. Government exists primarily to collect taxes and spend them on unworthy projects. Plainly, a 'drone /helicopter survey' with high-resolution video & thermal imaging, and mapping initiative is absolutely necessary at these locations. The stationary geodetic survey satellites are the best suited to perform such tasks and monitoring for the entire country! There is a lot of available technology that can be applied. Again and again, it emphasizes the fact that people will have to elect capable, technology-savvy, knowledgeable, leaders with problem-solving abilities, initiative, and vision; above all, a sense of duty to the elected office.
It's to be pointed out that in Kerala, every inch of land is cultivated and fully utilized! Therefore, it must be praised and promoted as a prototype for "sustainable habitat"!

It is imperative that the world adopts some of these proven methods to mitigate the depletion of natural resources caused by the reckless and voracious consumption that support wasteful and unsustainable lifestyles and traits of modern civilization. If not, the ramifications will be unimaginably unforgiving, damaging, and brutal to the point of extinction of the entire human civilization. Worse of it all, that would likely happen only after gradual and prolonged periods of sufferings; and miseries periodically inflicted by the 'controlling mechanisms' of nature.
Thereafter, the 'ending' may occur rather suddenly with no notice!

Chapter 7

The Epic Saga of great human Endeavor!
Reclamation of land from the vast lake
For Rice cultivation.
A glowing tribute to the forefathers for their fabulous legacy

The Vast & indomitable "Pazhukkanila", where the reclamation from the lake basin started

The region of Kuttanad in the South Western most state of Kerala is waterlogged, the lowest lying region in India stretching over 100,000 hectares, half of which submerged all year round. It's known as the "Rice bowl of the state for its lion's share of rice production. Water and Kuttanad are synonymous in character and significance. The inhabitants of Kuttanad of about 1.75 million live in small hamlets, often separated by waterways or paddy fields of large acreage.

Within the boundaries of the tiny villages themselves, there exist a number of streams, canals, and water-breaches. Each one is surrounded by paddy fields. Farming is the principal occupation and livelihood of the residents. A variety of products and cultivation of minor crops on a dwarfed living-dry-land supplement a modest extra income. The paddy fields here are mainly carved out of the 'reclaimed' lands lying on an average of 2.5 m below MSL (Mean Sea Level) protected from the frequent tidal surges, cyclonic storms, and monsoon floods by man-made earthen/mud barriers, locally known, simply as 'bund' consisting of a series of dikes, levees, embankments, and causeways.

In a move to broaden the state's tax revenue base, sections of the unused barren lake basin were auctioned off to bidders who were deemed able to pay taxes, initially as leaseholders. They were promised permanent ownership upon reclamation, and development for cultivation. Such a seemingly attractive deal was full of perils and ruins. The ownership clause was an integral part of the offer and absolutely necessary so that bidders could be enticed into the proposed deals. Except for some exotic varieties of fish, the lake offered nothing else. Twice every year, during the two monsoon seasons it became extremely turbulent, violent; became indomitable, caused floods, havoc, and destruction. There were few takers. Only a couple with extraordinary vision, doggedness, financial resources, and project management & technical expertise would attempt such a huge undertaking of an entire lifetime and beyond. The big ideas, in order of importance, were, tax revenue, mitigation of poverty, work for farmworkers, employment opportunities for skilled workers, business opportunities for moneylenders- banks, loan sharks, small industries, and last but not the least important, the avoidance of political unrest. A significant improvement in the state economy and prosperity was envisaged without the government having to find and invest any sum of money and other administrative resources, or technical know-how. In reality, it lacked 'all of the above'! It was a clever government strategic plan, the earliest of a massive 'public-private enterprise' scheme, that costs absolutely nothing at all for one of the participants! How wonderful! The sole role of the government was to facilitate and control without staking any resources. A handful of adventurers and entrepreneurs including my own ancestors as pioneers, teamed up to form alliances, joint ventures, and cooperative societies. Money was scarce. Banks wouldn't lend for barren land or non-income-producing assets. They have never done that in its history and will never ever. But some individual loan sharks were only too eager to exploit the situation. They also had the ulterior motives of bringing these dashing and adventurous people from another cast and beliefs eventually down, as they were seen as the firebrand of people who will soon usurp the power and influence from the established, elite social class. They stipulated predatory lending criteria; demanded exorbitant interest rates, with the additional pledge or lien on living home, prime Palm groves, gold & other assets worth many times exceeding the value of the loan amount, as collateral. The greed and deceit were at its feverish peaks! Without a doubt, it was a '*Lenders market*'; Candidly, a 'loan sharks' paradise; borrowers' hell!'.

In essence, these lenders outdid the greedy and self-proclaimed '*too big to fail*' Wall Street bankers who caused the Year 2008 worldwide stock market crash, and unabashedly earned hefty bonus packages immediately after using the "free" money generously provided by the Treasury (Reserve Bank) for rescuing the economy they wrecked and making many hapless homeowners instantly homeless! They are still reigning supreme as 'wealth-builders and visionaries on the Wall Street without any compunction, or consequences whatsoever.

After several prolonged struggles and setbacks, the entrepreneurs managed to raise enough funds, recruit manpower and devise schemes and techniques; finally embarked on this risky and precarious venture. Firstly, swaths of relatively shallow land were marked out from the vastness of the lake into several disparate (distinct size and contiguity) blocks with tall bamboo poles. The blocks otherwise locally known as "*Pata-Sekha-rams*" (Blocks of paddy fields) were of varying sizes, the largest being the '24,000' acres, in direct view from my childhood ancestral home. The mean average was only about half of that. Each block was separated by as much as 175 m of a buffer zone for the water to freely flow around its periphery. This scheme limits the water pressure to a somewhat lesser magnitude, giving the bund a chance to withstand the critical monsoon forces of thunderstorms, torrential rains, floods, undercurrents, tidal pressure, thrashing waves, and still able to survive. They hoped the water, as usual, will take a 'path of least resistance' along the designated, alternate pathways without challenging their '*bunds of mud*' (dikes-An embankment of mud & soil built to prevent floods). The canals, besides being used for transportation of men and materials, besides using them for extracting mud, dirt, and bulk materials for the embankments. As a first step, bamboo poles were erected at 5.0 m apart to identify, and carve out a suitable area of size. Then, they added a second row of bamboo poles for the measurement of the width. That would be the exact width of the dike/bund.

They brought homemade bamboo mesh, made like slats of '*window blind*' and tied them to the erected poles. The 'frame' was further reinforced by 'coconut/palm wood piles' in between. Having made these leaky compartments in the water, the next step was to compact them with mud extracted from the lake bed itself, dug from both sides of these barriers. Hundreds of people toiled over a span of more than two generations to build the earthen boundaries with their bare hands and primitive tools made by the village blacksmith in his modest smithy or foundry. They created tools out of scrap metal and iron; forged, hammered, bent, cut, and shaped on the anvil and fitted them with long, smooth wooden handles crafted by the carpenter.

As an interesting factoid: Today's industrial manufacturing is generally called "Subtractive manufacturing", as pieces for a finished product are cut and assembled from large sheets of material, as opposed to the "Additive manufacturing" (AM) in the latest '3 D printing' technology by depositing layer by layer. Understandably, it minimizes waste, enormously!

The resourceful, indigenous people with strong hearts and expansive lungs dived deep into the bottom of the lake, brought the "*Katta*", the extracted mass of solid mud-cake, scoop by scoop, heaved one by one into the large "*KettuVallam*" or work boats. They staggered on the way up, briefly collecting breath and balance on the multi-tiered bamboo stilts (an **H** shaped rudimentary scaffolding of a sort, the platform made of just a piece of bamboo cross-piece) temporarily planted at the spot. They worked from either side of the work boat to keep it in balance, preventing water from entering and capsizing the boat. The weight of the Katta feels manageable moving through the water (a hydraulic phenomenon), but requires much stronger effort to heave it out of the water, clearing the "gunnel" (the upper edges of the boat's hull, the body of the vessel) and dislodge into the work boat. They used both hands and shoulders in the process. Each scoop deepens the floor, making the next dive even deeper! When the depth became too deep, they transplanted the scaffolding to a nearby location;moved the boat, and repeated the process.

Later, they maneuvered the boats propelled by long and light bamboo poles to bring and deposit the mud in the 'frame partitions' that another crew had constructed previously on dry-land. It is to be mentioned and greatly commended that every item used was of natural, renewable, and locally available materials. It reduced the cost considerably, though during the initial development stages it proved no match for the ravaging floodwaters.

Manual dredging bring mud-cake from the bottom of the lake, a rudimentary step in the reclamation effort

The forefathers knew right from the outset that it would take a very long time and continuous maintenance effort to strengthen the dike well enough to equal the forces of water & floods. Cruelly enough, the great flood of year 1923 (Malayalam year '99) set the clock far back for them by several years. In reality, the time and events proved it to be an enormously arduous and treacherous path, fraught with many unforeseen adversities, levy breaches, and loss of entire crops. The farmers bore the brunt of the entire, untoward devastation, while the government amusingly watched from the state capital far away from the spot.

The landlords gave the workers prevailing fair wages and served them hot meals with rice, their favorite '*Chem-meen*' (krill/shrimp), boiled tapioca in flavored coconut paste when they returned fully exhausted from the grueling morning shift. The cooks at home looked out for the laborers, spotted them from as far away as the eyes can locate. They prepared the food items ready to be served hot as soon as they arrived. When the children of the workers came to see and greet their fathers, they were also invited to join in and enjoy a good meal and intimate family time. Most landlords treated them as committed people and dependents of the family-owned landholdings. The system bore striking resemblance to the landlord-'vassal' relationship prevailed in the feudal system, only that the folks were looked after with greater generosity; giving them a place to live; a small parcel around their dwellings to cultivate, and reap a harvest for themselves. As witnessed by my own family's dealings and interactions, some were more generous, meeting their family expenses (births, weddings & funerals), medical needs, supplying materials for thatching their huts, providing financial aid, and sponsoring their cultural and social events. When the floods inundated and ravaged their living huts, they were relocated to higher places; the most trusted and the loyal were allowed to stay at the landlord's premises, out-houses, barns, house verandhas, and wherever suitable under the dire circumstances. Obviously, the landlords benefited from the homage and fealty. By all accounts, it was a social, humane relationship; mutually beneficial, and generously altruistic.

Little by little, through the concerted efforts and arrangements, they built up tens of km long earthen barrier, encompassing many thousands of acres of newly arable land in each block. It was inconceivable to build them until it became a reality one by one in great astonishment to the local people as well as visitors from the highlands. Today, there may be several Thousands of such 'Pata-Sekha-rams' of various sizes in the greater Vembanad region. It must be reckoned as one of the man-made '*wonders of the time*' in ways, magnitude, usefulness, and human endeavor. Regretfully, the greatness and impact of this extraordinary feat on the economy and welfare of the local population are now almost entirely forgotten. There is no commemoration or event to celebrate and laud the great achievement of our ancestors. Every generation needs to be motivated and encouraged to excel themselves. It must be inculcated and ingrained effectively in the minds and history of the nation.

Descendants of the pioneers searched their attics and rummaged through the boxes and furniture drawers, hoping to find documents describing any of the plans/sketches or schemes the forefathers had devised and created. No one could find anything at all. They assumed that the plans were possibly drawn then and there in the mud with a stick.

Alas! When the floods came, everything was wiped off!

The local culture and practice seemed to be averse to documenting anything. Everyone, including carpenters, electricians, mechanics, and others do keep and figure out everything in their deep depths of mind. Others are left to find out for themselves, only after everything is completed! This is in contrast to the modern day project undertakings, where everything starts with a written plan, drawn sketch, or diagram; time-line schedules, Bill of materials, manpower & cost analysts, budget, funding source, etc. The great examples I cite and mention are actually from the time of Leonardo da Vinci and Pythagoras. One has to assume our ancestors used similar methods and techniques as the two masters did, though they didn't leave any traces for us! I have witnessed my maternal grandfather having several voluminous ledgers kept securely in his cabinet. He had personally updated each and every transaction by himself before going to bed. He had one for family affairs, recording important events, births, astrological sign, details of education, wedding, etc. one each for every category. I know this for a fact, as I was allowed to sleep in his large bed until I was 12. By the time he arrived, I was already deep asleep! Surely, it was in the habit and practice of the time to carefully record everything. Prior to the times, treated date palm leaves and a sharp needle were used for writing which of course could not withstand the test of times. It baffles me quite a bit as to what had happened to all the documentations of such a major undertaking; quite possibly, the government officials confiscated them in one of the foreclosure proceedings!

The Arabian Sea merged with the lake at several locations, as the water needed to flow between the reservoirs. The thrust from the tidal waves from the massive sea is enormous. The barrier embankments were quite frequently overcome and overflowed by the combined onslaught by the salty tides emanating from the sea, and the torrential monsoon floods. The dike/bunds were constantly under attack, tested to their limits. In other instances, breaches were caused by the lashing of storms and thrashing by humongous waves following thunderstorms. Emergency crews of task force kept vigil day and night with their scant resources and modest 'know-how', but with strength of grit, courage, and determination to stem the tide; bridge the breaches; save lives; despite encountering dangers and hardships. Besides, lightning strikes are dangerous common occurrences. The packs of dark clouds carry varying amounts of electric charges and '*potential differences*'. When they run into each other, the air gaps between them rapidly break down, discharging massive bolts of electricity. The suspended conductive particles in the atmosphere pass the current to earth and bodies in between. The lightnings can produce currents up to several thousands of amps, exceeding limits of fatality of all the living beings, when hit. In addition, they could heat up the immediately surrounding air to incinerating level. Bodies of water, trees, and other objects attract electricity in varying degrees, depending on its characteristics of conductivity, mass, and intensity of the bolts. Out in the lake, there is no place or shelter, barely exposing everyone. The indomitable, indigenous tribes who constituted the bulk of the labor forces worked in the forefront, bore the brunt of most of the perils, and dangers at work. In my faint memory of early childhood, I recall my father and grandfather coming home many times in the middle of the nights, drenched, shivering; bruised, shaken, hungry, and sick needing immediate medical care and intervention. The moans, groans, and commotions would wake up even a blissfully sleeping child! While the servants and my mother took care of them, I simply stood there crying inconsolably, failing to grasp the gist of all; thinking perhaps they were attacked by crocodiles, or robbers, and that they won't be here tomorrow to cradle, caress, or care for me.

The "*Pzhukka-Nila*", the innermost region of the lake, also close to our ancestral home, where the first reclamation project was started, was infested with poisonous snakes and freshwater crocodiles. As work uniform, the divers wore just '*thongs*', skimpier than those of the female beach volleyball players. It was permissible in the very remote and lonely location they do their work. Before diving into the waters, they rubbed their bodies with specially formulated balm of repellant, like the farmers treat plants for deer resistance. They were the '*unassuming & unsung*' heroes of the era and one of the protagonists in this great endeavor. The farming communities in Kuttanad owe enormous social debt, gratitude, and admiration for them. This is still the single most important historical venture in Keralam as it is directly related to the "*essentials*" of human life. It lifted millions of people out of poverty, starvation, destitution, deprivation, and famines. During the harvest time of about 5 weeks, the remote fields suddenly become alive; vibrant, brimming with joy and vitality. Laborers from far and wide would flock to the area and lay camps on the embankments like the migratory birds do before the Winter. They move their tents from field to field and reap the golden grains of paddy from the sprawling fields of Kuttanad, one at a time. Men & women, young, and old, labor in earnest and enthusiasm. It has been a great collective effort that will ensure their food sufficiency for at least the next 6 months. Indeed, it was a great sight to behold and join in the happiness of the fellow beings. Regularly at these times, our parents took us to see the great crowd, as everyone work in a festival atmosphere. We were very excited to see and talk to the new people who arrived here for the occasion; often fascinated by their accent, mannerisms, and cheer. Some would give us hand-made gifts shaped out of the reaped products, a blow whistle carved from the hollow stem or things brought from their native hamlets, or sweets they cooked. Every one can readily witness the remarkable trans-formative changes this project has brought about. It remains as the only place in India where cultivation is taking place below the MSL, and that too at a staggering depth of over 3 m below MSL!

This endeavor was the **human equivalent** of the epic reclamation of Keralam from the bottom of the sea by '*Parasuraman*' with his amazing magical and divine power!

This great legacy of ensuring the sustainability in '*food production & provisions*' by the ancestors is now forgotten and gravely mismanaged by the successive incompetent governments, but the people in the region still regard it as a remarkable feat. Irrespective of the merits, everything in the world and life fade in the passing of time. Reasons are aplenty. Perhaps due to the dynamics of nature, constant motion of the planets, needs, necessities, discoveries, technologies, and inventions that bring forth changes in importance in current priorities, lifestyles, attitudes, and perspectives.

On successful completion of such work of great magnitude and consequences, and due to the scarcity of similarly enduring undertakings, the skilled tribes had to disperse to various parts in search of work. It was quite heart-wrenching and deeply painful for my own family to see them depart, as it lacked the means and resources to support or keep them engaged, especially bearing the heavy burden of mounting debt and the constant harassment from the creditors. There were many tearful moments for us, as each dependent family bid goodbye, choosing their own time and occasion to pursue their individual destiny and livelihood. We were very much part of each other's lives, interacting on daily basis, sharing common struggles and destinies. As I recall these painful events at this time of writing, I am totally overcome with emotions of sorrow, pain and intense nostalgia. A strong bond and relationship cultivated over several generations were thus broken, although the descendants of the times still maintain occasional contacts at major '*life events*'. This gives us precious opportunities to share each other's life stories, and other mundane matters. It also helps us to reconnect missing puzzles and links from the past to our present lives, as well as getting an authentic account of events of the time; one filling the other making some sense. It felt like reconstructing a lost civilization of the past.

My earliest memory of seeing the VallamKali is when the same tribes celebrated their annual holy "*Makam*" festival, which my family regularly sponsored. As young fans, we rooted for our sponsored race boats and all the familiar faces paddling them. We're elated when they greeted us from their boats, making us feel that we were special indeed and important to them. I also remember that their arch-rival in the competition was the "Ara-ya" fishermen community. The fishermen were also closely well-known folks in our hamlet. They used to bring their catch of lobsters, prawns, shrimps, and mullets early in the morning from their nightly fishing expeditions braving extreme weather. It seemed that the most prolific catch was possible at night and under hostile weather conditions. Perhaps, the fish never sleep, always want to nourish themselves. Staying in the water and moving around cost energy too! Both these tribes were hard-working people, skillful, adventurous, and resourceful. Each served the community well, mutually dependent; earned much recognition and praise for it. The fishermen community became a displaced community at the advent of Pata-Sekha-rams due to large reduced areas of available fishing waters. But they were quickly absorbed and integrated into the farm labor. It was another great gesture and accommodation by our forefathers. Thy did always keep the interests and well-being of the fellow beings in their decision-making, and very next move. Obviously, the tribes had multiple skills, and capabilities that could be readily adapted and utilized in the agricultural sector. They were strong, hard-working, adept in water, resilient, and an adaptable group of people. Above all, at the moment, farming was better rewarding. When there was no cultivation, they could go back and pursue their favorite occupation of fishing, keeping their tradition, and earning an income in lean times. They also realized that, as food, rice was more essential than fish. Jokingly, they were reminded of their distant tumultuous past as '*pirates*' and having had several brave encounters with the Maharajah's marine fleet. The race boats of today were the ones they had to encounter oftentimes. Without them, we wouldn't have any Chundans, Vaippu, or Oadi race boats. We owe it to them; gladly, Of Course!

There existed some long-standing feuds between the two indigenous communities. My paternal grandfather was instrumental in brokering peace and amity by integrating them into his workforce. He treated them equally well and frequently rewarded them for their extraordinary skill and hard work. He was able to establish a 'community spirit' and notion of 'common good'. Likewise, he donated his coveted prime land to church and community to build school and play ground. Furthermore, as a legislator, he argued for their upliftment and 'social rights' authoring legislations. He sought the help and resources of his two well-educated sons in drafting them. A pioneering "Debt-relief act" is still considered as his main legacy to the farming communities, as the very first in history. Our family is greatly indebted to the two participating tribes for many reasons. I remember with a lot of gratitude and deep emotions that they rescued my infant brother from drowning. An old and trusted paddler of my grandfather volunteered to take me across the turbulent lake in a '*family houseboat*', braving waves and stormy weather, helping me to reach college at far away state capital at the very last hour of the last day of the 'closing of admissions'. On that journey, as soon as I saw a speeding Public water transport boat plying to the next town, in great excitement, I waved to them to stop, hastily hoped onto it, totally forgetting to pay off the kind, poor man who brought me punting and paddling all the way quite far from home in soaking and torrential rain. I am being tormented for the omission even to this day, knowing that it was a lean time due to floods; he had no money with him; he was exhausted; had to dredge all the way back, alone, and hungry. He could have even fallen sick due to my grievous fault. Gladly enough, continuously for the 6th generation running, his descendants are still with us in close daily association. I console myself that I could still make amends for the appalling failings in some small measure!

....Now, back to the story of the reclaimed paddy fields!

To survive the incessant onslaughts of nature, the levees were strengthened by planting coconut palms, banana plants & bamboo trees, as well as other vegetation which has deep, dense, and fibrous root systems. Keralam lies within the same latitudes (parallels) as Central American countries in the west, Thailand, Cambodia, Malaysia, the Philippines in the East; Somalia, Ethiopia, South Sudan, Chad, Nigeria, Cameroon, Benin, Togo, Guinea, Ivory Coast, Sierra Leone in Africa. Its location is on average 12° N of the equator on latitude & 76° E of Prime meridian on longitude. Christian missionaries brought new crops and produce from the tropical Latin American colonies of the Iberian countries. Most of the items adapted well to the region and climate; proved quite beneficial in multiple ways. The mutually serving colonialists and proselytizers did many good deeds, enabled mass education, did explorations, made inventions;

introduced machines, modern systems, and brought worlds much closer together. In their zealous pursuits though, they destroyed ancient civilizations, exploited the colonies; looted and plundered their wealth and gold; introduced their gods; committed genocides, atrocities of 'ethnic cleansing'; captured them as bonded slaves; brought them to build opulent palaces, places of worship, monuments, castles, cities, bridges; palaces & Parliament houses; mercilessly imposed their will on the natives; tortured and enslaved the population; spread many horrible diseases such as malaria, influenza, bubonic plague, STD, and genetic diseases among them; killing many millions, as a result. Lynching and scalping were other accepted ghastly practices. Horribly, there were even laws asking "kill & destroy hostile native tribes!" Repugnant, indeed!

Truths need to be told! Recount the actual history truthfully as once happened! Though, it hurts deeply, perhaps raising some ire, indignation, condemnation, and reprisals in the process!

To this day, the colonialists claim their explorers discovered new places and therefore have the '*eminent right of ownership*'. It is amusing that all these places were already inhabited and civilizations flourished since thousands of years prior to their arrival. The colonialists proved, "*Might have the right of land and property, as well as the ownership of righteousness*!". The world is still yearning to hear the actual truth (quite different from the 'manufactured truths'), sincere and credible form of apology; offer of reparations; heal the centuries' old scars in full parity and reconciliation. The vanquished natives of the land still wait for firm reassurances that their sacred lands will remain habitable for the generations to follow. Their lands won't be drilled, mined, pillaged; no oil pipelines shall run through their '*reservations*'. The dirty Canadian 'Tar-sands'-oil is, in effect, holding native communities in constant fear of hazardous oil leak, permanent and widespread contamination of their pure soil & water, and catastrophic explosions. Their legitimate and grave concerns that the ancestral lands will be scarred, desecrated, and will be made unfit for the posterity to live on have fallen so far on deaf ears. They are vilified and ridiculed; brutally brushed aside; physically assaulted, spurning their fervent protests and pleas.

Each passing day, irreparable damages are perpetrated by reckless actions and suppressing dissent with violent force and arrogant impunity!
The colonists owe humanity mountains of irreparable debt!

Here in Kuttanad, the period of cultivation is limited and determined by the seasonal intrusion of saline seawater into the lake and all the other connecting water bodies. The salinity destroys all plant crops. If we take a look at the seas and oceans, except for somc coral reefs of Calcium Carbonates, nothing else grows there. As they often say “oceans have become plastic-fantastic” destroying the reefs, also contaminating fish which gets into the food chain.

In the reclaimed fields of Kuttanad, the water had to be emptied before cultivation could begin in the next crop cycle. Initially, the task was carried out by sheer manpower, turning “*Chakram*” or giant '*water wheels*' fitted with flat wooden planks arranged like turbine fins. The wheel was suspended from columns that could be lowered in steps in relation to the receding water levels. A crew of two or three pedaled the wheels, day & night, sitting on multi-tiered stilts. On the lower level, where the load is at maximum, usually, a pair of twos had to be posted. As can be imagined, it was a laborious and arduous task. If my maternal grandfather had not demonstrated to us how it worked, by him sitting on the stilt and pedaling, we would never have known anything about it. He also let us try for ourselves on smaller ones. Our short legs, feather-weight, and underdeveloped muscles simply struggled to complete a full rotation! We are grateful to our '*Appachan*' for showing us how it used to be done in those days so that we could appreciate the many struggles they faced, and how they were overcome mainly by improvisation, grit, and perseverance! The contraptions gradually disappeared in the following years. It was truly an ‘evolutionary’ period in farming in submerged fields. There are not any other places in the world where below MSL cultivation continuously take place! With the advent of machinery, they were all replaced by a combination of kerosene oil engines and transmission belts driving an 'impeller'. The continuous and fast rotating action created a strong ‘*suction*’ force. The water was sucked up and lead through a large vertically placed cylinder, finally expelling it out into the lake via a horizontally placed wooden drainage contraption.

I recall playing 'hide 'n seek' among the number of engines in the family hanger. My younger brother, Tom's favorite hiding place was inside the cylinder of a humongous British-made engine, with piston and crankshaft removed for maintenance, but traces of spilled lubricant oil still sticking on the finely-machined inner side. He came out looking like the ‘chimney sweeper on the London Coo’ in the musical film “*Mary Poppins*”. Together, we tested our strengths on the '*Fly-wheels*' trying to turn each one, starting with the smallest and methodically moving up the scale to the biggest as the weightlifters practice.

The rudimentary methods of irrigation & emptying of water from the low-lying fields

Each time, we looked at our bulging muscles, gloating and admiring their strength and capability. We used a teamwork approach to overcome the weight of the wheel. I wanted to turn in the anti-clockwise direction, while he wanted just the opposite, which was for some reason the most difficult. Only in the mechanical engineering classes, it came to my knowledge that the compression cycle needed the greatest effort. It did also occur to us that if we succeed to produce an initial movement, the rest will become much easier. That was, of course, our first awareness of the heavy force called '*Inertia*'.

Inertia is a force to overcome not only solely related to the load, but also for anything associated with starting! This lesson was very useful in later years in life.

For obvious reasons, Keralam had built a number of dams. The hydroelectric projects began to produce electricity. When it became available, locations of the paddy fields obtained priority connections. It paved the way for high-powered electric pumps to be introduced in farming. Another important lesson was learned that technology makes many things easier; renders less efficient skilled-workers and things unwanted; instantly redundant. Sadly, the imported costly machines had to be simply discarded. Unceremoniously, they joined the mounting pile of junkyard waste somewhere, only benefiting a few enterprising scavengers for a short while. The oil / Internal combustion engines suffered the same fate. Many farmers like ours were left with no '*Return on Investment*' in such farm machinery and suffered huge investment loss for pioneering early modernization efforts. Tractors were practically banned because of objections from the government and farm-workers' unions. Suddenly, our imported Massey-Fergusson tractor had no field to till, even our own fields. All these negative forces coincided, conspired against the farming communities, leading them to a precipitous tumble towards ruin and bankruptcy. It was a cruel and hopeless situation indeed of immense ramifications extending to further generations!

About the same time, some unscrupulous lenders, in collusion with the aid of government bureaucrats and without any due notification to the borrowers, unilaterally amended terms of the loan agreement committing blatant forgeries. They altered and predated the stipulated 'tenure' of the loan by crude but clear 'overwriting' on the original document. The scribbling and the difference in the color of the ink itself amusingly exposed the conspicuous malfeasance quite clearly. It was so obvious and a glaring offense that the lower court wondered "why the *plaintiff* did not ask for criminal prosecution of the *defendant*". The superior class, in their hubris and nobility, developed a notion that all other people are ignorant and stupid! The government, judiciary, and the entire bureaucratic establishments were traditionally and overwhelmingly dominated by the same cast as the influential loan sharks. Nepotism and awareness of class and caste ran deep in the national culture in those times. So, this kind of forgery could be secretly carried out without much difficulty, notice, or fuss. All the vested interests converged; they conspired; bribery is also ingrained in the culture and part of any transaction. The rewards were too big, mightily enriching, and indeed tempting for all participants in the labyrinth of administration. With this simple act of falsification, they can also bring an ambitious 'outsider' to his knees. Both "Jealousy and greed are blind!". They initiated '*premature*' foreclosure proceedings which dragged on for decades along the entire legal course, reaching all the way up to the Supreme court of India. As to be expected, several appeals, twists, turns, as well as reviews ensued;

ultimately and predictably, resulted in the loan shark's favor. At one juncture, the attorney for the lender who pleaded in the early stage of this litigation was promoted to be a judge in the state's High court. He refused to recuse himself, although he was very likely to show prejudice against a party or for having a great personal interest in the favorable outcome for the defendant. This was contrary to the traditions and established norms of impartial jurisprudence. It also occurred against vehement objections from the appellant. In stubborn refusal, the judge did hear the case himself and judged in favor of his former client without any shame or qualm of conscience. His court considered the glaring *forgery* as 'trivial and did not materially violate' the loan agreement. In every book of law everywhere, the '*term*' of loan is a vital and *inviolable* part of the contract agreement. It was an obnoxious decision and a terrible miscarriage of justice, as this set an egregious and gravely unjust precedent in carrying out the law forward. Many lenders hastened to cite this ruling and exploit the situation without feeling any guilt or impropriety. It devastated the borrowers. The ramifications were quite severe and enduring, comparable to the excruciating experiences of many in the recent mortgage/housing crisis in the US, where Wall Street has been lavishly rewarded without any consequences whatsoever for the malfeasance, while many homeowners lost their dwellings. The systems, laws, governments, and establishments all over the world are hijacked by the wealthy and powerful few colluding with politicians and legislators. Bribery is defined as illegal small payments made to facilitate matters for everyday transactions in poor countries. When the same happens in rich countries, it involves huge sums of money. They are euphemistically described as 'lobbying fees, speaking fees, gratuity payments, service fees or cynically as the 'cost of doing business'! They attach some sort of 'accepted and expedited' way of doing things in a fast-paced business world. A famous biblical saying comes to mind: "Brother, let me take the **speck** out of **your eye**,' while you yourself fail to see the beam in your own **eye?".** Sadly, democracy does not work for the common good and common people. Democracy, by definition, is the rule by the majority. The overwhelming majority of people in any society or nation are common people. So, today's democracy is an 'oxymoron'!
A conundrum, riddle, and a contradiction, difficult to fathom and remedy!

All the tasks and travails were forgotten with the passing of time. No one actually cared how the things they see and experience today came into existence in the first place. They have been accepted as existing since the beginning of time. It is like the new immigrants coming to a developed country and finding everything 'ready for use'.

Notwithstanding the plight of the entrepreneurs, thanks to the massive transformation initiated by them, the region has gained a certain amount of self-sufficiency in food production. Appropriately, the region of Kuttanad has come to be known as the 'The bountiful *Rice Granary of Keralam*'. Beyond the paddy fields, the embankments created new space for living, accommodating a growing population; found space for planting coconut groves and other minor crops. In essence, with a series of similar developmental projects in and around the lake, the region has gradually been transformed. Remarkably, the newly created landscape of water bodies, waterways, strips of land, blocks of fertile paddy fields, growing a diversity of food, and cash produces plainly blends in with the nature. The vast and indomitable lake has been tamed to a manageable level; helped to build modern infrastructure bridging several landmasses of isolated strips, villages, and towns to the main land. It is a testimony of thoughtful vision, indefatigable tenacity, ardent aspiration, and the unwavering spirit of the times. The '*Kuttanaadiens*' (the inhabitants of Kutta-naadu) consider this venture as a monumental legacy, bequest, and an enduring gift from their modest but gallant forefathers. The succeeding generations are deeply indebted to them for the gigantic creation and the invaluable contribution. This must be considered as one of the greatest human efforts of the time in the history of Keralam.

Several dams and other hydroelectric projects were since constructed, as well as numerous major and durable projects. They were all completed using modern engineering technology, heavy machinery, public finances, foreign know-how, and other industrial resources. But, they cannot diminish or dwarf the significance and magnitude of the audacious reclamation project carried out by a group of private individuals employing many times more people over generations without all the abundance and readily available support resources of modern times.

According to mythology, **Keralam** was reclaimed by ***divine power*** from the bottom of the **sea**; while **Kuttanad** was reclaimed by sheer **manpower** from the depths of the **lake**!

This human endeavor endures as an epoch-making undertaking of enormous consequences for the whole region, as well as numerous enduring benefits to its people!

Chapter 8

Betrayal by the Government
Plight of Farmers

At this magnificent stage of development, the communist state government took over the land from the owners paying pittance or nothing by '*the stroke of a pen*' offering 30-year worthless '*paper bonds*'. The term paper bond is literally apt, meaning it wasn't worth the paper it was written on! The action was carried out without any recourse to adjudication or arbitration. "*The brave maketh; the cowards stealthily taketh away*!" They didn't feel any pain because it's all somebody else's possessions none of theirs! The ownership rights and laws of "*real & personal Property*" were ignored, totally! The entrepreneurs were suddenly left with tremendous debt obligations and little means to pay them off. Instantly, they became deprived of their valued resources, taking away their occupation livelihood. This action and measure doused the flourishing entrepreneurial spirit of the time, caused a steep downward spiral in the agricultural sector. India is still considered an agricultural nation with a need to feed over 1.3 Billion people, rather than an industrial nation. Therefore, the damage was severe and had many ramifications. Under the burden of inclement and debilitating debt, most of the pioneering families abandoned farming altogether and migrated to the midlands. They had to learn new skills and lifestyles for survival. Now, only small-scale subsistence farmers with limited capital resources and know-how engage in farming. They are unable to utilize improved machinery or scientific methods of cultivation. Inexplicably, the farmers and the region continue to face appalling neglect and scant consideration from the state government. Ironically, people who had never planted a seed or dug a hole in the soil, become ministers of Agriculture. Unpredictable weather patterns blamed on '*global warming*, debt burden, price erosion of grains & farm produce, coupled with galloping inflation create unsustainable conditions for farming communities around the world. In Kutta-naad, the traditional farmers have become averse to farming. Consequently, migrant workers have to be brought in, from thousands of km away, as far as the NE states of Assam and West Bengal. It speaks for itself!

Paddy-cultivation is manpower intensive. The harvest, reaping the fruits of labor is the only joyful aspect of farming. The farmers accept the fabulous bounty in great gratitude!

Another shenanigan by the government of the past, was equally disastrous to the farmers. At the same time, it was so hilarious to the media and the public that the farmers were required to obtain a mandatory certificate: “Permission to harvest" from the local Village officer of the respective jurisdiction, the lowest in the totem pole of the Revenue department of the state. Suddenly, the petty bureaucrats have gained clout and recognition. The cunning scoundrels wanted to make the best of this ‘god-sent’ opportunity. Near the harvest time, they used to go into virtual 'hiding', temporarily living in wives’ houses or some cheap hotels. The desperate farmers cannot find them! If the ‘absconding’ bureaucrat were to be found as a result of some detective works, the poor farmers will be told that according to the records, the owner was actually living in a different jurisdiction and this officer is not the one who should issue the certificate. In other instances, they will be told that the fields lie in a separate jurisdiction, always going back and forth in a cynically cyclical closed loop bureaucratic circus. In Kuttanad, invariably, the owner lives in a village; not in the lake where the paddy fields are located! The fields are low-lying and for paddy cultivation, not suited for living, as the dwelling will be submerged in floods during the annual monsoon season. It was totally dishonest, painful, and ridiculous; speaks itself for the efficiency and quality of the state government. Many farmers did not have time for these deceitful antics, or petty pranks. There is a lot at stake for them. They have a sacred responsibility of feeding humanity, just as the health care workers do for the sick and dying. Understandably, they were panicking, seeing their crop going to be destroyed by the delay or the imminent arrival of the rains; often decided to take matters into their own hands and went ahead with the harvest without the mandatory permit. The missing officer suddenly appeared with policemen brandishing batons or 'lathis', some carrying abandoned guns from the colonial barracks. They intimidated the ‘offending’ farmers; in many cases, arrested and humiliated them in front of everyone in a cynical act of shaming! Adding insult to injuries!

Harvest is not the absolute end, but the beginning of the ‘fag-end’ (The end or final part of something) of the season for farmers. There are at least three more outstanding jobs to be completed. The immediate one is to select a suitable and most productive batch of grains and prepare them as seeds for the next crop season. The selected heaps were spread on huge and supple grass mats; dry out under the sun; then sorted and stored in specially constructed and spotless barns or storage facilities devoid of moisture, humidity, mold, pests, and other deleterious elements. Now and then, the facilities were aerated by letting dry air in. Periodically, the samples are thoroughly examined. As children, we were often asked to do the task of opening the shutters or windows. We could feel the heat oozing out of the barn, lashing at our faces and chasing us, as we ran away. Any condensation could cause mold to build up. Only my maternal grandfather told us why it had to be done, none else. Having grown up, I resented and regretted those unforgivable omission from other elders.

A few weeks prior to the time of the next crop cycle, the seeds were packed in porous sacs woven from dried leaves of water cactus. In prior weeks of planting, the sacs of seeds were immersed in shallow water. They are frequently examined to see if all of them has germinated, showing sprouts. When they did, they were taken out and kept at room temperature, and set ready for planting! Well, there were no agro-industries that sold seeds like several in Western countries. Although there is a ministry of agricultural development, it acted simply as a ceremonious entity. Most often, the minister was just a politician from somewhere and knew little about farming. According to the official version or justification, as the state has an agrarian label, anyone in the state has some knowledge of farming. Well, I'm availing myself of the same qualification to say something purely based on my observation and experience in farming.

In other countries, some companies sell seeds after first acquiring a patent. Once a farmer buys seed from them and later makes their own seeds, it becomes a grave patent infringement, for which the poor farmers are taken to court and ruined for good. A major US entity is one such monster and hated in the farming communities around the world. The same company sold cotton seeds to Maharashtra farmers at a premium price, convincing them that the seeds were weed resistant, besides requiring minimum use of water. The gullible farmers were told that it was a miracle seed; believing in the miracle and sale's pitch, they spent a lot of their borrowed money. The crops were so bad that many didn't bother to harvest. The rest of the story is horribly tragic and heart-wrenching; some paid the ultimate price with their own lives!

A pompous and unscrupulous marketing campaign blitz with tragic results to the victims, but rich dividends to the perpetrators!

The farmers have to remove and clean the field clear of any remaining 'hay stumps' so the fields will be ready for the next cultivation. The final step is to irrigate the fields with sufficient water to hydrate the field well and protect the ground from the sweltering summer sun that could evaporate many soil nutrients. The monsoon floods will of course submerge the land, flush clean and bring in some nutrients from the higher grounds and heavens.

The pioneering entrepreneurs had been very generous, too; had some noble visions for their communities. They donated lands and resources to churches and temples to build places of worship, schools, hospitals, and community centers. These institutions helped to mellow the agitated souls and educate the minds. The organizations recruited educated, dedicated, and unselfish professionals to run them. Among other things, these sites mainly served as places of education, hope for the future, cultural & recreational centers promoting interaction, discussions, problem-solving avenues. People of all walks of life benefited in multiple ways. Again, the thoughtless government extended its vicious power to take over the schools and began to run them far poorer than mediocrity. They didn't know much about education because they didn't have enough. Again and again, the government unleashed its cynical penchant for wrecking things that are of great importance and agents for progress for the public. Apparently, they don't want anyone smarter than themselves! The egregious betrayals by the government constricted the paths to success for many. It spelled a severe decline in living standards and a catastrophic doom for the state. The democracy and government have devolved into an '*elected anarchy*' and a reckless wrecking ball!

The globally prevailing conditions for farmers are appalling; actually exacerbating. The 'climate change'-induced consequences of drought and consequent desertification of arable lands will certainly lead to world-wide famine and deaths. This unavoidable scenario is quite frightening, yet the world remains seemingly defiant and insolent. The closing of the narrow 'window of opportunity' to stem the tide, is gathering speed and acceleration. Therefore, collectively, the entire humanity should address the dire situation with immediate, drastic, and fitting remedies. Tragically, the dereliction and neglect with which this existential threat is treated by the governments of the industrial world is quite dreadful. This is in spite of the fact that food is the foremost item of 'essentials' for every life and its survival (In all faiths and religions, traditionally, humans begins their prayers with "Give us this day our daily bread…!, reminding of our most essential need"). Farmers are the ones who are providing it. The very neglect and scourge they face, should be counted as one of the contributory factors in the mass migration of 'climate refugees' across the globe and over the treacherous seas. The exodus we are witnessing today is the flight and struggle for sheer survival. The governments and organizations are preoccupied with only trade, business & defense matters. Some heads of governments are never home; they are traveling worldwide; pretend they are making business deals. That's where matters of Millions & Billions of money being talked about and dealt with; there are always plenty of chances for the decision-makers to 'harvest' some unaccounted cash and favors. The distressed farming communities are convinced that their causes and interests remain largely ignored. Tragically, some take their own precious lives. It touches only a few gentle and hapless hearts. Out of desperation, many flee their homes, native countries, relatives, and friends to an even brutal, unwelcome, and unknown world outside. They wander, stagger, and struggle in the exodus; become vulnerable and exploited. Even within the borders of a country, people migrate to urban areas without possessing the necessary skills to succeed, or barely survive. The ominous trend will continue until farmers are rehabilitated with firm assurances of property rights; necessary aid & protections; granting capital at generous and concessional rates; offering subsidies for fertilizers, electricity, and farm machinery; providing know-how to increase yields, and remedies against weed & pest control, crop failures, flood/drought, etc. Governments all around the world are extremely sensitive, sympathetic, and quick to respond in frequent stock market crashes as and when they occur. With urgent alacrity and actions, the treasuries keep their 'lending windows' open wide generously dishing out large sums of interest-free loans, besides assuring tax cuts for 'all business across the board' whether they caused the crisis, or whether they are affected or not! Farmers simply complain to God and hope some blessings will get dropped in their way.

The proverbial 'manna' is stuck in transit, somewhere between heaven and earth!

Incredibly,the world is only after the "chips", not morsels!

Save earth! Save lands; don't torch! Save farming! The farming and cultivation must be elevated to an essential service and noble status as benefiting the whole humanity!

A major factor contributing to inflation is the erosion of currency values and its purchasing power. A notoriously hostile neighboring country is gleefully helping the situation by feverishly printing Indian currency and smuggling it across the border. They use gulf expatriates and villagers from border states to carry out the illegal operations who unwittingly fall victims. An international ban on the export of essential currency-making materials such as paper, ink, and other materials is slapped on this country. Currently, a painfully excruciating, costly, and inevitable action to withdraw high denomination currencies is underway in India in order to flush out the counterfeit and "black" money.

The farmers shall be guaranteed living wages, fair prices for their produce; assistance with distribution & selling, as some of the urgent and mitigating measures. In between their preoccupations and indulgences, the authorities are asked to spare a moment, look out through their office windows of their edifices; see the sights open to view and reflect on it for a precious little moment. They will see some heart-wrenching realities of life of others, their fellow citizens.

Foresight and preventive measures are essential attributes for good managers and administrators. There are plenty of clear warning signs and brewing discontent among people. Ignoring the crisis may result in grave ramifications, such as refugee situations, social unrest, a spike in crimes, revolutions, and ensuing worldwide instability and chaos. Without proper attention and bold actions, our systems, and civilization are subject to peril and destruction.

The stark contrast between the luxurious life and status of the business executives & government officials on one side, and the common citizens on the other, are best understood in pictures

The mass migration of humanity from across vast areas of the country, between borders, and across the seas pose many hidden dangers of unknown proportions, hunger, periodic contagions, diseases; social /cultural conflicts, human trafficking, inhuman living situations; health hazards, and squalid sanitary conditions, to name a few.

It takes only a spark.

The ominous catastrophe is imminent. Folks! Shouldn't the world do something about it?

Chapter 9

The Magnificent Vembanad Kayal

The picturesque port city, Kochi

The pleasant and picturesque coastal city of Kochi, known as the 'Queen of the Arabian Sea' from the earliest colonial times, is situated at the delta. Thanks to the torrential rains during the annual Monsoon seasons, the region has an abundance of rivers running through and across with a myriad of rivers (a total of 44), 2 large lakes, a few lagoons most of them interconnected with naturally formed myriads of intricate canal networks providing long and continuous waterways. Most of the estuaries lead to Vembanad Kayal (Lake)- the longest in India. The lake meets the mighty Arabian Sea at Kochi. It is a cosmopolitan city, an education center with a number of colleges & institutions, a naval base, and an oil refinery. It is also a site of chemical industries; business and posh shopping centers. The location is naturally suited for shipbuilding and is a major shopping center. It is fittingly called 'Mumbai in the South'. The greater metropolis of Kochi encompasses residential towns in its sprawling suburbs, modern transport systems including Metro and waterways. Natural and man-made islands huddle together, connected by roads, causeways, and dikes. Regular boat services ferry goods and passengers to and fro. Several other boats and barges ply up and down further deep into the inlcts which extend far to 90 km stretch in one direction, and 60 km to the other. The multitudes of rivers continue the journey from there deep into the heartlands extending the water routes. The city has an airport, a naval base, which was, until recently, allowed to be sparingly used for domestic passenger flights. It had been a stunning experience to behold when landing at Kochi airport. I have heard several eyewitness accounts of visitors from out of state as well as foreign tourists of their pleasant experiences and excitements about it. As the plane enters the airspace, emerging from the clouds, and exiting the flight path that was hugging the coastal shores until then, the scenery suddenly changes, offering a variety of spectacular sights. The plane hovers in the skies in its descent, takes a sharp 'hair pin' turn; the passengers feel a momentary rush of blood and excitement; feel an uncanny and momentary loss of gravity. Spires of churches, Flag masts and sprawling court yards of a temple, dome of a mosque, and the bell tower of the Paradesi/Malabari synagogue promptly come into view heralding that you are arriving at a destination of secular tolerance and traditions. The plane flies low over palm groves; green & golden coconut bunches can be clearly spotted. The paddy fields with golden grains ripe for the harvest; tiny islands; lagoons of different lengths and breadths; meandering canals; an array of overpasses and bridges moving traffic and people are the sights to behold, in a unique, magnificent aerial 'bird's eye view'. The Bolgatty palace, a Dutch colonial relic now with a small golf course, sits at the edge of its eponymous tiny island spanning two sides of its corner looking across the water towards the bustling city. The Visitors feel that they are welcomed in the homeland with its scenic splendor, splendid charm, and reassuring calm.

Kochi, the port city, modern, cosmopolitan, bustling, yet sublimely serene and buoyantly beautiful

The passengers lean over and admiringly look out through the tiny 'peep windows' of the aircraft. They see the view of the Arabian Sea on one side; its gently arriving waves leisurely striking the highly reflective white 'sugar sands' on the beach; shimmering waters of the serene lake on the other; ships maneuvering to dock at the wharf; cranes busily loading and unloading; a newly launched tanker getting ready for its maiden voyage at the shipyard; a large vessel equipped with masts, rotating radar antenna and mounting superstructures are seen with navy men in white & blue uniforms in 'guard of honor' salute. They guard the shores and outer seas. It reminded me that I once coveted and unsuccessfully tried to get a Short Service Commission in the Indian Navy while studying at a professional college. I could have been one among them! The road and train traffic move briskly by; pedestrians walk along the 'Marine drive' boulevard and the surrounding promenades. Some are seen seated on the parapet wall along the promenade watching across to Wellington island and Fort Kochi as the water traffic of many kinds leisurely pass by as usual. The waterfront Subhash Bose Park in Ernakulam in front of my alma mater, Maharajah's college and across the student's dorm brings youthful memories of the past. Domes & towers of educational institutions peek over the tall green trees, the edifices attract curious attention, their campuses bordering the brink of busy roads and edge of water bodies. Sacred Heart college at Thevara, is one such prestigious catholic college. The children lean onto the windows; the scenes and sights they are witnessing spread glee and excitement on their faces. They had been looking to spot their cousins, grandparents, aunts, and uncles who might have come from the hamlets miles far away to escort them home. But, momentarily forget about them in the scenic imagery. The bustle and hustle of life are warmly felt. The air passengers will shortly become one among them after the plane has safely landed at Wellington Island soon.

Travelers recount the landing at this airport as one of the most magnificent and memorable sights in flights!

The region of Kuttanad lies about 55 Km further South East of Kochi. It is practically waterlogged. The traditional transportation and cargo movements were by boats of various sizes constructed for custom purposes. Everyone must own and know how to row a boat, just like the land-based inhabitants are used to their bicycles. The love of water and water sports is innate in the spirit of Kuttanad.

The many ways of making honest living the hard-working, self-reliant and resourceful villagers, commonly pursue

Nowhere is it more evident than in the enthusiasm for VallamKali-the regatta of the boat races. It is ironic that people are generally poor swimmers for lack of good swimming techniques, although they can float and survive in tides and water for a long time. Most are 'autodidacts' who learn skills by themselves or by instinct; by force of necessity without assistance from any teaching or coaching resources. They are not consciously cognizant of the hydraulic and hydrodynamic properties of water, its pressure, or buoyancy. Fishing is a common activity. Fishermen and birds fish alongside. The bluish, brown beaked Kingfisher finds a perch, looks around; suddenly swoops low over the water, makes a swift dip, and comes out holding the fish in a single continuous motion, goes back to its perch, beats the prey against it before swallowing (They don't want to swallow a live fish jumping inside stomach). The cacophony of quacks rises in the air, as the shepherd in his flimsy boat cleverly balances himself and herds them together with his punting/poling-pole. They provide families with eggs and fowl meat, low in cholesterol. The egg is rich in vitamins & minerals; the yolk contains the bulk of it, except for the protein and riboflavin in the white. If children get a 'cackle berry' with double yolk, they think it means good fortune and prosperity. Most families have hatchery. This person is now taking the ducklings for feeding, just like the shepherds take the sheep to the pastures for grazing. As they float in the water, they busily dabble with their beaks; find plenty of edible vegetation and food particles among the algae in the water;busily gobble them up in great merriment.

His little brother maneuvers another tiny boat, skillfully throws a fishing net in the manner of a 'Discus thrower' maintaining critical balance on it. He catches an assortment of small fish with slender bones. They are quite delicious and crunchy; crumble easily in the mouth with a gentle bite when finely fried. A housewife comes out of her house and throws a small nylon net she made with a float attached in the middle, into the shallow waters; toss a handful of leftover food as bait and go back inside. She returns to the spot sometime later, and lifts the net with a light bamboo pole in a swift swirl, and whirl motion; catches an assortment of small fish, enough for supper for the family. Nearby is a narrow but strong stream flowing into the canal. Schools of small fish “Pallathi" (chromide) with silver and golden spots & stripes and protruding eyes swim against the current. Big fish from the river lingers around the mouth of the stream, hoping to make a meal out of them; they find the water in the upstream too shallow and leave the scene wasting no more time. Water buffaloes grace nearby; little birds are seen sitting on its back, establishing a vantage position for catching fish. The buffaloes are not too bothered; make no attempt to jerk them off and continue what it has been doing. They are herbivorous! Some birds eat parasites, leeches & lice right off their backs. The four-legged animal is not capable of scratching its back. So, they come to settle into a mutually beneficial arrangement. Water lilies float; they bear lilac flowers. Beetles and butterflies shuttle back and forth from flower to flower sucking its nectar, also adding a buzz to the ambiance. Bees, flies, and wasps do the same. Bees have two sets of wings, while the others have only one. The honeybees and bumblebees are excellent pollinators. They live in colonies and are matriarchal. They need pollen and nectar from flowers for their high-energy activities and survival. The bees can perceive the ultraviolet light spectrum, and use the 'dithering' [*computer graphics:* The use of dot patterns to approximate colors not available in the pallet] ultraviolet pattern as its 'nectar guide'. They carry pollen from flower to flower, aiding pollination; pollination leads to fertilization; fertilization results in reproduction. It is important to remember that good pollination can improve the quality and quantity of agricultural produce by substantial margins.

This continuous process preserves plant & animal life, quite another wonder of nature, that's seldom noticed except by agriculturists!

The Amphibians swim and splash around; make frequent hurried trips to land and back. Frogs of green squat on broad floating green lotus leaves, fully camouflaged. Their presence is felt by the sporadic crying "mack-mack". The Mother frog stacks 'frog-ling' (baby frogs) on its back. Tadpoles live underwater. They breathe through the gills. They are still in the evolutionary process of metamorphosis. A turtle climbs onto a driftwood; slowly retracts its neck, like the piston in a locomotive, as it starts to move. The little turtles try the same, they slip off back into the water. Turtles are ectothermic, aka: cold-blooded amphibians- simply means their body temperature adjusts itself to the temperature of the surroundings. This characteristic gives the ectotherms the ability to adapt to different habitats and conditions (similar as the Chameleon adapts its color). In contrast, mammals and birds are endotherms or warm-blooded animals; have to maintain a constant body temperature, independent of the environment. The rodents frequently come out of the den, dive, and scurry around looking for prey. An otter rushes out of its burrow near the water's edge, dives deep, and stays submerged for a very long time. They have webbed feet which seem like a folded paddle and have sharp claws. It can be a vicious predator, brave and bold. They don't back away from anything, as a pack they are seen even attacking ferocious alligators. It seldom lends itself to full view; yields a fleeting glance when it suddenly emerges from water; stealthily swim back to its holt. One can rarely see what it is catching. The socially organized Pelicans with their large beak and neck-pouch fly in as flocks; skim the surface and scoops up fish along with water in one swift maneuver; drains the water from the beak before swallowing. The 'wood-note' and the drumming rhythm of the pileated male woodpecker can be heard coming from a tree on the banks. It claws onto the side and diligently carves out holes on coconut trees for its roost and as a 'love nest'. In contrast, the woodpeckers in the Grand Canyon National Park in the US drill thousands of small holes on tree trunks to hoard acorns (fruit of an oak/pine trees) as future provisions. The squirrels like them too, and often come around stealing from them. It is quite amazing that the woodpeckers have the ability to do the pecking actions without suffering any brain damage. On that count at least, it's much superior to the human capability! The falcon hovers high up in the skies, spotting chicks on the ground. They are long-sighted! Swoops down and pounces; the chic shrills in panic. The mother hen and the housewife join in screams to foil the attempt in vain. The falcon swiftly flies away, clutching the prey within the sharp claws of its feet, the chic still screaming in fear. A flock of noisy black crows scavenges the carcasses some carnivorous birds had left behind and keep the surroundings clean and sanitized; while the quiet Coucal, the burgundy - blackbird with brightly colored 'spot' eye, forage the area. A bunch of frugivorous/fructivorous (fruit-eating) birds share jack fruits and mangoes on trees planted on the strips on the banks while the insectivorous Kakka-Thampuratti, the blacker than black 'Drongo' devour all the insects it can find helping farmers. It has a single color; has shiny feathers tucked neatly together, looking like a non-stick 'Teflon' or waterproof material. It is always active regardless, sun shining or rain falling. A pair of gentle Myna birds perch on the stems of banana branches facing each other, swallowing insects. The sparrows hop from one tree to another, constantly changing positions, merrily chattering; seem restless and suffering from ADS 'Attention Deficiency Syndrome'.

Together, they make a sonorous symphony.

Around here, nothing stays still or quiet; neither the current, nor the wind; people, flies, or bees!

The biosphere support and maintains all life-forms, and the tropical creatures!

Boats for a variety of custom uses: Houseboat, Work boat, Cargo, Tourist, Travel, Luxury & Sightseeing cruise

'*House-boats*' ply the route now and then, carrying honeymooners, company executives on seminars, and tourists from out of state. Many international Eco-tourists as well as nostalgic people from the former colonial occupiers of England, Netherlands, France & Portugal are among them. Some are seen pointing to various sights and nudge companions to view the same. They haven't seen such sights before. They live in the midst of hustle and bustle in big cities; lead hectic routines; most of the time confined within the four walls of their Hi-rise flats or looking at walls of other buildings through the glass windows of their own multi-storied office building. Occasionally, they go out to the well-manicured parks where plants and hedges are planted in perfect symmetry and configuration, trimmed and edged, cut and groomed; on weekends, to zoos where animals are kept in cages. Most of them spend the work-hours looking at computer screens, 44 cm diagonal; other times intently looking at even smaller screens, 11 cm in size of smartphones. It is their memory, brain, pathfinder; keys to the house, bank accounts; where music, photos, video, and vital personal information are stored; the all-important pieces of software applications or 'apps, and a lot more. Life comes to a screeching halt without it. What a pity! They dabble with selected flexible fingertips; swiping, stroking & pinching on the screen with gentle dexterity, sending messages, texting, and tweeting. They walk around talking to themselves; some dance; lose their 'presence of mind'; vigorously gesticulate; become oblivious of the dangerous traffic around. Some grow long hair concealing the earphones that are hanging from their earlobes eliciting an inquiring look from onlookers. Good Samaritans just want to make sure that all fellow citizens are okay! It is all manifestations & trappings of the so-called Cyber living. For many, everything happens in “Cyberspace” which has so many components and specialties. The response and feedback are instantaneous, in real-time. No one person, knows exactly how it all comes together. It's the new lifestyle, ethos, and larger-than-life itself. Face-to-face communication is a thing of the past, even between parents and children; between close family members.

Me and my cellphone; life cannot go on without it! Perhaps, only a catastrophic '*phase reversal*' of the earth’s magnetic poles or impending climate disasters alone are capable of stopping the trend.

That's the world the participants of the High-tech life normally see and are most familiar with; hardly imagine the serenity and beauty of the rustic settings as they see it now really exists. This is where the essentials and basics of life take place; the ecosystems develop; where food is grown, supporting life and humanity. Here is the heart of the world is situated; the center of oxygen production; the brain and command control of the climate reside. Life may go on without “Apps” & “chips” but certainly will not without food! Now, they see a natural world, rather amazing, and mesmerizing ; quite distinct from theirs!

Along the waterfront stand a row of richly painted posh bungalows of tropical architecture designs with patio and terrace offering unobstructed view of the flowing river. Beautiful flower garden and manicured hedges with neatly planted bushes provide a pleasing appearance whole year round. Boats are seen moored at the boat slips built on the banks or the leading canals.

Bird species of the Lowlands of Kuttanad indicate a vibrant biosphere; they go where fish is

Beyond the aesthetically built masonry wall fences that surround the bungalows, some brick columns are visible through the dense bushes and trees. They carry birdhouses with pigeonholes in them. The 'band-tailed gray pigeons quickly find them and make them their own homes. They are gregarious birds, sociable; like to live in a community. They don't seem to have prey or predators; eat only nuts, seeds, and fruits. A beautiful “Vezhambel", the great Indian Hornbill (also the state bird), distinguished by its curved beak & colorful head-plumage, which was feared of facing extinction, descends from the Hi-ranges, perch slanting on the side of a tree, haplessly looking skywards crying for rain. That's what they usually do. Apparently, it cannot slurp and swallow at the same time; the water has to drain down the throat by gravity! According to ornithologists, it lives the entire life in the same roost, is monogamous having the same mate life long, just like humans used to be in the long past. Birds are seen with camouflages, and in contrasts with the surroundings. There are flocks of different kinds, sizes, shapes, chroma, and tint: of feet, beaks, bellies, caps, cheeks, and chine. They add beauty, color, and visual diversity to nature. During flying and landing, they spread the wings fully wide, revealing the entire plumage in vivid colors. Each has its characteristic style of flight and glide. The big birds glide gracefully after flapping its wings a few times; the smaller ones need to flap almost continuously except for landing maneuver. Each bird possesses unique hunting techniques typical of their flocks. It is quite enchanting to watch their behavior. It stimulates a sense of wonder, charm, and amazement. Birds of the same feather, flock together. They hum, chirp, call, trill, troll, twiddle, twaddle, and dance turning the area into a sonorous zone. The males sing to attract females; the females usually chatter in response. They communicate in what seems to be a telegraphic pattern of short and long sounds, or in computer terminology in codes of different word lengths (of 8 Bits, 16 bits, etc.). Birds have syrinx membranes which are like stretched skin on a drum, or like the tympanic membrane of human ears. The skin vibrates when air passes through them, producing sound waves. The syrinx is divided between each of the lungs and is controlled by muscles to produce different sounds. A songbird on the tree branch sings to attract a mate. No response is heard. It flies to a few different locations and repeats the same song; soon seen ushering a mate and disappearing into the woods. It seems that all living beings seek companionship and love, a strong and common instinct of nature! Some birds are domineering; use their distinct sound to claim a territory. There are birds whose sounds can travel several kms; the whooping crane of North America is one of them. They must be real loud and penetrating! There are swaths of sparsely populated regions in the US and that's where they are commonly found; perhaps the voice is seldom heard by humans and therefore don't feel disturbed. Bird's hearing is much superior to that of humans in terms of alterations in pitch, tone, rhythm as they can resolve sound components in small-time and frequency units. While a musician struggles to distinguish $1/16^{th}$ notes from the $1/8^{th}$, birds can resolve $1/128^{th}$ notes, by a better factor of 8. Astonishing! Imagine! The frequency difference of music notes of the adjacent semitones of middle A & A♯ is roughly 26 Hz. Many musicians find it difficult to differentiate between them. Birds are thought to be able to decipher a difference of just 6 Hz. They could probably hear the fundamental frequency and its many even and odd harmonics. The essential point is that they understand each other. It is amazing that they know their nests, routes, paths, and migratory routes without the aid of "GPS" or other external devices. During their migration, they cover fields, cities, water bodies, and hills cleverly manipulating wind, moisture, draft, and other natural forces. Some birds even try to cross the Himalayan mountain range, braving the extremely frigid winter weather and rarefied air. Sometimes, they have to retreat and make fresh attempts. They seldom give up easily. If a single bird is found to be in dire straits, they know their task is going to be insurmountable or dangerous; wisely and instinctively abandon the foolhardy attempt. It's an important trait to emulate! It must be concluded that they have functional social, practical systems to live their lives and survive conditions. While flying, they form different intelligent formations, apparently to evenly distribute the adverse wind tunnel and draft effects among themselves. Penguins are often seen periodically rotating their respective positions in their 'huddle' during severe winter blizzards. This is done to distribute the brunt of the forces and share the brutalities of survival equally. There is 'equality' amongst them, which humans could certainly emulate!

It's a socialistic and empathetic culture that is very much lacking in the human specie!

Farmers plant and weed in the paddy fields in knee-deep irrigation water, while the egrets & storks with their extra-long bills walk around; hunt for fish and worms in the vicinity. In the muddied water, because of the depletion of oxygen concentration, the fish surface periodically giving its predators a better chance of spotting and catching them. Some urchins apparently skip school for the day; stealthily move around with "Ottaal", a bamboo-built 'truncated cone". They plant them in the shallow muddy waters; trap a few fish; catch them one by one, by hand, approaching from the access hole at the top; deposit them in the net tied around the waist. Periodically, an adult person comes around, collects the catch and carries it away to his Vallam, and empties them into a container. The live fish command premium price in the market and luxury hotels. They are delicious too!

A family in a Vallam, cut and reap "kaitha" leaves, from waterside cactus plants and another variety of tall grass plants, olive in color, with a sickle; gather it in bunches and load it into the Vallam. They will take the bundles home, dry them out under the sun and weave them into soft, pliable "pullu-paya" or grass sleeping mats.

Only a hard-working lifestyle can ensure a bare survival and well-being of farmers, almost everywhere in the world

They are well suited for the climatic conditions as it is a heat insulator, and can be cleaned with a soaked towel. Many of the village folks like to sleep flat on the floor of their huts without a pillow or cushion. They prop the head up with a rolled up shoulder scarf/cotton towel. The 'grass mat' provides a bit of cushion between the body and the hard floor; supports the back and neck without causing discomfort. After use, it can be rolled up and put in a corner behind the swivel door, until time for its next use. It is nature & health-friendly, as well as space-efficient; a ubiquitous item in a village household. Importantly, this provides an occupation to practice in the lean, dreary, and miserable monsoon season, earning much-needed cash.

At a different location nearby, divers dive to collect clamshells. They usually dive deep and stay down for a very long time underwater. Once, on a family trip in a VallamBoat, as a child, I started crying inconsolably as divers failed to come out of the water at the time I expected. I imagined that they have drowned and perished. I will not see them any more! Sometime later, when they emerged scooping up a basket-full of shells with clams inside them, I started jumping and clapping, almost falling overboard. My father laughed uncontrollably. My mother explained to me that the shells contain the shell meat she often cooked, which I liked very much. "Your father liked it too!". She said. One can witness all around an earnest spirit and natural instinct to survive. It is also the most honest way of earning one's living, reaping only from the sustainable resources of bountiful nature. The incident helped me to cultivate an appreciation of the remarkable skills, talents, and exceptional capabilities people and creatures possess. Alas, the world lauds only the sports people who score a 'come-from behind' win, the records set, their stamina, courage, and character; ignores the talents and efforts of the ordinary people! The clam meat is an exotic item and a delicacy for the habitual drinkers of Kerala, much like the Black Sea Beluga 'Caviar' served as an 'hors d'oeuvre', or appetizer for the very affluent, the connoisseur, gourmand, and the gluttonous of the world.

Innovations & inventions happen in everyday life; many more than in laboratories and research institutes!

Watching the actions and behavior of the species in the surrounding nature, one must conclude that "Wisdom" has many components. It may begin with sensory powers, perception, instinct, reaction, processing of information (logic, comparison, analysis, etc.), deduction, inference, comprehension, application, iterations, and corroboration. The elephants in Sri Lanka sensed the December 26, 2004, Indian Ocean Tsunami minutes before the Tsunami arrived ashore. The tusker led their herds to comparatively safer inland and higher ground, interestingly, well before humans reacted. They can pick up sounds as low as 5Hz. Their large feet that are always planted on the ground, felt the vibrations. So also the mountain goats grazing near the most active volcano in Europe, Mount Etna. All the goats respond and lead themselves to safer locations. It is no wonder mankind continuously learns and acquires wisdom from what is open to view in nature. They usefully apply the valuable lessons in life. "Bio-mimicry" as explained in several places in this book is an example of adopting this knowledge and putting them to good use.

Folks, we must look around and perceive; not just see, and erase!

Chapter 10

Those days of the forgotten past

Cargo transport of yester years, propelled by wind power, manpower & tide power alone; **NO fossil fuel burnt!**

Today's houseboats are modified and customized versions of KettuVallams used for cargo transport in the years past. Its hull was made up of jointed planks 'stitched' together with coir ropes run through tiny holes, binding the stuffed “Cha-Kiri” or coir fiber fillings on the inside to prevent leakage. It was an effective method of boat building and held up well for many years with occasional repairs. They were fitted with loading compartments that could be removed or readjusted as needed. The KettuVallams were transformed into "Oadu-Vai / PayaVallam" when fitted with a sail. The use of wind power originated in the ancient lands and islands, not 'invented' by the sea faring conquerors in the 'age of discovery' [an arrogant and preposterous misnomer suggesting that "a place is considered existing and inhabited only after the Europeans has seen it, and set their foot on it]

The PayaVallams could carry heavy loads to distant places. They were propelled by a single mainsail. Unlike the jibe of a yacht, it is rectangular, fastened to a detachable mast erected close to the bow; two bamboo booms tied to the sail, one horizontal at the bottom end, the other diagonally across from top to bottom gives it a shape, strength, stability, and flexibility. The sails were woven from plant leaves and controlled by ropes made out of coir fiber. It can also be shifted from one side to the other in the act of 'jibbing', as the heavy mast rests on a rounded cup-base. The big sail can be folded and rolled up by detaching the bottom ends of the two stabilizing bars and secured onto the mast at night for safety or when no wind is blowing. The design was simple and innovative for simple and quick maneuvering by one or two people. They have 2 removable thatched curved canopy shades of semi cylindrical / truncated cylindrical shape to protect the cargo from sun and rain. One fits the front half and the other the rear. The middle get a covered sliding entryway.

The navigation was done by a manually operated big, broad steering paddle by a steerer at the stern. They maintained an excellent safety record. When the wind became still or when going through small canals, the Vallam was pushed off with a long bamboo pole against the canal floor by punters or rafters from each end. This action is called 'punting or poling'. The one at the front plants the pole near the bow and walks on the broader gunnel (brim) towards the middle as far as he could go, pushing on the pole; then letting it float on the surface of the water while he walks back to his original starting position to repeat the action again and again. The punter at the stern did the same, only that he started from a point as close to the middle as possible and pushed towards the stern. They work from opposite sides of the boat, avoiding clashing of the poles and for easier steering. At any given moment, one pole is doing the punting assuring a continuous propellant force. The method is very efficient and effective. The repetitive action in such a rhythmic and synchronized pace propels the heavy cargo boat forward. When they come close to each other, one talks and the other listens, the role alternates at the next turn, as in telephone talks on analog lines of the past; conversed with each other taking turns. They tried to cover the most distance at night, when the weather was cooler. Shimmering night skies reflecting on the waters and a simple hurricane lamp at the bow as 'headlight' led them on their way. They sang melancholic tunes of solitude, sorrow, and struggles; miseries of being away from sweet homes and loved ones, often less mindful of people living along the banks and sound asleep at the time. In the quiet of the night, sounds feel louder and travels farther. People were used to hearing them and actually enjoyed listening to the 'boatman's genre' of gentle and nostalgic tunes. It gave a distinct ambiance to the night, which otherwise was eerily quiet and uninteresting. The ballads they sang, were simple, sung from the heart and souls of ordinary people with their genuine natural feelings. The tunes were actually soothing and soporific, inducing sleep. Over the years, they have developed unique themes and genres. They described the nightly scenes, the starry skies they see, the shooting stars across the skies and in front of them; the different phases of moon- new, crescent, waxing crescent, waxing Gibbous, Full, waning Gibbous, and Waning Crescent in keen observation. They told apocryphal (fictitious/made-up) stories of monsters in the water, the mermaids; stunningly beautiful nymphs inhabiting the nearby woods tempting them and the beauty of the green pasture lands; pretty maidens living in the huts on the banks of the canal junctions, seeing queer visions & illusions in the darkness; places, and festivals they see across the banks on the waterways and the embellished versions of their own nocturnal adventures. Their song and sound also served as nightly 'locational bearings' to the oncoming KettuVallams helping to avoid an inadvertent collision in the narrow canals. They were the most 'distantly traveled' folks at that time; their stories genuine and believable. When they see their colleagues making the return trip, they become excited; passed money and exchanged messages to be delivered to their waiting families, as to how they were doing and how far they were along in their journey, estimated day of arrival at home, and so on.

Message delivered! Postage: not necessary; Bank clearing fees: 0.00!

The boatmen make their own food on the boat itself. They carry with them a teapot, portable stove, and an 'urn' after the E. European style 'Samovar' to keep the tea hot a while longer over the burning coconut shells. Furthermore, they used slow-burning and smoldering rice chaff for heat on cold nights. Along the rivers and over the lakes, they unfurl the large sail, also woven out of fully renewable water cactus material ('Pullu-Paya') and let the wind power propel the boat. It's a good time to cook some food. The boatman at the bow usually performs that duty, while the one at the stern steers the boat along the right course. Under the hot and humid conditions, the tired steerer usually succumbs to a 'siesta' and simply lets the boat drift in the wind that is always present over the vast bodies of water. He had quite cleverly and skillfully secured the steering paddle in a knot, preventing it from slipping off and getting lost in the vast lake that used to be turbulent and crocodile-infested in those times. They don't carry more than the necessary change of clothes and hate to get wet and soaked.

Whenever possible, they take short-cuts and safer passages through the narrow canals connecting large bodies of water of rivers, the confluence of rivers and that of lake and canals; see sparsely populated dwellings of thatched huts where families live a modest lifestyle and existence. Along the route, they watch children engaging in playing simple neighborhood games with amusement; remind them of their own children; delightfully imagine if they might be doing the same at this time; felt missing them so much in deeply agonizing homesickness. They knew their wives were toiling alone, besides managing rigorous chores and naughty sons. Occasionally, one felt remorse for the sudden outburst he had towards his wife the other day. He takes comfort from knowing that their daughters do always offer helping hands, as they witness their mother's toils upfront and feel her pains; he imagines his little daughter running out of the house to greet him when he finally arrives home; the teenage son, demure and sulking, keeping a distance with a frown on his face for having had to do certain duties mother had given him in father's absence. His immature mind, holds a tiny bit of grudge that he missed a few good times with his friends. Mothers assign pieces of work to sons to make them grow up fast and feel responsible in life. When they get lost in thought and brooding, the homesickness begins to ache. "Life has to be lived; the needs of livelihood had to be met!". They murmur to themselves and reconcile their plights with a deep sigh of sorrow and helplessness.

The dusk marks the culmination of the day's hectic activities. The boatmen see people getting back home in a hurry. Immediately, a brief flurry of other activities begins in every home. A chant in great humility and supplication emanates from a waterside hut "Rama, Rama pahi mam!" (Lord Rama, Have pity on us; help us to meet our basic daily needs !)... sung by a boy and a girl squatting in front of a 'deepam' (oil lamp), palms folded as in "Namasthe" sign, and in deep reverence.

Meanwhile, the "Lord's prayer" emanates from the neighboring home where the assembled are seen kneeling with folded hands. A loud prayer call echoes from the top of a minaret from a distance, from the other side of the bank. A moment of solemn holiness pervades all around. The oarsmen rest their paddles; stay still in respectful homage and say their own short prayers as they would, if they were at home with their own families.

The boaters spot a faint reflection of a lamp in the water; then the source itself, a lamp post beams a bright light far away across the lake. That is the lighthouse towering over the coconut palm groves at the small 'foreshore'. Village of Karumady, about two-third-way between Kottayam and Kollam towns. They were the two business centers at that time, and connected by 120 Km long continuous water route. Originally, the light tower was installed there by the British colonial Administration after a series of fatal boat accidents involving their own men. Soon, it became a tourist attraction. Now with fond and nostalgic memory, I recall my maternal grandfather taking me to see it and visit the other attraction at the nearby location, the idol of the young, meditating Shri Buddha, reverentially and affectionately called "Karumady Kuttan". The dark granite idol has the torso split exactly in half, but miraculously the head remaining intact. The idol is a relic from the 11th century and only relatively recently recovered from the bottom of the heavily traveled cargo route. The speculation is rife that it may have fallen off from a cargo boat in one of the afore-mentioned accidents. Colonialists coveted ancient artifacts from exotic countries. With the proliferation of Billionaires in the world, its value can only appreciate. They prefer it over gold which is mainly regraded as 'hedge against inflation'. To own one of them and proudly displaying it make people look rich and adventurous among peers.
Besides, individually, he/she owns a piece of relic from the ancient civilization!

Now, the idol remains consecrated inside a "Sthupa" of a mini pagoda at the bank of a long winding waterway-canal watching over the boatmen and passengers. The site has since became a Buddhist pilgrim center and being kept under the protection and patronage of the Kerala Archaeological Society.

This location is a familiar place to the frequent travelers. The lighthouse was intended to guide the boats to the 'neck' of a rather narrow canal leading out of the lake in dark, stormy nights. That's also where boatmen look forward to congregate and rest. Surely, they will meet a few good men there. There are some card-game bets of the past to settle with some scoundrels. Once they arrive, they begin to chat about the tide & currents, the 'toll men' and policemen; the going rate of bribe payoffs the excise officers demand at the Thottappalli 'canal locks', [canal locks are used to ensure that the vessel with large displacement of water does not get stuck at the bed of a shallow canal pass way connecting water bodies at different depths] and the premiere release of the latest cinema in the nearby 'talkies' theater. They play spirited card games, staking moderate bets; after which, catch a fully stretched-out sleep on a different surface other than the hard-wood of their boat, tolerating the loud snoring of the fatigued fellow boatmen beside.

Well, this is one of the many events or lifestyles of the forgotten past, and none make any mention of it any more. But humanity that lived at the time period benefited enormously from the toils of these brave men. They were independent teams of 'small-business entrepreneurs' who never achieved wealth; only earned a moderate premium for their labor who leased the big boats and provided their own manpower and labor. At the end of the trip, they divided their share of the 'charges' received from the traders after deducting the rent owed to the boat owners. Some earned extra money for the loading and unloading of the goods. Sometimes, they find cheaper goods to take back home. During their one-day rest, as a ritual, they would watch the release of the latest movie so that they can boast about it in their village or can share the story with wife in a kind of romantic setting of her taste and liking.

The cargo boatmen are certainly vital links and contributors in the evolution to modern times. These folks provided vital services of transporting essential goods and materials, helping business and economy function and grow, besides providing a livelihood for their families in a very honest, enterprising, and sacrificing way. Their work is now supplanted by truckers who transport goods much faster by roads on lorries or by rail by the Southern Railway Services, an arm of the Indian Railways, which is the fourth-longest and the highest in passenger transport in the world.

The cargo boats of the forgotten era or the KettuVallams got a graceful transformation into modern-day "Houseboats" of the Kerala tourism industry.

With no pretension of its original 'not-so-glorious status as cargo-utility boats, the houseboats are refurbished into luxurious, pleasure boats to take the tourists for sightseeing and a whole-day ride along the many narrow, winding, long and large, vast and pristine waterways, and extensive canal networks in Kuttanad. They are often seen plying up, down, and around or anchoring at scenic spots. Plenty of videos and audiovisual entertainments are provided on demand while enjoying the unique ecosystem, flora, fauna, fountains, and sceneries of the region. Fresh air and cool breeze are the added bonus. The tourists coming from the large metropolis and overseas often gasp at the most beautiful sunrise and sunset scenes from the comfort on the deck and balconies of these houseboats.

The tourists from the Arab countries are particularly thrilled to watch the different phases of the moon, and the brighter stars on a 'no-moon night' in dust-free clear nights in a gently rocking houseboat at a lonely location on the vast lake, just like they spend the nights during the desert camping expeditions. By the way, the humanity must be thankful to the moon for slowing down the earth's rotation with its gravitational pull, thereby keeping days considerably longer to its 24 hours! The Arab tourists, also enjoy and feel dismayed at the torrential monsoon rains of days and nights despite the dangers of thunder & lightning. In the ecstasy, often, they are seen standing outside while raining down to get soaked and drenched to their great merriment. It must be more enjoyable than any shower-baths in the 5-star hotel suites they are accustomed to staying, or their own palaces, and mansions. Intentionally, some choose the heavy rainy season to visit Keralam. One can easily see they surely and sorely miss the rains in their own countries. Nowadays, due to the climate change, the torrential rains happen in even dramatic fashion by 'cloud explosion' and not by 'melting' of the heavy clouds!

Bravo! Ye, brave cargo-boatmen of yester years, we salute you wholeheartedly!

Your contributions were too vital, and consequential to be forgotten.
It will be chronicled as one of the great services and contributions to human advancement, and civilization.

Therefore rightfully, you've earned a creditable place in the history of Keralam!

Chapter 11

The "Ulbavam", Genesis of God's own Country"

According to Mythology, Keralam itself was reclaimed from the depths of the Arabian Sea. This probably is another factor contributing to the greater affinity and bond of Keralites to water.

Yet another piece from the mythology is even more interesting and staggering. It explains the "Ul-BhaVam", the genesis. The earth existed hundreds of millions of years, in historic periods or episodes called "Yugam", without any life form. The conditions were similar to the other known planets in the solar system which are full of rocks, granite mountains, dust clouds, deserts that continuously emit gaseous substances. Many cataclysmic events and upheavals had been constantly occurring that brought about ensuing and consequential transformations. At one stage, a series of monumental and sustained chain of 'cause & effect' sequences took place. That was indeed the most significant and colossal of all. It was an age of prolonged and severe solar activities known as a "Yugam of solar flares".

The sun began to emit plasma and magma saturated with minerals, ions, and gases into space in much more than usual quantities and frequencies. The plasma, composed mostly of electrons, protons, and ions, and the magma, of molten rock, mineral, and organic matter were profusely spewed around. This was the result of the combination of activities of the sun enabling the fusion of hydrogen, deuterium, and helium, in exothermic (releasing energy in the form of heat and/or light) reactions. The phenomena will actually continue with varying intensity into eternity, or until the death of the sun. Embedded in them, were the strong magnetic fields. When the moving magnetic fields collide with conducting materials such as metallic particles, electricity is produced; the electricity, in turn, produces its own magnetic field when the current passes through them at high frequencies. A phenomenon of mutually stimulating and enduring energy creation was thus initiated. It propagated itself as the 'electromagnetic wavefront further and further at greater speed. This self-sustaining wavefront has many properties such as energy, wavelengths, electromagnetic fields of mass, 'attraction & repulsion' to name a few. It manifests and felt in many distinct forms and forces. Light is one of them, which is electromagnetic radiation within a tiny portion of this humongous spectrum. Today, mankind utilizes this phenomenon and range of frequencies as radio waves for satellite, terrestrial communications, radio telescope, industrial applications, and lower frequencies for ultrasound and infrared based applications. Thus, different forms of energy, such as heat, light, sound, chemical, nuclear, etc. came into being. These energy sources produced distinct effects. Difference in temperature levels produces heat, boils water, which in turn produces vapor; vapor gathers energy; rises up, and forms clouds. The clouds gather moisture that make them heavy. When the temperature cools down, they condense and make rains. Differently, energy can be harnessed and concentrated. When that's done, we have beams of light. In essence, Laser beam is intense, concentrated fine beam of light. Light has its on characteristics. It passes through different objects differently causing various effects. If it cannot pass through, it may get reflected, deflected, or partially pass through. Each of this situation can be applied for everyday application. So also all the different forms of energy. In effect, the mass of these 'composite cloud' formations permeated into all directions and infinities at the speed of light; reached the tiny planet earth in its epic journey. Some stopped at various levels above the earth; others descended to the surface. The floating clouds of magnetic materials gave the earth its protective layers of troposphere, stratosphere, ionosphere, and magnetosphere in that order from the center of the earth extending outwards. In addition, there are interplanetary forces and energy fields. All of them possess different, chemical, temperature, and magnetic energy properties. They filter ultraviolet radiations harmful to life-forms, as well as the excessive heat from the sun, and from among many other undiscovered hazardous elements. Together, they formed the earth's protective shield or space mantle!

Millions of years later, another colossal cosmic event transpired,...
The most momentous of all events in the history of the universe!....

Chapter 12

A PRIMAL CHAOS BEFORE CREATION!

It was caused by the alignment of several large planets from the vast cosmic and galactic systems in a literally unique occurrence. Humongous geomagnetic storms from the simultaneously occurring solar flares released 'coronal mass ejection from sunspots' that swelled in size as they traveled outwards; twisted & flipped, vigorously & continuously; lit up the skies with spectacular auroras; churned the earth's magnetic field lines those span between the North & South poles and affecting anything far, around, and in between. The combined forces had a gigantic, *strong, 'strangle-hold'* of anything in the path. The gravity plummeted many folds into negative numbers to the order of 10^{-6} g. That created tremendous suctional forces which caused the water from the abyss of the oceans to become agitated, turbulent, and rise; also, lifted dust and heavy rocks to orbital heights. It was the proverbial "*Big Bang*"! The four elements of matter Earth, Water, Air & Fire combined in various combinations and proportions. Each of them contain pure elements, nonmetals, alkali metals, alkaline earth & rare earth elements. Water jets shot up into space, breaching all the protective layers of earth, the tropo-, strato-, meso-, thermo-, and iono-spheres, and over 100 kms above the sea level. This is compared to the undersea volcanos shooting water jets to a height of 20-30 Kms on the average. The water contains pure elements; they interacted with the cosmic, magnetic, and ionic particles and formed new compounds. Being lifted closer to the hot and bright sun, they were melted, fused, and underwent further transformations in the space. They gained different masses, density, and properties. New substances were created. In the perpetual rotation along their distinct orbital paths, the planets gradually drifted away into different directions ending the 'celestial chaos', restoring normalcy and balance. Eventually, gravity returned! The particles that remained suspended all this time in the space, gradually descended to the earth. Over the next '*Million-year cosmic cycle*' the particles underwent further transformations of chemical, physical, electromagnetic, and atomic mutations, formations, and metamorphosis. This, in turn, set out another sequence of cause and effects phenomena with 'feedbacks, regeneration, and amplification'. The process culminated when a semblance of earthly balance and equilibrium came into being. The activities produced alkalies, acids, minerals, metals, and compounds. Of the greatest significance were the Carbon compounds, especially the organic carbon compounds. The compound is found in almost all living organisms, in cells and bodies as protein, fat, carbohydrates, etc. Carbon has the exceptional ability to bond with one another, and other atoms, forming various chains, rings, and three-dimensional structures. It is ideally suited to build life forms. Under the influence of varying conditions, organic & inorganic materials; proteins & carbohydrates, amino & nucleic acids were formed. [In a nutshell, Nucleic acid contains chains of nucleotides in the cell, such as DNA, RNA that can store genetic codes / information; Nucleotides, in turn are Nitrogen bases of ATGC: Adenine, Thymine, Guanine & Cytosine, made up of Carbon, Nitrogen, Hydrogen, and Oxygen in different proportions. The various ATGC sequences contained in a strand of DNA are instruction codes for the organisms in the living process, to make proteins in order to develop, survive and reproduce. Proteins are made by different combinations of 20 different amino acids linked together in the form of chains created to perform vital functions for sustaining life]. All the 'recipe' of ingredients for lifeforms thus came into existence one by one over a lengthy period of time. Ultimately, this created the 'DNA' molecule, which is the basis of life and species as we know it today. The '*biosphere*' came into existence where it was conducive for plants and vegetation to grow and multiply themselves. Perhaps, some lifeforms existed already, but perished in its infancy as there was no ecosystem to support living. Eventually, over millinea, an ecosystem developed in its natural course of metamorphosis and mutations that could now support and sustain many biosystems. The existence of environment became contributory to the formation and birth of many new species. They could live, multiply, proliferate and diversify into many forms and subspecies, a case of cause & effects; actions, reactions & interactions; also, multiplication & proliferation that result in abundance of quality, and quantity!

Some sages and *Maharshis claim* it all came into being, right here at their home country of Keralam; hence the name: "**God's own Country**"!

Now, having a living environment, it has to be maintained and sustained to allow life to go on. The many requirements and activities needed for the purpose are met by many seen and unseen phenomena. Some of them take place continuously; others recurring periodically; still others on 'as needed' basis. Believe it or not, one such activity is carried out by the volcanoes located around the earth, both terrestrial and under seas. There are dozens of them active everyday in various intensities. They store tremendous amount of energy as thermal, physical, chemical, potential, and kinetic. They contain minerals with many chemical combinations; nutrients with phosphate, sulfate, nitrates, carbonates, water, and many other elements. The volcanoes, in fact, erupt after sustained activities after moisture has built up excessive water pressure within. The lava of molten magma help support many species and life forms. They make the soil fertile. One must conclude that among the noxious gases they spew out, there are useful ones too!

The world is certainly a busy, constantly active, powerful, and wonderful place. It's beyond human comprehension!

Bizarre Solar flares & sun's intense activities led up to the genesis, the creation, and evolution

Of the planet earth and life on it!

According to yet another folklore, Keralam itself was formed from the bottom of the sea when the immortal Rishi, (sage) Parasuraman, a celebrated warrior, an avatar of Vishnu, and a prominent figure in the epics of Mahabharata & Ramayana, also the 'Privileged Sevens to whom the epic Vedas or divine doctrines were revealed, identical to Moses in Hebrew, hurled his magical Axe on a lengthy leash into the Arabian Sea. The Axe hit and broke a tectonic plate. This plate has been rubbing against another huge mass and building up pressure for thousands of years. A 'breaking point' had reached; only needed a "trigger" in the right direction to untangle itself. It pointed towards the Southern most strip of the subcontinent of India!

A privileged divine message was revealed to Lord Parasuraman as to what is going to happen imminently to the humanity. In his infinite wisdom, the lord knew, if left alone, it will cause earthquakes and Tsunamis. Tsunami, a Japanese term for "Harbor Waves" sets out series of humongous waves and turbulence when earthquakes befall in the seas. The earthquakes, in this case "ocean quakes" are primarily caused by "tectonic movements" or volcanic eruptions. Unlike the frequently occurring ocean waves generated by wind and storms, the tsunamis consist of multiple waves that move at much faster speed and ferocity packing powerful currents arriving at the shore in quick succession. Without a hesitation, the lord decided to act swiftly. He used his only tools he had at that time. They were an Axe hanging from his shoulder, tucked on to the "clavicle" (collarbone), and a long leash in his hand which he carried with him always. He hurled the Axe on the leash high and wide with all his might in the "hammer throw" style. It was the 'longest throw' ever in history of mankind and in historical significance! He sensed and calculated the directional force and oriented the rope in the exact 'line of the least resistance'. Patiently and craftfully, the mighty lord maneuvered the leash to carve out and cut loose the chunk of mass. The breaking up of the piece set off a relatively mild tsunami which actually helped the large mass float on the surface of the water. Furthermore, the ocean waves from the so-created tsunami helped the floating piece approach and collided into the subcontinent forcefully, fusing into it with crustal melting and violent tectonic activity. The resulting tectonic and orogenetic forces (the forces that create, continents, oceans, mountains, and earth's lithosphere) created the sprawling mountain range called SouthWestern Ghats, as well as the High /Mid /Lowlands of coastal Keralam with precipitous slopes and gradients stretching from the 'fusing-together-point' of hi-ranges in the inner Eastern region to the outer low-lying Western Malabar coast along the Arabian Sea! With this deliberate act, Lord Parasuraman saved the world, and simultaneously reclaimed a new piece of land from the depths of the ocean!

Keralam was thus formed, giving credence to the common saying "Keralam, God's own country!"

In essence, these many mythologies were not just folklores, fantasies, or figments of wild imagination, they were actually a way of spreading scholarly wisdom to the public at large without incriminating themselves from the wrath and fury of gods, orthodoxes & fanatics of religions, believers in creeds of the time; kings & tyrants, and the despots. Throughout in human history, iconoclasm, heresy, unorthodoxy, and impiety have been severely punished. They were treated like 'enemy of the republic', or committing treason against a country in the modern era. Before the age of mass education, story telling was the most appropriate way to disseminate information. Importantly, the authorship could not be positively traced or verified to a person or the exact time. Among the vast population, they appear gradually, but spread fast. Those days, the learned people of the time liked to freely share the information they gathered or known, to as many people as quickly as possible. Story-telling & 'Word of mouth' were the reigning 'social media' format and parlance of the day.

There was no patent or copyright restrictions in those days!

The sages were altruistic and charitable. Without having to learn or teach the scientific theories for any of their discoveries, they wanted to direct the energy and thoughts of the humanity towards the real causes; made people aware that the occurrences are largely from natural causes. Until then, people simply believed in mere superstitions, or considered them as expressions of displeasure by the creator. Mythologies gave a different perspective; helped people to imagine and think about viable solutions leading to taking necessary actions with immediacy and seriousness.

In spite of having a dubious origin and mystical theme, there are always certain uncanny insights and profound knowledge ingrained in mythologies!

Chapter 13

The deep ravine which has thus come into existence is called "*Paa-ThaaLam*", an abysmal dungeon of perpetual doom and misery. We were told by parents that it's where the sinners are condemned to eternal damnation. The gruesomely ghastly '*macabre dance*' and playing of the '*Devil's trill sonata*' by skeletons are two simple introductory forms of torture. Furthermore, it is only an interim holding place, they reminded, before being thrown into the blazing hell among many fire-spitting imps, devils, and demons. By the age of 11, I felt I committed enough sins to be doomed to that damned, miserable place. Sadly, my younger brother who was much more mischievous, violent, and had been constantly giving me a '*living hell*' would certainly join me there, adding to my misery. Actually, I wished that place only for the people I hated at that time. I was bad! The children listen to the entire story in rapt attention, missing no word of it. Our Primary school classmates also confirmed hearing similar stories from their own parents too. So, it must be real! They must have visited this place without telling us! We concluded. Life starts miserably already! It was quite frightening! We made our own wild conclusions from there on. Oh,..That's why the dead were not returning. They go either to hell or heaven; most of the people we knew will go to the former. We started making a list, and soon it became very lengthy. The other list was virtually empty. No wonder, no locals had so far been worthy of Sainthood, while there were plenty with Latin & European sounding names whom we could not naturally relate to, in our lives. They have it all! Look, heaven and earth are theirs! Certainly, it spiked our jealousies. It added another grievous mishap and mystery to life, inevitably, also an uneasy feeling of '*inferiority complex*'.

We imagined if we could somehow go to Europe; from there it would be a facile and an automatic passage to paradise! When I finally reached there, considered to be the 'promised land,' immediately, it became quite clear to me that the people there, are destined to the straight path to hell bypassing the "*purgatory*'. At the same time, my trip to heaven seemed a bit brighter; such a marvelous place could not be and should not be kept empty!

The same folklore has been told over generations to the impressionable and captivating grandchildren as if the elders have seen it happening in front of their eyes. Sensing fear on the little, innocent faces, in the end, they politely termed it as mere fiction, adding that it would affect only real bad people. "There is a thing called 'deliverance' meaning atonement for the sins. But one must constantly strive to earn that by doing good deeds, being virtuous, do penance, reciting prayers over and over, etc.". Those words from our grandparents, of course, didn't impress us very much. We did not want to work hard at those 'difficult-to-achieve' remedies, although it may lead to eventual salvation. However, it offered a measure of momentary solace and a precarious reprieve. Actually, they feared that questions will be asked when children grow up. They will demand pieces of evidence to support their tall tales, ultimately losing the whole of their credibility with it. Cleverly, they foresaw the possibility and predicament the religions will ultimately be facing, when people start believing in truths, scientific theories and other findings.

Well, for now, let's hope that no force nor our sins shall split the adjunct land of Keralam asunder, and revert or submerge it back to its original place where it surfaced from. Certainly, there is a specific reason to dread that possibility. Something similar did indeed happen at least once in history, according to Plato. The mythical island of "*Atlantis*" sank deep into the bottom of the sea. It was believed to have existed somewhere in the Atlantic between England & Portugal. It's haunting to think about what really happened to the people living on that fateful island.

Horribly and tragically, it remains lost forever!

In our small world, people like to hear myths, mystery, intrigue, fiction, science, '*godly and heavenly stories*' in many fascinating combinations. The dynasties of rulers & emperors, religions & cults glorify them; elevate them to dizzying heights, evoking their divine knowledge and '*infallible*' privileges and rights. They thrust the myths of deities and demigods down upon the gullible populace with the sole purpose of controlling their minds, in spite of causing great fear and making lives more miserable than what they already are.

It's no coincidence that most '*unbelievable and 'bizarre*' miracles happen out of sight and in lonely places such as mountaintops, out in the high seas, or abandoned caves our own fore-fathers once lived!

In catechism classes, we were told confessions, penance and good deeds will earn '*bonus points*' ('indulgences', in ecclesiastical terms) which will commute sentences in their count and severity. We were told about 'plea bargains', 'blood money', presidential pardons and sentence commutations in certain countries. In all cases, the defendants have to cite some 'reasons for clemency'. So, we tried to accumulate these bonus points at a frenetic pace. But soon, will fall victims to some sinful deeds, and outbursts. Our hard-earned points will quickly vanish. Later in life, it occurred to us that credit card companies utilize the same heavenly principles. Visa, Master card, American Express & Discovery, all bestow bonus points and '*redeem* reward' facilities to their cardholders for the sins of 'borrowed money' by charging exorbitant interest rates. Thank god! There is mitigation for every sin. Indulgences can be obtained by '*alms giving*' (donations to charity) as well. Regularly, on Fridays a set of unfortunate poor would come to the gates; our grandparents gave us coins to give it to them. We scrambled for the coins to earn more 'free' bonus points. Again my brother Tom won out all the melee. Seeing this, my grandmother would secretly take me aside and would give me another coin. I would run after the people and give them the coin without Tom ever seeing it.

We have been wondering why we were living in the most obscure, wretched, and '*difficult-to-access*' corner of the village while the prime land on the bank of the "Kodoor" river with all the conveniences and great value had been donated to the church & community, something I fail to comprehend and reconcile to, even today. Our family must have committed something terribly evil, to warrant such a huge sacrifice and had to make amends for it in a very exceptional way! Or, it might have been done for some sublime causes to ensure the salvation of our collective souls. We were totally confused! "When one mystery is fathomed, somehow, another pops up. Perhaps, that's how a sense of balance is arrived at. Life itself is a discovery, experience, and adjustment", according to a very old saying. The unofficial village "*guru*" respectfully called CV for C.V Ramakrishnan Nair, who later joined an '*Ashram*' founded by yogi

Vivekananda Swamy, would frequently invoke old sayings of wisdom, philosophy, and spiritual matters to sweeten and embellish the local rumors he heard recently. Occasionally, when he visits the village, like the benevolent old king Maveli, he visits every home and seeks out each person by name to talk to. He knows everyone and everything that concerns our hamlet. People trust the body count he gives out over the government census registry and far more accurate than the 'Voter's list'.

When I was probably Eleven years old, I remember vividly, it had been raining continuously for a week. Water was rising up to the '*verandah*'. We had not seen any of the men of the house for a while. We were restive and quite bored watching only dark clouds, frogs, snakes, and the same school of fish that we couldn't catch. Not only that, but we had no opportunity to have some fun or play. Often, we pretended fearlessness at the loudest thunder or facing up to the intense lightning, earning scoldings from mothers. We were actually marooned and confined inside our home. My mother, her in-laws, and my cousins lived in the same house in an extended family set up of the time. One day, at noontime, before lunch, all the household was assembled in great haste and told us that we need to pray; pray very, very hard. We were baffled; usually, it takes place at nightfall. "The Rain is not stopping; water is rising fast, now cresting, the tidal waves are threatening the embankments of the paddy fields. They may soon be breached or be totally overflowed. Any of these scenarios will destroy this year's crops. Our lives will be ruined. I want each one of you to kneel down, spread your arms wide like Jesus was put on the cross, and recite the "*Litany of All Saints*"; ask them to intercede and plead for forbearance!". Our eternally stern, stolid, and strict paternal grandmother ordered in a solemn, despairing, and trembling voice. "It is too long, I'm hungry; I can't do it" my little brother protested. Others also seemed not too enthusiastic about it. There are too many people and places for the petition to pass through. Perhaps, we were subtly familiar with all the "*red tape*" and bribes in our system and all the different places we have to go to get a routine thing finally done. That could be the case in heavens too; naturally, we assumed! Life on earth should likely be similar to the one, as life itself came from heaven! We put our simple known logic together to make a little sense.

Without further ado, we started the Litany with utmost humility and devotion displayed never before, kneeling on the 'needles-like' rough coir carpets, hands stretched out wide, enduring the aches and pains on the weak shoulders and at the kneecaps devoid of adequate cartilage for cushion. Who did actually introduce this torture method for a simple prayer to the kind and benevolent God? Involuntarily and incoherently, I murmured. Perhaps the colonialists or their missionaries meted out the same torture to the indigenous natives, and they found it quite effective and sadistic enough to please themselves. We started with the "Lord's Prayer" and then on to the Litany; my mother led the prayer as the '*lector*' as she knew the long list of Saints by heart right from her convent run boarding schools; the rest of us trolled the 'refrain' "*Lord hear our prayer*!" as fast as we could without paying attention to which saint was called up on now to intercede or what we were actually saying. Our goal was mainly twofold. First and foremost was to finish the litany ASAP and end the unrelenting pain in the joints; secondly, we had to send the petition sooner before anything bad should happen. There was no time to waste! Instantly, we understood the value of time of split seconds! We reached almost at the end of the Litany, with only a very few saints left to finish the list. Then,... a distressed man appeared by the window. He was drenched and shivering in the rain; his thick black hair torn and disheveled by the lashing thunderstorm. Obviously, he was hungry and tired; probably traveled like that on his entire way. My mother who was leading the prayer suddenly stopped and got up to inquire, causing an abrupt silence. He was the loyal and trusted messenger sent by my father from the remote paddy fields that now lay hopelessly submerged in the surging lake. The man sat there crying loudly and in spasmodic fits of overwhelming sorrow not being able to utter a single word.

The horrible message was delivered 'without words'!

We stood around in great fear and perplexity; '*heard*' the eerie silence, saw stony faces, all our beloved ones standing frozen in time, right there, as in a 'tableau', distraught and staring into the emptiness extending far away to the fields in the lake. The whole energy was drained from our body. There was an unceremonious end to the holy chant. We realized that our prayers were not heard. Furthermore, the fate of my dad was also an additional horrifying thought to worry about. We suffered a huge disillusionment; in addition to the sufferings of the aches and pains; piled up only more troubles; licked our wounds; gravely resented the ignominy and the ingratitude of the Saints; cried inconsolably for long without understanding the exact reason for it.

A crucially defining and testing moment of my faith & beliefs at the early age of 11!

As children, we were dictated solely by instincts, similar to animal's behavior, exercising no cognitive judgment or reasoning capacity. The incident was quite disturbing; we wanted to fathom the mystery. When we approached mothers, they turned away, struggling to suppress grief and tears; said nothing at all. In our society, minors are not allowed to speak up, unless they are spoken to. Male elders seldom talk to little children except in the way of scoldings or spewing commands and instructions for running errands. They demanded reverence and implicit obedience at all times. There was a great wall, greater than the '*Great Wall of China*' in between. They did not overtly express love or kindness, nor show or teach anything. An exception was seen among casts who practiced traditional skills. They took youngsters into their folds, as apprentices of the traditional trades, mentoring and equipping them with skills early in life to earn a living.

They did care for their progeny; did what they could, like the wild animals train their cubs for hunting! Regretfully, the same didn't happen in our case!

One of my paternal uncles was a versatile and talented person. He was artistic, a handyman mechanic; had his own tools of spanners, wrenches, pliers, and set of carpentry & masonry tools. Our family had a rich collection of engines for various purposes of farming and irrigation. He would take these engines apart in our own hangar or courtyard, clean, lubricate and reassemble with ease, dexterity, and skill. He would run the engines so loud. That would catch the attention of everyone in the neighborhood. Immediately, they would come and watch the moving parts of the engine with curiosity and astonishment. When the children huddle around to watch the 'miracle of the machines', he would stop what he was doing and walk away without saying anything. Occasionally, he would take out his ornamentally carved flute he made in his own design and with his tools and play it. His favorite tune was "Ana-Thala-yolam Venna TharaamEda, Anantha, Sri Krishna, Oaadi Vaata!" (I shall give you the purest curd on earth, as big a size as that of an elephant's head, if you come running back to me now, oh, my handsome prince, Sri Krishna!). When we ran back into the house and room, interrupting our serious games, where the music actually originated, he would have quickly put the instrument back in the Mahogany credenza he built for himself; pretended nothing has ever happened, and whatever may have happened had nothing to do with him! He enjoyed playing pranks. We failed to understand the '*double meaning*' of the music and his action! Lord Sri. Krishna was an accomplished flutist; enjoyed playing it frequently. The pure white silky curd, extracted from the coagulated cow milk, served by the beautiful cowgirls, was the only other thing he wanted more! We wondered about our uncle's actions; if it were an act of creating mystery, a show-off, or simply a way of attracting attention to himself. Hilariously, he was gloating inside. Obviously, he was a fan of Lord Krishna and had copied certain *other* mischiefs and traits of the lord.

I rued that such a wealth of knowledge, expertise, and talents were simply wasted without sharing or benefiting any of his immediate relatives when I haplessly struggled in carpentry, smithy & foundry workshops, and machine drawing. I was clumsy and awkward in my execution! Certainly, I could have fared better as an engineer and as a fast learner! However, his son, my cousin, seems to have inherited some skills via genetics and DNA. He could assemble radios, satellite systems, install rainwater collection in cisterns, erect irrigation piping networks, recycling systems, etc. He built for himself a gazebo solely from vines, creepers, foliage, flora, flowers; added a fountain as in the mystical paradise built for 'Sakunthala,' the beautiful maiden in the mythical romantic novel "Sankunthalam" by the prolific literary genius Kalidasa. It is a cool place, 'naturally air-conditioned'. Besides, he maintains a variety of groves, nursery & fish hatchery, and a host of other fascinating creations of his own. He is the most talked-about guy; has a large fan base as well, reaching the status as the 'most interesting' person in the village.

One of the merits of an extended family system is the opportunity for interaction between so many members in close relations and age groups. The members are trustworthy, protective, benevolent, and caring. One can learn from each other, form ideas and plans; collaborate, cultivate team spirit, build leadership qualities; can identify themselves in common causes; confront adversities as a unit; develop a sharing, socialistic, unselfish, and tolerant attitude. It is always reassuring to know that other human beings are also like yourself; their needs and desires are identical; goals, aspirations, and issues converge; have similar thoughts, emotions; success & failures; frustrations & disappointments that are common in lives. Yet, somehow, each individual's experience and personal destiny come out quite differently. One understands the paradigm early in life: "United we stand; divided we fall!"

There lived an elderly man known among the villagers, nicknamed "TangoThoma" living in our immediate neighborhood. He loved to talk, played chess games with anybody who knew the moves; was witty, casual, held no pretenses of the generation gap. He used to take afternoon drinks following his habitual siesta. He felt reinvigorated after this ritual. In his mind, new thoughts would spring up; along with a spurt of renewed energy. He would then look out for any passerby to talk to and spend time with. He claimed he was used to periodically seeing "visions" and had been hearing "*callings*". Only little children would believe him! He was a habitual troublemaker at home. For that reason, he was eventually relocated to an isolated section of the barn in the premises, near an old stable with cattle and a haystack. He has told us many interesting stories before- events of the past, thrilling adventures he performed in his adolescence; the fights he fought with thugs and thieves; his conquests of many kinds; the number of favors and miracles received from his divine patrons the '*All Souls*', mainly his own great ancestors, in times of dire danger; occasionally displaying passion & devotion; at times anger & frustration in many melodramatics, and contrasting episodes. He became a heretic and had freely confessed to it.

Hesitantly, we told him what had happened at home that day and patiently waited to hear his '*wise*' comments. He pondered for a moment; we assumed he was not interested in our misfortune; but, continued after a suspense-filled long pause. "Listen, children, you have to understand Malayalam is not understood in Latin countries-Italy, Spain, Portugal, and other European countries. Heaven is colonized and patronized by the saints from these countries. They heard you but did not understand... Do you understand me?". He contorted his face; paused for an answer. We felt like crying again. "All these sacrifices were for nothing?. It's not fair!" we thought, but kept a restrained reticence about it. Perhaps, words didn't come out. We didn't dare to say anything blasphemous under the grave circumstances. That, sure is hell here on earth, and after death!

TangoThoma was pensive for a while; watched our reactions; narrowed his sight, bringing the eyeballs to the extreme ends in a critical expression, and continued. "In the Syro-Malabar diocese of Changanacherry, we clamored for our own bishop; instead they appointed the bishop of Kandy, Ceylon as ours too, remotely shepherding from there. He was the closest available clergy with some drops of European blood and ancestry. If the Ceylonese would accept him, surely these lowly '*Malabarbarians*' (a racial epithet) will! The Curia in Rome opined and decreed. When the newly designated bishop arrived to assume leadership of the diocese, the parishioners revolted, remonstrated, and rejected the '*fiat*'. Reluctantly, fearing yet another schism in the church, and to pacify the irate believers, they decided, reluctantly of course, to install Mar. Thomas Kurialacherry, the son of our own land as bishop. By the way, folks, he is also my uncle, on my maternal side! He will get me to paradise; will take care of me to the life-end and beyond. Actually, all Syro-Malabar Christians are related, believed to be converted by the apostle Mar. Thomas Slihah himself, from 12 chosen Nampoothiri (Brahmin) families, the privileged priestly upper cast. My great...great fore-father was one of them. Initially, they married only among themselves (endogamy-marrying within a group) and multiplied fast, hearing the 'word' of the Lord! So, our Syro-Malabar Christianity precedes a few Hundred years before Roman Catholicism, which begins only with the Emperor Constantine. Apparently, his mother was a Christian. Until that time, the Christians were persecuted in the Roman Empire! Could anyone imagine?

Naturally, we became suddenly very proud of our ancient heritage, pedigree, and pioneering status in the religion. Our parents will love to hear this story, if they didn't know it yet. We were impressed with venerable old Mr. Tango Thoma for the very first time. Instantly, in our minds, he became a wise man, totally ignoring his branding as "perpetually drunken" and a habitual trouble-maker!

Now, look at you, ...you guys are doomed! Ha.. Ha.. haa!". He burst into fits of ridiculing laughter. He sensed our waxing interest and increased attention.

Tango Thoma continued vigorously:

"Well, before, they ruled Malayalam was not good enough for liturgy, or in other words, it is not God's language. In my younger days, we had been praying in unknown languages, followed foreign gestures instead of our own "*mudras*"(signs & gestures). The church did not recognize Malayalam as a derivative of Sanskrit, which itself had been Gods' own language since the beginning of time! What are you telling me?...yeh..ha... Mr. T became very emotional, agitated, belligerent, and loud. Perhaps, slowly, the alcohol found its way to the brains. We were totally embarrassed and wanted to run away; waited for an opportune moment to do so. It was getting dark too; the time for the regular nightly prayers was soon upon us. Now,... You miserable chaps, You would know why you cannot find me at the church. Well, then, it is getting dark, mothers are calling, go home in peace. Aiyi, boys, ...wait! He called us back,...you could still be rewarded in heaven; when your prayers finally reach there and get translated!"

Later, he murmured into our ears, "don't babble out these things to anyone; you will get into a bunch of unwanted troubles. I want a church burial, amidst the skeletons of my own ancestors". Ha ha haaa..! His laughter lengthened and became hysterical.

"Yes sir,...sorry,...No, Sir,... we don't want any trouble;... none at all!"

We gave him a courteous bow in a Namasthe sign. We ran to our house sprinting as in a 100 m dash in school games.

Baffled, humiliated, and dejected, we returned home at the end of his tirade. The bold revelations helped us to define what racism was, and that it has a divine aspect too. Sainthood is based on how many other souls were proselytized, converted, and baptized, rather than the piety with which an individual lived out his life. Implacable and greedy people always want to possess and control the best things in life for themselves and even the whole of paradise! Secretly, I decided to learn one of these languages- the language of the Saints; weighed my options. I didn't want to learn Italian, because initially, they killed Christians and drank too much wine, an alcoholic beverage, but was declared it holy thereby usurping the noble place of pure coconut water of our land. Columbus, an Italian explorer 'discovered' America. It facilitated the way for other colonialists who perpetuated atrocities and ethnic cleansing on the native Indian tribes. I didn't know anything at all about them but really felt sorry for them as they share our name "Indians". I was equally abhorrent towards learning French because they charged colonies cash and gold for vacating the countries they violently occupied- even the poorest island nation of Haiti is still paying annual "*ransom*'! They left the Francophone states plundered, ruined and in strifes. Nor did I want to learn Spanish, because they were the last Europeans to convert to Christianity, much later than the *Syro-Malabarians* or even other European nations. All have much less Christian heritage than us. The Spanish were the most brutal and fanatics of all; stole gold from South American countries after slaughtering Kings and indigenous tribes; also not Portuguese, because they were partners under the patronage of the church along with Spain; committed the same or worse atrocities and brutalities. Through the mediation and arbitration of the church, received Brazil as their own colonial empire-state, while the Spanish were awarded the rest of South America and Central America.

The Indian defense minister V. K. Krishna Menon, a Malayalee, proclaimed “we wiped out 'the last vestige of colonialism' from India” after liberating Goa from the Portuguese occupation. According to history, they have the dubious distinction of being the first colonialists to arrive in India; also spread horrible diseases in the country. Joining forces with the Spanish conquistadors, they decimated the great “Aztec” civilization of Mexico, the Inca civilization of Peru which flourished in the North-Western coast of South America. They made no attempt to explore and resurrect the “Mayan” civilization that thrived for 700 years, the longest in history, in Guatemala. The extent of its glory is still being discovered with the aid of GPS/geodetic/laser scanning techniques. All these long years, the remnants of the great civilization had been hidden in the invasive, marauding, and spreading tropical jungles. What a horrible and unforgivable destruction have these colonialists and invaders inflicted on humanity! I didn't like English either, because they massacred peaceful protesters, women, and children who attended a freedom struggle rally within an enclosed fortress in Jallian-Wahla-Bagh, in Greater India. The colonial troops blocked the few entrances and opened fire indiscriminately massacring many hundreds of people. Until today, the colonialist perpetrator never even cared to apologize. Instead, deplorably, the Parliament House honored the brigadier Dyer who ordered the killings. The British, also stole the largest and coveted Kohinoor diamond. Her majesty wouldn't give it back; shamelessly keeping it without feeling any qualms about it; although stopped wearing it on queen’s headgear in public parades and processions. Apparently, they are asking for “provenance” from the true owner, the deceased Maharajah of India for proof!

It is praise-worthy that the people of India are forgiving in spite of all the atrocities committed against them! They choose peace, harmony, and good relations for the good of future generations. The enduring Indian mantra of diplomacy (even to the Chinese) is "Friendship to all, malice to none!"

Well, I chose German, mainly because my grandfather had some cordial contacts and spiritual dealings with German missionaries whom he sponsored and protected. Besides, they didn't do anything particularly wrong to India. After all, we share the same Indo-Germanic languages. They are rumored to have migrated from the Indus Valley Plains in ancient India following the decline of Moen-Jo-Daro/Harappa civilizations. I was told they brought 'Schokolade' and 'Kuchen" (chocolates & cake bakings) as gifts when they came to visit my mother's family. They were such delicious variety and not found in shops locally. They were original German! Everyone often felt a longing for the taste. I wonder how did they keep them from melting in the tropical heat! But my main goal was to get to heaven and lead a trouble-free life from there, and then onto eternity. From there, I can sponsor all my loved ones, like the so-called "*chain migration*' as in the US. How wonderful that would be! My brother Tom, who soon after his troublesome childhood years became my closest, passed away a few years ago. He, and three other departed younger brothers thereafter, must be now doing this work in heaven for the rest of us!

My aunt in the convent sent me a children's book of angels and paradise on the occasion of my First Holy Communion at the age of 7. It was my first illustrated book in original color. I liked all the pictures except the hideous snake on the apple tree tempting Eve. However, it gave me all the reasons to hurry up and learn a '*paradisaical*' language. Quickly, I learned some words to be uttered at the gorgeous gates of heaven: "Wunderbar, Viele Grussen lieber Engel, Lassen Sie mich bitte r'ein gehen!" (How wonderful to greet you dear angel, please let me go in...!)
I didn't want to be turned away because of a misunderstanding in a vital greeting protocol!

Or in the worst case scenario, to the Satan: "Entschuldigen Sie bitte, Lass mich bitte nicht in der gleichen Kammer mit Herr. Adolf Hitler einsperren! (Pardon me!... Please do not lock me up in the same cell with Adolf!). After all, Adolf caused deaths and torture of millions of people. Perhaps instinctively, it occurred to me that the torture in that particular cell would be worse than those at Guantanamo. A simple investigation suggests that even there, only a single page from the '*Book of Torture*' by the colonialists, which was originally performed for the '*divine glory and winning souls*' was applied. They were in fact legacies from Roman emperor Nero, who enjoyed applying these ghastly methods against the early Christians. Over the years, it has served as the "*Modus Operandi*" for many subsequent totalitarian regimes and despots, extending to our present times and spy agencies of democratic regimes with the additions of 'poisoning' and 'nuking' dissenters and political rivals. In the early colonial times, not only kingdoms were annihilated, but also the resourceful and ancient civilizations. As mentioned earlier, Aztec, Inca, Mayan, and other civilizations were brutally destroyed. The subjects were killed off after subjecting them to severe torture for not readily converting to the "new faith'. For historical purposes, some replicas, and models of these gruesome torture devices & contraptions are preserved in a few museums in Central America. I happened to hear trustworthy accounts from citizens of those regions and members of the indigenous tribes, where history is re-told in stories by ancestors from the ancient times.

Many sorts of '*crimes against humanity*' are being committed today in the names of religion, faiths, ideological and dictatorial pursuits, as well as for creating monolithic nations. Only a hapless few, all of them from the so-called '*Third World countries*' are subjected to trial & punishment in the International Court of Justice. The leaders from the powerful and influential countries remain privileged above law, live as noblemen, unscathed, their reputations and standings untarnished, proving to the world a good media & PR campaign can burnish any heinous acts; consider atrocities & crimes just a trivial aberration, a valid justification for the unfortunate transgression; ultimately rendering it legally acceptable, and culminating in absolute vindication. It makes all the torture and atrocities vanish and forgiving! "What is done, is done with!" The never-ending Iraq and Afghan wars are a few of such examples. As we have learned, battlefield victories are only temporary. We are shown again and again:

There are no winners in wars; only casualties of many kinds!

Complicity, abetting, and aiding is a crime no matter by whom, or for what reasons. They are punishable crimes with proportional and commensurate with severity!

Well, it didn't take long for me to find out that Malayalam is equally popular in hell as German, English, French, Italian, Spanish, and Portuguese.

What a waste of time! No one told me earlier!

Many eminent thinkers and anthropologists wonder if our lives on this planet would have been better if the ancient civilizations still co-existed and were allowed to flourish. Life and nature with its essential resources of fresh air, water, and other pristine natural resources would have lasted much longer, serving humanity for several more generations.

Chapter 14

THE JOYFUL FESTIVAL SEASON

Soon, there will be races in this river; here, there, everywhere, local, regional, and mega races, all with enthusiastic participation from people. This spirit transcends caste, creed, faith, politics, wealth, social ranking, and other divisions notwithstanding. Lately, the interest in regatta has spread to the regions of AshtaMudi Kayal, the ('octal-pronged' lake) which is the second largest in Keralam. It's a lagoon and considered a wetland area under international conventions and definitions. The surrounding population largely consists of fishing communities. Within a short period of time, they have proven their extraordinary skills in rowing and paddling, and worthy contenders as well as enthusiasts in the races. There is a widespread sense of participation, competition, common goal, community spirit, sportsmanship, friendship, and joy among all people. It exemplifies a feeling that we all belong to the same, one and only 'living planet', our beloved 'blue planet', the only planet that is known to support life in the solar system! There are ample reasons to be alive and live happily together! In the village, people lead simple lives. Their lifestyle and beliefs are traditional; they want to provide for their families by maximizing available natural and renewable resources, which is mainly the house they live in and the land & fields they cultivate. They share their knowledge, skills, and manpower; borrow and exchange items in casual manner as opposed to making business deals or rentals. They celebrate feasts, festivals, and ceremonies in great communal spirit and cooperation.

The harvest season has just concluded. All the important work has been completed. The bustle & hustles have come to a screeching halt. A well-deserved rest period is beckoninh them. It is the right time to take part in some fun & games, feasts, festivities, and social interactions within the community. It's largely realized in VallamKali. Each "Kara" (hamlet) has its own race boat or simply 'Vallam", just like the European countries have football clubs in every town and attach their identify, fame, and prestige to the success of the team. All festivities and competitions must conclude before the onset of the ravaging monsoon season. Surely, that's a big gamble and stern challenge!

Chapter 15

Early preparation before the Vallam Kali

The 'Kaaranavan, *'the raceboat* steward/custodian', and the "*Ashaari*", the carpenter get together. They assess essential maintenance items to be performed on the race boat and start on it with great alacrity and meticulous attention. At first, a special formulation, a grease recipe, which is actually a concoction of fish oil, egg yolks, bee wax, charcoal, wood turpentine, this oil and that, all well beaten together into a dark thin film of coating, is compounded locally. This is intended as a lubricant or water-repellent, applied to smoothen the surface below the chine or the waterline. It reduces the '*hydraulic drag*' as well as sealing any hairline cracks or fissures that might impede smooth gliding motion through the water. With the help of enthusiastic volunteer manpower and co-operation from the villagers, the boat is turned upside down on stilts. Carefully, they apply the grease to the underside of the boat in the same orientation as the wood grains are, with similar precision as the Ayurvedic masseurs massage the body along the contour of veins, ligaments, and muscles. My brother, who was mischievously whimsical and aptly nicknamed '*naughty Tom*' who was 8 years old then, was never shy of action. He brought out uncle's shoe shining kit to do the job, uninvited, for which he got loud and swift admonishment from the elders. He was given a stern '*restraining order*' from entering the "*Valla-Pura*", the boat hangar. The carpenter made a toy for him- a "*Pamparam*" or '*spinning top*', as an appeasement to win him over and to influence his strict compliance. Children seldom had toys of their own. Therefore, obviously, it drew some envy from his peers. Contrary to the norm, we were quite fortunate to have some rarely seen toys, a toy piano, and a cork gun which our paternal uncle, Anthonichan '*Appa-ppan*' brought for us after his post graduation studies in Law, from the elite Aligarh University. My maternal uncle, 'Chacko Chayan' had joined INA (Nethaji Subhash Chandra Bose's army) and served in a foreign country during World War II. In fact, he ran away from home to join the non-official army and caused enormous agony and anguish to his parents and siblings. When he finally returned, he had brought a gramophone, a binocular, and several sets of prestigious uniforms. In gestures of appeasement and affection, he gifted the gramophone and the binoculars, all rare items, to his only sister, my mother, whom he missed for several years and missed her wedding. My brother and I were the unintended beneficiaries and soon became '*spoiled brats*'. He would dress himself in the uniforms, that were too large for him and beat us all up with the baton, claiming he has the power and authority; like 'If he were a police officer!' Thank God, he didn't have guns like the US police officers!

Traditions and customs are respected and properly followed everywhere and every time, 'as it were in the beginning and till the end'. God, parents, elders, and teachers are revered in 'filial piety' in the same order, as in most other oriental cultures. Besides people and positions, certain animate, inanimate things, as well as certain shapes, objects & figures are also revered in sections of the community. According to the implicit 'code of conduct'; disrespecting any, will invariably invite wrath of gods, holy men, and fanatics in the form of banishment, ostracism, or 'fatwa'. It is all considered blasphemous; seen as hurting people's feelings and vilifying age-old beliefs. Even, Science, facts, or disguised references are considered disrespectful of someone, a religion or a belief. Words should be carefully spoken! Somehow people automatically know "the do's & don't s!"

It is a grave offense, far beyond simple misdemeanor!

Chapter 16

Making of a Chundan vallam tribute to the Ashari the carpenter

My maternal grandfather was a great fan of boat races. He had sponsored local races and held executive positions in various committees of larger ones. It was quite fascinating to hear from him how these race boats were originally made. He told us that his older brother-in-law was the *Kaaranavan* and a leading paddler of a racing boat. He had accompanied the team of master carpenter and craftsmen to the tree forest. Everything takes shape and form in the sagacious, and resourceful mind of the master carpenter, who selects the right wood for the job. He likes to use natural wood native to the prevailing weather conditions; travels to the deep, dense forest in the Western Ghats, locally known as the upper Hi -ranges, where Teak, 'Aanjili' and Indian Mahogany trees grow strong, tall, straight, and vigorous. Teak is a close-grained hardwood. Aanjili is the preferred wood to build a race boat. The evergreen tree can grow upto 40 m long and 4.5 m in girth. Like teak, it is very durable and comparable, but likely less costly. The wood lends itself to accurate work with tight-fitting joints, and facile carving. Wood with irregular grains like that of Maple is not suited for the purpose here. The colorful hazel brown, black striped Sri Lankan Calamander wood also receives much consideration. It's also close-grained, perhaps a little too hard and heavy. The master carpenter picks and chooses the standing trees with the similar precision and specifications applied by the superlative 'luthiers' like Antonio Stradivari and Andrea Guarnieri in selecting wood for their exceptional violins. Strictly speaking, all search and find stumps of similar tree once grew in the vicinity; strenuously compare forestry records. They would meticulously examine the cross-sections and the microscopic cellular structure of wood, including annual rings, rays, grain patterns of the trunk. The annual rings appear like concentric bands, while the rays extend radially outwards almost like a straight line indent or mark. It tells the tale of its growth and can be considered identical to the age-marks on human forehead. A transverse cut of the wood shows the cross-section of the wood with annual rings, revealing the age of the tree. The vertical cut through a tree trunk shows the lines forming the inner structure of the wood, which are grains. The oval-shaped figures interrupting the grain patterns are the knots indicating where the branches grew out from the trunk of a tree. Knots weaken a wood's structure, and in races, cause unwanted friction and drag, slowing its motion in water.

A tree climber uses a rope tied to his waist to measure the length, which is later used to control the actual felling of the tree. Actually, he takes just the end of the rope, carefully arranged in loop that is free to unfold without getting tangled. His partner on the ground feeds the rope reducing the weight and help the climber with a facile pulling of the rope. The climber hangs the loop on successive branches on his way up. The action needs good technique, planning, strength, balance, and dexterity (they must have learned the technique from watching animal that live on trees!). These people are professional loggers with traditional skills, and have been mainly doing this kind of work for generations. It's a very physically demanding and dangerous work. There's no way to bring heavy machineries deep into the dense jungle without damaging many plants and trees. The danger is even greater, that they are not used to wearing safety harness, hooks or safety nets. The climber keeps a simple, soft cotton sling, about a half of the diameter of the tree; hold it between his feet for traction around the outer tree trunk; uses hands for leverage and hopes up in 'monkey strides' (factually stated, and not in any derogatory sense!) to climb. It's an audacious task; requires a great deal of agility, concentration, body balance, skill, and precision. He pauses and hold onto the few strong branches on his climb (Remember that the selected trees haven't got many big branches!). At the end, he makes a wedge at the apex by chipping a small piece of the bark to make a 'foot -hold'; hold on to strong branches; balance himself, then securely perch himself; survey the ground below and the surroundings where the tree may fall.

At your service!

Elephant & Mahout, inseparable beings; work in great understanding, friendship & partnership

The climber seldom uses any restraining devices except for some ropes or rudimentary tools. One has to conclude that in occupational skills, these workers use only traditional, handmade tools and techniques. They seem to be averse to any mechanical, safer, and modern methods or hoists. Apparently, they believe it needs different understanding and technique. It messes up many other learned skills and acquired capabilities. Any missteps or error in the execution would be even more dangerous. It's quite scary to watch them performing the many obviously hazardous tasks; but at the same time, equally surprising to see how they accomplish each one with seemingly facile dexterity and favorable outcome. The climber had to do the routine three times, each time tying a rope at three close by heights. The initial climb is the hardest for obvious reasons. The members of his support crew tie the other ends of the ropes onto anchor hooks pitched firmly into the ground at 120° apart, or simply wound around bigger trees that are found suitable. A bigger rope is wound around in a few loops at the bottom end and restrained against a group of stronger trees to prevent the end from swaying, pitching, or jumping off out of control. I like to compare this web of ropes to the 'spider cams' used in the live broadcast of sporting events. Here, the load is much more cumbersome, and many times heavier! The weight of the tree is distributed from the bottom to the top in its entire length, not concentrated at one point. The restraining leverage applied by the ropes have to be well-balanced too. Now, the setup is complete for a controlled felling of the tree at a precise spot on the ground. The crew is mindful of the perils involved. Worse, the mind-computations can go wrong under different load, force, and leverage conditions. They take necessary precautions and locate safe path and space to get out of the way, if anything should go wrong; maintain a serious disposition; remain alert and laconic in communication with another. Each seems to know the right sequence, timing, and individual role in the entire task.

All are 'cautiously optimistic'!

It has been a common practice in our community that the minors are sent to accompany the workers on coconut harvest trips, which happens every 45 days or so. The purpose was to 'watch & supervise'. It was intended to prevent losses due to scattering of the 'felled' coconuts disappearing among the thick bushes or any intentional pilfering by onlookers or tenants. In their naiveté and innocence, children are more likely to confront and question the offenders. Quite dutifully, the elders used to inculcate three main cautionary and important instructions:

1. to keep an extra-safe distance away from the tree for own safety; watch and avoid being hit by scattering coconuts
2. to not speak to the climber nor ask questions distracting him while he does his climbs and perform dangerous work.
3. To be respectful and thankful for his skills and service

At the tree-top, here in the forest, the climber trims up the rest of the branches neatly with a sharp sickle-blade carried with him, dangling from the prominent 'depression' on his clavicle or collarbone; the sharp edge is by 'design and make' is situated safely away from rubbing his skin on his back. He never bothers to cover or shield the sharp edge. To suggest anything may irritate him and roil his stern disposition. Others keep quite; quietly let him do his thing! Sure enough, he finishes his task beautifully!

Indeed! man's ingenuity and improvisation manifest themselves in many unusually creative ways. Yet, as prudent precaution, People on the ground must be constantly aware of the catastrophic event of the tool slipping off, and quietly falling from height!

On the ground, the wood cutters use Axe to hew at the bottom of the tree to break and bring it down slowly. They work in tandem or alternate; pausing and examining the progress and the 'lean' they want to create. With great care and precision, the tree is felled without causing damage to itself or others, in a methodical manner, controlling the slack of the four ropes and sometimes using other trees as lean, prop, and support. It is progressively slanted, as the puppeteer manipulate the action and movement of his puppet. Finally, they let the tree gently fall to the ground covered with soft and thick haystack arranged the entire length of the tree for 'soft landing'. It's quite astonishing to see the tree land exactly in the middle of the 'bed of hay' with a few harmless gentle bounces, as planned. It was like an infant merrily jumping on a cushioned bed or sofa!

It must be added that the forest conservation rangers/officers certify each step and issue licence to cut the mature trees in specific locations, after several bureaucratic protocols of documentation and processing. They will observe the entire operation to assure that there is no damage to other trees or the saplings they have planted. They impose other relevant and necessary compliance and requirements. Significantly, it is the 'Amazon' rain forests of the subcontinent consisting of abundant varieties of biodiversity, the home where thousands of species of tropical, medicinal plants & vegetation grow, as well as animals inhabit in its wildlife habitat! There are many animals such as elephants who don't view kindly of anyone encroaching and damaging their home territory. It is supremely important not to agitate or antagonize them. In other words, this work has to be carried out in extreme secrecy form the tusker elephant herds!

The remaining branches of the felled tree are neatly trimmed and dressed up to a long, thick, straight piece as a well-tidied timber. After critically examining the wood several times, the eagerly waiting carpenter takes all the necessary measurements of length & diameter. He forms a sound mental plan of his subsequent work at the boat hanger. Then, he goes on to identify, mark, and certify each log. They are then moved down the slopes to the mouth of the 'Pampa' river. This operation is carried out by an experienced 'foot-mahout' commanding trained team of tuskers. They use a chain harness for skidding the logs. Several small, trimmed, and polished round logs are used as 'rollers'. Another tusker is tasked to transplant them progressively ahead of the entire trek that may be K ms long, needing several hours. At narrow paths and spaces, the tuskers skillfully slide the heavy timber. It's quite fascinating to watch how the mahout and the tuskers communicate and coordinate the activities. Instinctively, both understand the role of the other fully and well at any moment and phase of their common task. The elephants are strong, intelligent (having 260 Billion neurons, compared to the maximum of 90 Billion for humans'!), and sensitive animals. Thus, the Mahout and the elephant, the two most intelligent animals on earth collaborate to accomplish this rigorous piece of work in coordinated & synchronized joint venture!

As a reward for the nice work done, the mighty tuskers get sumptuous meal, rich with a variety of supple plant stems, green leaves; delicious jack fruit, and banana, following a luxurious massage bath at the shallow end of the river. It is also given lavish playtime in the water and receive pampering rubs & scrubs. Later, the mahout and the animal get to rest at a large open space. Soon, both succumb to a well-earned siesta in the soothing evening breeze under the shade of a sprawling tree situated nearby. The tuskers start snoring in hissing noises, sometimes twitching its trunk apparently trying to reach and embrace the master sleeping close by. The man doesn't feel bothered at all! Actually elephants have remarkable body balance like athletes and gymnasts. Its uses its massive four feet and keep the weight within the gravitational center. They could sleep standing up. That's what the wild elephants usually do. Tamed ones are accorded extra comfort. They are trained to lay flat on the ground. Humans like them a lot. If a big and strong enough bed could be made in 'super-quadruple-extra large sizes, the elephants will probably get one too!

At this time, at the slops of the river bank, a different crew tie up the pile of timber logs tightly together making a floating scaffold & platform to transport it through the water. The oarsmen cleverly negotiate a favorable combination of time, tide, and flow of the current using a combination of punting, poling, pulling, and steering in the journey. Mostly, they are maneuvered through the middle of the river. The strong current may sometimes become too hard to control, making the lanky load accidentally bump against the banks on the turns causing damage to the cargo and danger to the crew. The diminutively resilient crew, live, cook and sleep on the hard logs, not showing any apparent discomfort under the changing weather conditions during the journey. [As a mature person, now I realize that all human beings feel pains and aches, just like me or anyone else. But they suffer and tolerate hardships for lack of other choices. The selfish, luxury-spoiled world has become averse to thinking about other people's plights and miseries].

Only during the heavy thunderstorms, they park and seek shelter at river side tea-shops, and wait it out. Depending on the length of the adverse weather conditions, they may order some hot food, sip a cup of warm tea, and may take a short nap. Usually, it takes several days and nights to bring the slow moving load of logs to its eventual destination.

The waiting master carpenter and the Karanavan, along with many volunteers joyfully welcome its arrival. Here again, a local elephant accompanied by its mahout pull out and drag each piece as carefully as possible and place it on rollers at locations where the carpenter has already marked. At this time, the designated hanger is wide open and without a roof to accommodate the large animal. As each piece is brought in, the carpenter hurriedly take another round of measurements and scribbles them on its 'face'. He has proprietary 'cryptic' sets of notations and units that no one else can read or decipher. The elephant and the mahout patiently waits for successive instructions. The logs are set to dry and acclimatize for at least 6 months before work on them can begin. Meanwhile, each timber is turned over, checked and mildly treated to prevent rotting, also as protection from ants, termites, and other pests. The logs are useful and coveted items, they must be also secured by proper means. Usually, chains and padlocks are used for the purpose. They are covered with a rainproof 'tarpaulin' canvas sheet. Few weeks prior to the start of work, sawyer jacks are called in to cut the tree trunk into logs per the strict dimensions provided by the carpenter. Soon after, the community volunteers erect a thatched hanger.

The boat making may begin anytime! Predominantly, it consists of a series of intricate carpentry work in great precision. Nobody has seen any plan or drawing of the final thing. Everything is shrouded in a veil of mystery as is the case in most everything in our community. There are some rumors and speculations coming out of some people who pretend to have the privileged 'inside scoop' as to the total length, paddling capacity, height, and weight of the chundan vallam. Nobody is bothering to verify the data anytime, for sure. By habit and behavior, people are casual, not precise! Apparently, the traditional philosophy is that life on earth is a transitory event; meet them when they arrive; it needs to be lived as such with whatever means one possess at that time!

The master carpenter seldom uses any sketches, drawings, or written calculations. He uses only hand tools, hand saws, an assortment of chisels, mallets, planes, hand drills, etc. which predictably, date back to the first person in the family who entered the trade. He is also averse to using clamps or vise that may damage the wood pieces, instead uses his toe as a "*claw*" as the falcon holds its prey. He applies his own weight to hold the pieces firmly in place, partially squatting with his feet on it. No one can sit that long in many 'yoga-like' poses without suffering muscle cramps or fatigue! Strangely, he uses a long, solid measuring stick with subdivision markings that he carries everywhere he goes, including on flights [he is allowed to take it with him as a 'hand carry'! Somehow, he manages to keep it in such a way not hitting or poking other passengers upon boarding or disembarking] when traveling to Gulf-countries for work assignment. The stick conforms neither to the metric nor to the standard units for the obvious reason that his trade practices were based on the tenets and dimensions specified in the Vedic Architecture developed by Lord Viswakarma, the master architect. That was a couple of millenniums prior to the relatively recent introduction and standardization of universal measuring units! The bulk of the work is carried out by carving and chipping methods. The pieces are assembled only by joints or custom-made wooden pegs or pins without any adhesives; nails, screws, staples, or the likes. He loathes chemicals. Apparently, he cannot stand its obnoxious smell, which causes allergic reactions and asthmatic conditions in people. The master carpenter utilizes assorted cogged joints- Finger joints for corners; for firmer joints the *Dovetail*; *Mortise & Tenon* to join horizontal and vertical pieces; *Tongue & groove* to join planks; *Mitre joint* for squares; *Bridle joints* for rafters; "*Cross lap*" for intersecting pieces and a number of other joints of his own invention and custom applications. One can hardly see him sort or arrange the pieces, but seems to find the right pieces promptly when he needs them.

I have had the good fortune and privilege to work and interact with many groups of professionals in my career. That includes Scientists, technologists, doctors, and lab professionals. Many genius people have had founded companies & corporations, developed amazing techniques & technologies. After a short while, having earned a rich status of some high degrees, they quit or retire; have other people run their enterprises. They go and 'enjoy' life; often lose purpose; indulge in some 'nonsense', sometimes ruining reputation. As can be seen in the images below, to be honest, I have yet to witness anybody working with more concentration, seriousness, purpose, and precision throughout the whole day, and even the entire lifetime than this group of artisan workers.

The master carpenter meticulously and artfully handcrafts a ChundanVallam in the same traditional way as in the 1400s

A few other carpenters, all of them belonging to his own immediate family, assist the master. This artisan community does not have any documentation, "How to do" or "Carpentry for Dummies" books. He needs none. He learned it first hand from his father and grandfather who learned from theirs and going back to preceding generations. All of them were preeminent masters themselves of the trade. They know everything that has to be known! This trait and trend are still prevalent in other skilled professions in the community at large, as well. The knowledge, skills, and techniques are transferred from one generation to the other by familiarity, family apprenticeship, and perhaps via "genetic" transmission. This is arguably the origin and essence of the caste system in the whole country. In fact, it's a world-wide phenomena, but India gets the whole blame for its negative implications! [In the historic period of Medieval ages in Europe, It was a usual practice to assign or accept surnames based on one's occupation, as examples: Smith, Butcher, Butler, Bailey, Miller, Abbot, Potter, Sailer, Knight, Thatcher, Tailor/Taylor; Hoffmeister, Baumgartner, Bauer, Becker, Muller, Cantor, Cohen, Mahler, Kaufmann, Krammer… (Ger.)…]

If anyone sneaks into his workplace and observes his work, one might think the carpenter is possibly building a door, an item of furniture, or probably a roof truss at other times. It is quite hard to make out any sense, scope, or actual purpose. If you ask him, he responds with a sullen silence and a wry look. When it is time to assemble, every section is found wrapped as unrecognizable figures in sacs made of Jute fiber and stored among hideous-looking scarecrow figures hung in the way. As timid and fearful as I was, never dared to sneak in, to examine in spite of my obsessive curiosity. He works with rapt concentration; seldom talks while working and seems annoyed at any distractions, including natural occurrences of rain, thunder, and lightning. When he is called for lunch, he shows no hurry; bangs some more with his mallet to finish 'that' particular piece he is working on. Discontinuity seems to be his worst enemy! He is extremely superstitious; consciously avoids any evil spell or omen of any kind. He is the most laconic of all the skilled workers. I do often wonder why! I have sat in front of them curiously watching their work of shaping and fitting, and chipping off some more wood to get to the precise fit. He never looks away; he is in a world of his own, alone-he & his work piece! You'll never see him throw out any item because of misfit or damage. That's how precise his work is! He keeps any unused wood material for any future use. Later on, at college, I have done some carpentry workshop as part of the curriculum. Almost all joints I tried to make were misfits and was totally useless for my assigned project. It made me feel 'small'! It used to torment my ego and damage self-esteem. About 25 years ago, it became necessary for me to undertake a formidable renovation task to a section of our ancestral home. It was all made of teak wood with carvings and glazed ornamental work. The carpenter was very hesitant as the 10 m wall was carrying heavy top-load. But we accomplished that task successfully. The opposite result would have been quite damaging and disastrous. It helped me redeem myself to some degree, although it was not an 'ego-trip', but an essential job to do. It stands strong still showing a pleasing appearance and good usage!

At the end of the day, the villagers come with bags and sacs; sweep and collect saw dust & wood chips, wood shavings from sawing, sanding, milling, planing, and routing. They take it home for use as heating fuel. "Wasteful behavior is a punishable offense by God and your fate!" That's the lesson taught in our society since early remembrance. [The trait becomes a habit that lives with you as long as you live. It made me really miserly too. Once, as an apprentice in Quality control engineering, I detected a microscopic fault in a measurement and alignment of a critical component. When it was reported to my supervisor, the entire production lot was discarded into a dumpster without any hesitation. It bothered me deeply. I decided QC is not a good fit for my mental health!]

Usually, the carpenter works until he cannot see any more and goes straight home; resumes work at dawn. Mothers wake up school-age children, hearing the sound of the mallet striking the wood. They scold the children who are late in returning home in the evening from the playground by the churchyard, or marketplace, saying "by this time, even the carpenter has come home!" After months of diligent team work, the race boat has gradually taken shape. The carpenters finished their task. Until this time, others did not have any idea of its form, shape, or size.

Preeminent village leaders and stakeholders are brought in, to see the boat and obtain final approval for the launch. They inspect the work with a magnifying glass at every joint as Sherlock Holmes looked for fingerprints on the murder weapon. They 'see and perceive', 'look and search'; knock on the wood and listen to the echo as physicians tap on the stomachs and tap limb-joints; sight the alignment, linearity & symmetry squinting their eyes obliquely at different angles, axis, and elevations imitating the wily crow that is setting itself for stealing food. Basically, it is a critical sight examination using all the human senses looking for any suspect joints or imperfection. Instead, they find all the joints, staves properly aligned; the ribs strong and stable, seamlessly watertight, and oriented towards the back, as rightfully desired. There will be neither ultrasonic imaging nor machine tests performed, solely, the visual alignment, and listening tests. They certify the work; unanimously approve it for launch by a resounding "voice shout'. After applying oil, the boat is 'smoke & heat treated' using frankincense, charcoal from coconut shells, and rice chaff to get rid of moisture and harden the wood, and also to prevent rotting.

The millennial generation is proactive, contrary to the lackadaisical temperament of the older generation. Demonstrating good judgment and prudence, the committee has recruited a Coach, paddling crew, captain, and other active participants. Everyone is waiting for the finish of the race boat to get on. In the meantime, several design contests have been underway for several important items needed for the boat: the naming, its calligraphic fonts to be used, uniforms for the paddlers, helmsmen & Rhythm makers; gilded bow & stern moldings, carvings on the prominent and visible parts, etched decorations, the iconic and all important insignia, paddles, hats, arm bands, souvenirs, so on and so forth. Anyone could participate; so there is strong interest. But school children and other student groups are the most interested. They seldom find a chance to earn some money or prize! The chosen winner can win cash prizes and citation from the organization. Obviously, they are rewarding and prestigious!

Quickly, the anticipated news spread throughout the community. Technically, the preparations for the launching can be underway at any time. The day, date, and time are actually determined by a lot of factors. Some are traditional; some astrological; some celestial, waiting for certain planets to align in an auspicious configuration. Others are the business considerations and media events organized by influential groups and sponsoring patrons.

It's a long, drawn-out process, like the 'bill' in congress, senate, and 'on the president's desk'!

The ultimate limiting time factor is the upcoming date of the Nehru Trophy boat race. The team of carpenters has done their job and moved on with the next one somewhere else. Apparently, they don't take any vacation in their lifetime.

Their work is continuous, life-long, and all the time!

Once, a colleague of mine gave me a book to read. It was titled " The richest man in Babylon". He was a blacksmith, a similar occupation as a carpenter; both hard-working, determined and destined for success!

Unbeknownst this fact, I sought help from one of the classmates /dorm-mates at Maharajah's college, when I had difficulties with Mathematics. Thanks to his deep understanding and ability to explain mathematical concepts, both of us scored full marks in the qualifying final Exams. He was a son of a traditional Blacksmith! At one time, before the Industrial Revolution, masons, carpenters and Smiths were considered as the most hard-working, skilled, precision and artisan work groups.

Chapter 17

The ceremonious launching of the ChundanVallam

People gather for the launching ceremony. They are eager and anxious, but reserve all judgment and comments until after seeing their Chundan in the water, as to how it looks, floats, the personality it projects. The ultra-conservatives will rather wait until they see the thing in motion. They want to touch, feel, push and rock a bit this way and that, before voicing their final verdict! Everyone feels they have a right to judgment, as they have used boats all their lives.

Entry to the 'Boat hangar" is restricted. The boat is covered and wrapped with stitched-together jute sacs. It is camouflaged like the new car model for the year before its 'unveiling'. They use abstract figures of evil spirits and demons to scare the naughty children away, who may be tempted to pull out some petty pranks. While passing by the hanger, the curious folks peep through any holes they can find on any side. The young boys climb trees and take positions for a good 'bird's eye' view. Eleven strong men, reminding of the ancient swordsmen in "Kalari-Payatt", dressed in loin-clothes of "KasavuMundu" with a gilded scarf around the bare chest in an iconic knot, carrying sword and shield, one at the stern end and five each on either side of the boat, stand solemn guard, as sentries at a Roman citadel. The boat is a treasure! True, it cannot be stolen by packing in a bundle, and that too in front of all the adoring fans. The ancient customs shall be faithfully observed whole across the board. Purposely, there is no guard posted at the 'prow' (the front-end of a vessel) to signify a free, unhindered pathway forward. A young person carries a silver bucket containing a perfumed solution in a colloidal state. It is aromatic sandalwood paste. He stirs the paste with an oblong dumbbell-shaped stirrer as he stops in front of each guard. The guards look solemn and stern and stand like the "Pontifical Swiss Guards' at the Vatican. Each dips his heavy palm in the bucket and apply the paste across his forehead, the 'Kuri', or 'thilak' with a dab leaving parallel marks looking like a 4 lane highway; then a few minor strokes in three places- left, right, middle, just above the chest buttons with the thumb. The brain, heart, and lungs are now protected from evil spirits! The frontal cortex is kept cool; the heart gets a brief massage, chimes a steady rhythm as a well-ticking metronome; lungs expands and contracts normally, attains the pace of steady breathing, so also the pulse! All vital signs of life are now well taken care of. The left-over balm is shared by the onlookers. When they return home, the family will spot the bodily decorations and ascertain the fact that they have a respectable standing in the community; will certainly feel proud of them!

The "Santhi", the spiritual leader blesses the boat smashing a tender coconut against the "gunnel at amidships" of the boat, spilling all the natural colorless, mineral juice onto it, to the last drop, making onlookers salivate involuntarily. He leads a procession around the boat, reciting solemn Vedic chants, frequently varying the cadence. Everyone thinks the others understand the full meaning of it except himself. The priest also carries a chalice, shaped like Aladdin's lamp which continuously spews holy smoke or "Dhoomum", an integral part of the rites in all religions. There are several instances of imitations and plagiarism in all religious beliefs and practices. He pauses at critical positions, solemnly swirls the holder in the 'figure of 8', making swirling smoke columns like the smokers making smoke rings in playful manner. The prayers are thus transmitted to all the gods in the surroundings in equal measure. Little girls with long braided ponytails dressed in matching two-piece silk skirts and long-sleeved blouses, adorned in fragrant jasmine flower-tops, follow the holy man scattering fresh flower petals much like the flower girls do in a wedding ceremony.

Boys, dressed like young warriors dressed in 'Mundu' which is drawn between the legs and tucked tight behind like a tiny tail, and wearing 'cock's comb' (it's a secret fact that the color and beauty of the comb is an attracting factor in mating!) styled head-dress rivaling that of a peacock, sprinkle perfumed rose water onto the boat and the assemblage. Every section of society has a role to play. So, each is recognized and represented. All conventions, traditional ceremonies, and rituals are faithfully carried out. The astrologer has pre-determined the most auspicious day, by date to a specific time, accurate by 'Neram' (time) 'Vinashika' (minute)- performing the so-called determination, the "Sakunam". The carpenter is greeted in traditional and ceremonious fashion, too. He wears a long white 'Mundu' reaching up to the floor; a fine white 'Juba' and a long "Kavani", a scarf with a broad, colored border proudly representing his caste and trade, hang from his shoulder in precisely equal lengths on each side, subtly making a solemn statement of accuracy and precision in his trade and traits.

The whole community participates in the ritualistic launching of the Chundan in one of the many roles on offer

White is a symbol of purity. It absorbs no light, nor heat; therefore comfortable to wear in hot and humid conditions. The carpenter receives a gift of traditional offerings- a bundle of 5 betel leaves, 1 well-ripened reddish-yellow arecanut (betelnut) bead, 3 dried tobacco leaves, a pint of "Chunnambu", a pungent white paste of slake lime, a stack of Rs.1111 in 11 Hundred rupees and 11 single rupee denominations. Probably, it is to mean that more of the same should come to make pairs. Importantly, all the numbers must be odd!

He also receives a gold-plated, calligraphic monogrammed chest with a comfortable swivel handle, called "Che-llam" to hand-carry the ingredients of 'betel chewing' wherever he goes to work. It is much more than a commercially available 'chewing gum'. This is his only addiction. People say it improves concentration. It is quite amusing to watch how this concoction is prepared for consumption. The green betel leaf is spread on the palm as looking into a hand-held mirror; the tiny veins of the leaf are delicately pinched and peeling off without tear; the paste is then spread with fingers in few light rapid downward dabs. The item is then tilted to a horizontal position as the mobile phone in 'landscape' display mode. The sliced pieces of the arecanut beads are then spread on it, followed by placing the tobacco leaves in transverse order. It is tightly rolled, bundled, and folded into a pliable 'pinch'; then held between the thumb and the index finger, directed towards the throat before depositing it deep into the wide-open mouth, as in yawning; the fingers barely escaping a painful bite. Hands are washed clean with water out of a "Kindi", a container made of Bell-metal with a spout curved upwards and shaped like a dog's tail. The 'chewers' do this routine in no time with an individual and fascinating artistic style, and manner.

In North India, similarly prepared ready-made packs are sold at kiosks as instant "PanMasala" in assorted and aromatic flavors. The chewed crimson-colored juice becomes splattered all over the lips, making the smear of a 'lipstick' unnecessary. The chewable potion is cleverly intended for the exercise of the jaws to aid in the circulation of blood to the brain that provides added stimulation for the neurons. In effect, it is similar to the exercise of walking to stimulate the whole body. The betel-chewing is a common addiction among writers, poets, accountants, and others who perform sedentary work. The Western equivalent is the 'chewing gum', a commercial product arguably disguised to avoid 'patent' dispute with the Kerala Ayurveda foundation. 'Sherlock Holmes' did habitually use thick tobacco smoke for increased concentration and intense contemplation, somehow avoiding asphyxiation! Don't you ever try it! It is carcinogenic; potentially causing lung cancer. We tend to believe and imitate what our foreigner counterparts say or do, without raising a doubt or an eye-brow; definitely a destructive attitude!

By now, all the packages of blessings have been asked for the victories, safety of the paddlers, honor & pride; reward & prestige, as well as good luck & goodwill for the entire community. Thus, the ceremonies come to a conclusion with a great smile and loud jubilant applause from all around the hanger.

The stage is set for the launching, but has to wait for another few precious moments for the "Sakunam", the determined most auspicious time ceremoniously and gloriously to arrive!

And as we surely must wait!.........

Betel chewing is a pastime, habit, and addiction for many local folks; you'll also know where they live!

It has been a common sight to behold in a village while passing by a house; that the housewife sits on the floor legs fully stretched leaning against the front-patio column; meticulously prepares the wrap and hands it over to her lazy husband who is languishing in the sprawling 'armchair' (Charu-Kasera) with both feet elevated and propped up on the two side arms of the Eazy chair; a "Vishari" (a hand-operated fan with a bamboo handle) woven out of "Ramachham" (fragrant root fibers of a water-side plant, woven like a car radiator to trap cool air) resting on his over-sized belly; an ornamentally shaped brass 'spittoon' (spit bowl) placed at the stern; today's newspaper within reach casually placed on the arm of a side table. She makes another one for herself. Both enjoy the 'chew' and immerse themselves into the latest "sruthi" or the 'rumor mill' / gossip making the rounds of the day- a rural, no-cost, 'non-print /paperless', ad-free (unlike the tabloid- style journalism of Daily Mirror, Blitz, etc.) social media format. The "Muru-kkaan" is a widely used mild narcotic, maybe an opioid, addictive (rumored to cause oral cancer), chewed like a 'chewing gum' of ingredients from the surrounding environment. The chewed contents develop into a crimson color, indicating that its juice is fully extracted. It is to be finally expectorated by hawking and spitting. The usual method is to spit it artfully, through the two widely spread out middle fingers which are depressed hard against the lips, and into a spittoon, if one is readily available or anywhere convenient in the absence of one. Imagine the discomfort and dilemma when either the nose, mouth, or bladder is full! They must be expelled from the respective organs.

Moments later, two distinct sounds of snoring can be heard, one in 'fortissimo' (a musical term for Very loud, marked on the music sheet as 'ff'), leading the tune; the other 'pianissimo' (a musical term for very soft, marked as 'pp') as a muffled background accompaniment. Periodically, the two solos blend into a dissonant duet. The dissonance interrupts the siesta for a moment or two, but quickly resumes when the consonance returns. No sooner had they fallen into the most blissful climax of the siesta, than they are awakened by a group of noisy school children passing by after the end of the school-day. The man in the armchair curses, refuses to budge. The housewife quickly spots the chicken feasting on the paddy grains left out on the straw mats for drying under the sun. She shrieks and throws stones at them. That's the known and quickest way to disperse miscreants! She also noticed the cunning crows had pilfered off some pieces of the salted fish hung out for sun parching. None is seen or heard in the neighborhood or in the vicinity; perhaps gone to drink water after tasting salt. Therefore, the act of revenge must wait! "Everything has a cost, even enjoying a little earned rest and comfort", she grumbles.

Housewives all over the world are in charge of nourishment, nutrition, and alimentation affairs for the entire family. It is understandable that they are constantly worried about it. These are provisions preserved for the lean and gloomy days of monsoon season, when it rains continuously for days and nights, preventing people from doing any productive work. Any amount of spoils will result in excruciating scarcity. Being just a few degrees north of the equator, it prevails here the 'equinox' more or less every day of the year. That is, the day and night are two equal duration of 12 hour periods. The evening ushers in a spell of slight moderation in temperature as the blazing sun descends from its peak. The change in air temperature fans a little breeze. That is soothing as it evaporates the perspiration from the body. Soon, the cows, cattle, and buffaloes will return from the pastures; wait around the stables. They need fluid; have to be fed with "Kaadi", the leftover juice from the staple food of boiled rice prepared for lunch. It's a tried & trusted recipe for the increased production of milk. The households consume the bland, but tasty solid white rice, devoid of all drained nutrients! The ducks head towards home by dusk, crossing rivers and canals, quacking all the way. Instinctively, they know the route back; go directly to its hatchery, situated beside the chicken's. Amazingly, like humans, all the animals know their way home. Thankfully, ducks are well-fed by themselves dabbling in the shallow-water vegetation; it needs only one-time feeding. There may be many nutrients in the floating algae. Only ducks seem to know that. It must be delicious to their tongue and taste! So people simply lead them to eat, while they eat the eggs they lay. Quite smart!

The prayer time arrives soon after nightfall; The country church bell tolls. It seems to be daylight sensitive; rings between 6: 45 and 07: 00 pm without fail. Before that, everyone must have taken bath and kept themselves clean, fresh and ready. The family huddles together and chants “Naa-mam” a supplication to god in Hindu beliefs; a rosary, or ‘Angeles’ by Christians; an “Isa” prayer in Muslim faith; others follow their respective rituals. They thank god for the blessings, bounty and good fortunes they have received today and ask for the same or more in the next day. The collective sound of prayers, mellow calls of beseechments, and humble solicitations fill the air and resonate throughout the village. The prevailing moist air spreads the sound faster, further, and wider.
Without a doubt, it is the most quite, and beautiful time in the village!

A brief period is devoted to family time. After a light supper, the children engage themselves in homework, some struggle with it. Math, Physics, Chemistry, and English seem to be the hardest. The parents lament they forgot even the “Pythagoras Theorem", let alone logarithms & exponentials; formulas & equations; atomic numbers & valency, Maxwell's & Faraday's laws, because of abject disuse. They don't use any of these things in their work nor will they ever again; all the efforts were just to pass the exams and get a degree in college, above all, a good marriage alliance. Grudgingly, they practice the golden rules, the Rules of the three 'R'- Refrain, Restrain, and Reticence. The younger ones shuttle back and forth between older siblings; decide to save the unsolved ‘mysteries’ pending help from the brighter classmates before classes start the next morning. The whole village is asleep by 09: 30 pm, excepting the obnoxious drunkards and the nocturnes. They are the people and creatures who regularly rule and torment the nights uncontested.

In a typical village household there are many tasks and chores to do, keeps everyone busy and engaged

The last passenger water transport service seems to be the loudest of all and passes by my village in a way announcing to the villagers that it's too late now, and "Thou shalt now hit the bed immediately!" It's the last event of the day; nothing more to look forward to until the very early morning. Anyone you have been waiting for should have come by that boat; no point in waiting any further. Housewives know that as an absolute fact! All animals are quiet at night until the rooster announces the break of the dawn gleefully and loudly without fail. In the morning, with a glimpse of the earliest rays of the rising sun, the rooster crows. He crows; crows again & again; louder & louder, announcing the end of the night and the rest-period. People respond. They get up, perform all the routine morning chores and clean up. Even prior to it, Hindu women proceed to the temple premises; take bath in the communal pond & bathhouse there

enjoying the privacy under the dimmest light avoiding penetrating & prying eyes; dress up neatly; pay homage at the temple on their way back to their homes. The Christian believers hurry themselves to churches to attend Holy mass, while Muslims flock to the mosque for their 'Fajr' prayers. People want to start the unpredictable day in a spiritual and disciplined way, leaving nothing to chance and make it a good day overall. My grandfather jokingly observed "it is the best time to elicit some good things from others: borrow money, seeds, or tools; ask a favor, help, forgiveness, or a good deed; concession, settlement of a dispute; a favorable business deal; a marriage proposal, etc. The window of opportunity is stingily tiny and transient. It will close before you know it!"
Spirituality decays exponentially with time, like the discharging of an 'electrolytic capacitor'!

The eggs have to be collected early in the morning. That is usually done soon after the hand-milking of the cows. A duck lays more than 100 eggs/year; they are nutritious although high in fats, contain vitamins A, D, B_{12}. B_{12} is needed for regulating metabolism, aiding in the formation of red cells, also bolstering the nervous system, and the development of the brain, (these nutrients are produced by bacteria, available only from animals, therefore an alarming deficiency in 'Vegans') as well as Omega-3, which helps to lower cholesterol. It seems to be having contrasting effects of providing excess fats, but contains something to reduce the ill effects! The eggs from the ducks are bigger, fetch a premium price. Supposedly, they are useful to prevent 'diverticulitis' causing intestinal pain.

Infants and growing children need cowmilk and dairy products as supplements. There is always good demand for it. Adolescent children are entrusted with the task of delivering the products in the neighborhood. They like the chore that gives them an opportunity to pilfer a few coins to buy sweets from the neighborhood store. The store owner wouldn't dare to betray their loyal customers. He likes their business! Besides, most parents trust their children. They are too little to know how to cheat; and too sweet to commit any misdeeds; they are my children!

Today, in the serene hamlet of AttuValli, at the 'VallaPura', the boat hangar, preparations are steadily underway at a feverish pace. The dignitaries are arriving. The scent of burning 'Chandana-Thiri', the sandalwood incense wicks, permeates in the surroundings; announces something important is taking place in the vicinity. The ambrosial aroma pleases the gods. The words "incense owns a Deity nigh" in the Xmas carol "We three kings" mean that incense is a sign of belief in the real presence of God. To every step and turn, there is a prayer, a ritual, and a ceremony. It reminds me of the Rabbi in the movie "*Fiddler on the Roof*", the musical, telling the tailor "there is a blessing for everything, even for a new sewing machine!" The "Chenda-Melam" group (drummers with Chenda, TomTom, Mrudangam, Madhal, & cymbals), the loudest of all, lead by Bhasi Vaidhyar and the complementary "Pancha-Vadhyam" ensemble of Vijayan Vilakkumatam (The band of the Five categories of musical instruments- Wood, reed, string, brass, Harmonium keyboard) arrive and alternate playing a mix of the familiar ballads and folks music. They are dressed in kilted loin clothes, bare-chested and a Kavani in a knot around the waist for support; simply stand around like street musicians and play their instruments. People squat on their feet on the ground needing no folding chairs; others spread their 'headscarf' on the ground and sit on them, on their 'gluteus maximus', legs crisscrossed into each other like the figure of 'X '; some stand, others casually lean on to numerous coconut palms nearby, overhanging deep into the water (please avoid doing that if you feel drowsy!); village urchins climb onto the mango trees and find their own 'gallery' seats on its many branches. Some in the audience have brought their preschool children; caress them in their laps. They are perhaps babysitting today, as the housewives went to visit their parents' families on 'Vishu' holy day tomorrow or gone for work in the rice fields. People are generally music savvy. Ragam, Tanam, Pallavi, and complex Rhythm variations are quite familiar to them. They have heard them all since childhood at various temple festivals and memorized them well.

To begin, the leader of the group set the mood by playing the melody of the 'Ragam' [There is specific ragam suited to a particular emotion, sentiment, themes, time of day, and occasion]. Another instrumentalist plays his improvisation or interpretation set by the ragam as 'Thanam' in 'Refrain'. 'Pallavi' has three parts and played by 3 different instrumentalists- one plays the melody of the treble vocal part; another plays the rhythm section or Bass; the 3rd plays the rhythm in rich variations altering tempo, dynamics and other attributes. In the course of the musical performance, suddenly, one comes forward and surprises his own group by fielding his secretly kept improvisation. Un-dauntingly, one of the others pick up on it and fluently plays it back note by note on his instrument adding his own embellishments at the end; the next does the same and passes his improvisation on to yet another, after playing his unique ornamentations; others do the same, or interpret the same musical phrase based on his own musical sense. Soon it evolves into a challenging competition; it escalates; becomes louder and dramatic with euphoric head shakes and body movements for emphasis, drama, and expression. The music swells in delectable listening pleasures. The audience takes sides and tease them on. Transitory tiffs and spats prop up among the performers; ego gets hurt. They make instant and opportunistic alliances, factions, and cliques to defend standing, and reputation, or taking a revenge, but quickly disbands itself at the next instance amusing the savvy listeners. Everyone enjoys each episode, movement, and the moment. They are kept wondering “now, what’s next?”. The drummers vary the beat and intonation mixing their improvisation with other instrumentalists, challenging the soloists to match their skills. The instrumentalists return the favor with added difficulty, complexity, and flourish. Each one attempts to outdo the other; tries to catch another one napping. It’s reminiscent of the most embarrassing moment experienced by the talented and famous Kunjan Nambiar on the sacred “Mizhavu”, a copper metal, kettle drum that is allowed to be played only by the high-caste Nambiars! Sacred persons are now extinct and the poor Mizhavu faced the same fate. Its place is currently usurped by a clay pot baked at the right temperature for good resonance in a neighborhood kiln, or one conveniently chosen from the many kitchen utensils mother-in-law had bought from the local pottery market. In this musical episode, the performers try to trap another musician missing a note or a musical 'rest'. The sentimental winner momentarily captures all the attention and the upper hand. He gloats and taunts others in a demure, but belittling chuckle. Eventually, all end with a conciliatory gesture of goodwill, sportsmanship in a group ensemble playing a popular musical flourish with a resounding ending.

Well, in early music, the vocal text was thought to have originated from poetry; melody from the manner of speaking, heard from different dialects; the rhythm from the subtle sensation of time, dictated by the division of time needed for speaking words or a system of ‘versification’ [telling in verses], phrases, or rhymes in a sentence. It is certainly linked to the human breathing cycle. Well, everything in nature has a 'time period', a cycle, hence a rhythm and frequency. They are all naturally interrelated! Western music was largely derived from the ancient Greek and Roman cultures. Pythagoras discovered the 'consonants' from which he tabulated intervals based on mathematical proportions, such as, the octave has twice the frequency, or 2:1; the Fifths 3:2 and so on. An Octave was further subdivided into 12 intervals (melodic or harmonic divisions). It must be noted that the 'note intervals' are not quite equal proportions. They are mainly musically consonant or melodious.

The music, divine and different; performed by a few, enjoyed by everyone!

The religions immediately recognized the power of music and gradually introduced it into chanting of rituals, liturgies, and ceremonies, creating devotion, inspiration, interest, and attention. "It elevates the soul"; laymen would say it elevates spirits. The effect was phenomenal, which prompted enthusiastic acceptance in Psalms and hymns. The famous 'Gregorian chant' that originated in the monasteries and cloisters was the fore-runner of all the later developments. Then later, spread to local churches. Apparently, the believers were fascinated with it; thronged places of worship, notably the churches and temples to the great dismay and delight of the hierarchy!

The churches were also the places the classical Western music flourished from. It had a captivating audience! They would listen, feel, participate and enjoy in a warm and 'charismatic' way. The music was further enhanced by instruments. They built huge, artistic church organs with 8', 16' & 32' organ pipes and full-length keyboard, several of them polyphonic. Its figure, the geometrically proportional vertical pipes, the multi-layer keyboards, pedals, the draw knobs, cabinetry, architecture are imposing, and awe-inspiring individual components. The church organ has the largest frequency and dynamic ranges of all the musical instruments; its sound swells and fills the space, vibrates, and produces various frequency & amplitude sensitive resonance and reverberations. The indescribable 'cathedral effect' is one of the most awesome. One has to be inside one, to enjoy the effect fully- another good reason to go for Sunday masses, when the instrument comes alive in its full glory and musical grandeur. Such an instrument was found nowhere else, nor affordable to individuals! Wolfgang Amadeus Mozart remarked: "The organ is the King of Instruments". (I would toast to that by adding 'the typical Indian percussion ensemble of Tabla & the various other drums are the ultimate rhythm makers in the whole of music; The Veena produces the most delectable, throbbing, pulsating "twang' with a finger slide!"). The 'Kappel Meister' (Choirmaster) and the organists passionately spent very long hours experimenting, discovering, and composing wonderful music which others could share and adapt to other instruments. Most of the classical composers owe their skills, fame, and opportunities to the churches.

Similarly, Indian music, dance, and various artforms flourished from the temples in almost identical fashion but much ahead of the Western musical renaissance. Many of the divas and deities themselves were magnificent and accomplished virtuosos or enthusiasts. The sculptures and carvings on the temple "Gopuram", the towering facades, portray Lord Krishna with the flute (wind instrument), Goddess Saraswathi with Veena (string instrument) Lord Nataraja in dancing postures, Lord Vishnu blowing spiral shell or the shell trumpet. It signifies that the music is divine, soothing, and sacred. The devotees wanted to impress and please the gods to earn favors and blessings. Obviously, that belief became an overwhelming incentive to practice and play music. The musical tradition at temples dates as far back as the Vedic times. The Flute, NadaSwaram (Oooth), and the various drums are temple musical instruments as old as the musical history itself. They used to be kept in the temples until time for use. According to the Hindu scriptures and Upanishads, 'Gandharva Sangeeth' (the heavenly music) consists of vocal, choral, instrumental music as well as dance in delectable and inseparable combinations.

Although the Hindu religion was tolerant about music, the Judeo-Christian religions initially denounced music for entertainment; branded it as 'pagan music' promoting emotions of passion, love, lust, and voluptuous feelings. Obviously, they did not want to share the legacy and beauty of the music! Even today, there are certain religions and communities in parts of the world that find playing music at eateries, restaurants, and parties as 'haraam' or objectionable. That's not the case here! It's the people's music, spontaneous, fluent, invigorating & enthralling!

No festival nor event takes place in Keralam without the ChendaMelam, it's loud, times the pulse & pace
A 'pacemaker' of a sort!

The crowd is swelling!

The musicians see that there are many to play for and be entertained; they will generate louder applause and personal admiration. The drummer, the loudest of all, and most easily excited as well, balances his stance. He relishes the moment and the ambiance. He holds his massive drum suspended from shoulders and waist; his sizable belly flaps provide adequate grip, feel, cushion, and reverberation. In a surprise acrobatic move, he lifts his left leg up and supports it onto his own right knee at right angles, in “Na-Ta-Raja” dancing pose. He holds his drumstick like a flute close to his mouth. He alternates his stance in quick succession; tries to emulate Michael Jackson’s “moonwalk”. The whiskers of his mustache resonate and vibrate with the beat; looks around inquiringly, in a manner of asking, “how do you like this, fellows?”. He challenges others who looked stunned and ‘awe-struck’.

No one can match his feat [daring act] or footwork!

He thrusts and rolls his eyeballs in triumph and self-admiration. The crowd applaud and cheer him on, prompting him to do more crazy things. He spreads his stance, placing both hands on his bald head during the musical “rest”. The audience thinks he is tired; done for the day; expects nothing more from him. Suddenly, he tosses his drumming stick backward in the air, as if he doesn’t need it any more. The amused children wrestle to catch it as the bridesmaids go after the 'bridal bouquet'. A serious-looking elderly man scolds and promptly disperse them. Thank God! The children still respect and obey elders! A lesson taught early at home has still some validity and longevity spared in them. The boys ran away with the unexpected booty. The drummer is unmindful of the melee behind him; goes on playing complex combinations of triplets, quadruplets & sextuplets at a vigorous pace and delightful sequence, using only his capped/padded fingers; using his heavy palm for the accent; looks around triumphantly. He spanks on one side of the drum with both hands, then changes to the other, producing different sound effects. As he performs, he jerks his neck widely back and forth; showing his 'giggling' flabby neck layers; his chubby cheeks elongate sideways in opposite directions, as if they were water balloons tossed around in a swimming pool.

He owns the stage; steals the show,... for now!

Meanwhile, a bus unloads adolescent boys coming from the Government college campus in the nearest town. The noise and the gathering catch their attention; they are bemused; find themselves in no mood to rush home and start homework. They survey the area; assess the surroundings; occupy a strategic spot, a clear ‘line of sight’ to the segregated area reserved for the sorority [social organization of women students at college] and village women.

Immediately, they spot some familiar faces, recognized by their uniform of modest dressing, white blouses with long sleeves extending past the elbow, neatly pleated, folded, pinned blue Sarees with three white stripes of borders. The blouses are fashionably designed with embroidered & grommeted 'air holes' providing ample air circulation and comfort. They are from the Maanikya Mahila convent college in another town in the opposite direction. They have seen each other at the same bus stop from opposite sides of the road every day. The young ladies have a strict dress code and well-groomed appearance enforced by the Rev. Sr. Maria Theres Scholastica, the principal, so gentle and caring as Mother Teresa, but stern and imposing as former Prime minister Indira Gandhi, insists that they stand out, with impeccable manners, conduct, poise, posture, and expression. "The world must appreciate beauty and character; treat all women with dignity and honor". She has passionately stated in her commencement keynote address, which began by reciting excerpts from a musical poem "*Oh, Eves!*" by *Aurora Bobbyalis*.

"First, He created Adam!

He was sad, lonely, lost 'n distraught
Confused, and confounded
So, He created the most beautiful Eve
Sent her to earth to be his mate
The most important creations of all!

Daughters of Eve! Most loving,
And kindest in human breed
Entrusted with performing the
Most essentials and hardest duties of life
Bearing children, and caring
Affectionately and tenderly
Motherhood bestowed upon you
The most rewarding and noble status
Amongst all others!

Bringing up offspring and children
Youth, adolescents, and grownups
Grooming dutiful citizens

With high thoughts and qualities
Passionate, and protective always,
Sweet and caring always

Feeding, nourishing, nurturing
Providing, and guiding
Spreading faith, hope 'n charity
They never leave a child, nor home
Cherish them in their hearts
Forever!

Your lives are most rewarding
Make lives worth living
Life doesn't exist without you,
Nor can it go on without you!
Oh, Sweet Eves!

Mothers bear the brunt, out in front
Seldom complaining
Spontaneous 'n natural
Make sacrifices without bragging
Takes, as it comes; gives as it's due
Make lives pleasant for everyone
Unmistakably, showing the world
Life's worth living!

They are not strangers;
They are mothers, wives, daughters,
Sisters, aunts, and cousins
Find them among all families
Neighborhoods and communities
All the whole world across!

May we ask ourselves?
Do we credit them enough?

Do we dignify them, properly?
Appreciate as much as they deserve?
In soul-searching manner!

Everyone seeks you and beckons you
When troubles strike
Squabbles, quarrels threaten
The sorrowing, sighing,
Sick, and hurting call

Trusting that help is at hand
To bring soothe, healing, and comfort
It's a foregone conclusion
But mostly taken for granted!

Making a home, doing the chores
Providing the essentials and nourishments
Toiling at home and work
Very late at night
From the early mornings
Oh! Eves! Your work is never finished!

We brim with awe and love for you
You are the hearts and souls of all lives!
Lend a deeper meaning
To everything surrounding it!

You may not know it
'Cause you seldom hear it
And never ask for it!
When others don't express it
Then and there!
Truthfully, deep in the hearts and minds
The world depends on you and admires you
All 're filled with love and affection for each of you

You! Are the joy of all lives and its origin
No matter whether you're born for
Pretty pageants, movie roles, or modeling career
Or none of it, at all!

The world owes women much
Each woman is precious
Worthy of great admiration
They make every house a home
Everyone yearns to return to
The most beautiful place on earth!
The world loves you!
Oh, sweet Eves!"

Rev. Sr. continued, reflecting on and elucidating the identical sentiment. "Pay no attention to the blustery and misogynistic comments and vilifications you hear on Television by the president of a powerful foreign country. Woe and shame upon him! Also from the party politicians, people, pastors, fanatics, and the Talibans who echo the malicious ridicule, and condone them! My dear ladies! Let it not affect your self-esteem, nor demean you in any bit. It is not the world-view, at all! It is not even the barbarian's view. It is far darker and sinister than that. "The onus is on the whole humanity to vehemently denounce and rebuke the pronouncements outright. In the rest of the world's culture, it is the duty and responsibility of people, especially men and fathers, to protect and honor the dignity of their female relatives, and loved ones. Let's wait and see what happens here. Each of you is a blessed individual, created by God, and carefully chosen to perform noble and responsible duties, tasks, roles, and errands. She said in a firm voice, ferocious rage, giving a stern reprimand to the habitual offenders and constantly failing cowardly minds.

"Women play the greatest, noblest, most significant, dedicated, and the most sacrificing roles in life. Next to our good Lord, they are the closest to everything in life that matters. They are daughters, sisters, wives, mothers, aunts, breadwinners, makers of food, besides being employees, employers, entrepreneurs, and all the other things humans do, perform, and accomplish. They do everything and accomplish much more! You experience and endure various physical, emotional transformations in life, adapt and accommodate well to all adversities and conditions; give and sustain life; nurse, feed, raise; nurture & nourish; provide comfort & love, mend & heal; look after the sick, old, and the dying in the family; the bigger ones take care of the younger siblings; educate them on growing up; clean and wash; maintain a healthy family. There is no relationship that comes into being in a family nor humanity without women playing a biological and important part in it. As mothers, females provide half of nuclear DNA, and a hundred percent of mitochondrial DNA. They enrich at least three families, one their own; the family they are born into, and the other they join; quickly become a vital part of each. Instinctively, you learn your role & importance; duties & responsibilities. You render meaning and civilization to life and the world!

Women, the most resourceful of all beings, the chosen ones, by God Himself!

The women dedicate their lives to providing the essential things and services humanity needs. That is why God has chosen you as the most trusted beings to carry out the important duties and tasks in life and to fulfill His command "prosper & procreate, and live on!". My dear ladies, God knows very well your toils and struggles in carrying out His mandate and will ultimately reward you with the best thing He can offer, and we all desire". Ladies, you are blessed and must be proud of yourselves; feel remarkable, precious and the chosen ones by God Himself!". Your mere presence make people polite and civilized whether it's your own family, school, gatherings, public places, or societies you live in.

You make every house a home, the most beautiful place on earth; yes indeed, my dear ones!

The students have great respect for Rev. Sr. Scholastica. She often chides and cautions them without being too obtrusive. The dress uniform is designed by reputed fashion designers. "The college uniform is much more than a dress; it carries with it your collective & individual identity along with dignified statements of yourselves. You are highlighted in the crowd, in the daily commute in buses, in parades, in celebrations in any gatherings. You win respect and appreciation from the public. It adds to your personal safety. It is egalitarian in the sense that it does not discriminate on the grounds of wealth, religious affiliations, style, fashion, or peer pressures; designed for style, modesty, comfort, and weather conditions. Your own esteemed student council selected from several competitive presentations and by 'online' voting by fellow students. The college administration is so grateful that you have accepted and implemented our suggestion. The uniform represents your alma mater, a unique culture, discipline, and distinction. My dear ladies, be vigilant and do not fall easy victims to treacherous temptations; unscrupulous and predatory people with evil intentions; impressionable, whimsical fads of the modern times, or trending social media trends. Your dignity and self-respect are more important than fans and followers. Naturality, simplicity, and modesty are vastly superior to gaudiness & tackiness, and the 'likes & thumbs up' you get. Health, cleanliness & tidiness are more important than artificial bodily decorations or chemical treatments on the skin; dis-figuration of faces. Let nobody, thing, or force shall corrupt, defame or diminish you". She said in a firm and solemn voice, penetrating into the minds and souls of the young students. She and her carefully selected staff inculcated in them, among many such noble teachings: the value systems, code of conduct, discipline, punctuality, honesty, kindness, humane empathy, success in life, and above all, hope, the audacity to hope. It was not just textbooks and lab experiments, but what matters is how each student develops into a remarkable person to the fullest of one's potential. As you earnestly pursue them, you'll find out that they are remarkably expansive!"

As Principal, she has been always an indefatigable and inspirational stalwart in character building, and bringing out excellence in students!

The fact is loudly and clearly demonstrated in the published academic excellence and job placement lists!

Beauty, modesty, style, character, kindness, dignity, and charm are synonymous with women's character

People around here are now clearly witnessing and feeling the confidence, dignity, and exuberance they display. They look very attractive and beautiful. They are the "Kan-monies", the darlings of their parents; torchbearers of their communities. Their wide dark eyes are accented with black "KunMashi", the eyeliner & mascara, drawn in the shape of the iconic Lotus flower petal. They have natural shiny, silky hair, blacker than the feathers of the "keechi" bird reaching below the knee, and longer than the longest & fattest "PataValanga" (the term my grandmother had chosen to proudly complement the hair of her own 'KanMani', princess, her only daughter, my mother. Botanical garden and orchards were her favorite places to spend time with; agricultural terms her favorite conversational simile & phrases!), the longest hanging item in the vegetable garden, the 'Kerala snake-gourd'!

Women sport various hairstyles of their own, besides the traditional ones; or alternate them periodically for convenience, better look or simply imitating celebrity idols. Some keep their hair braided; double tailed and double braided; some propped high up in thick fluffy knots as the apex of a ChundanVallam, bedecked in colored ribbons in 'butterfly' knot. Some of them bring the tail- end of the hair forward and fiddle with it, or use it as a hand-holder; some others just let it loose, falling and flowing, the strands fluttering in the wind reflecting the 'soon to be setting' sunlight, or as an accent to the better side of their pretty faces. Without a 'hairpiece or wig, and unlike the presenters on Television networks, most have a bundle of hair, larger than the "Ana-Thala", head of a baby elephant. Adorned in them are fresh yellow, white, blue, or red flowers of hibiscus, rose or marigold stuck behind the head designed as a 'forbearance' to deflect unwanted stares. Perhaps for identical reasons, some wear flowers in the front towards the better-looking side near the ear, drawing all attention to the expensively fashionable golden ear-ring bought recently and is currently wearing. Many others wear jasmine garlands over the full length of the hair, the other end just falling on the forehead- casually. The cutaneous (of their skin) color varies in charming shades of 'gray scale', of black and brown depending on the amount of 'melanin' each person has. It provides natural protection from the sun in the form of sunblock and sunscreen; giving insulation and temperature regulation to adapt well to the hot and humid conditions. The eyes are wide and shapely to adjust to the bright sunshine outdoors. Likewise, the nose is naturally formed to adapt, allowing inhalation of adequate fresh air to keep the body healthy, strong, and humming with vitality. The protective 'mother nature' has unique ways to shape and form all its creations in the most suitable and adaptable ways and styles!

The breeze disperses the aromatic fragrance of the flowers and the hair oil in the air, announcing women folks are present in the vicinity. In this society, women are considered beautiful when they possess the qualities of virtues, charm & cheer, gentleness, kindness, and excitement. All the ladies wear "pottu", the uniquely Indian beauty 'dot-spot' on the forehead, centrally placed, the color matching the pieces of dress or fairness of the skin, the shape matching the geometry of the face. Some sports an ornamental nasal piercing too with a golden needle and a tiny bud of pearl, at the tip; may replace the tip with another color or shape indicating a sentiment, an occasion, or a cryptic message- either that they are married, engaged, widowed, or at the moment, awaiting!

They possess aesthetically groomed and shaped flexible figures from the daily practice of yoga, "Bharatha Natyam" dance poses, exercise, school sports, a number of physical labor, and chores called upon to carry out at home. Everyone realizes that health, energy, and strength are essential throughout life and are made aware of them at an early age. Things are mainly done and practiced for these important purposes.

Not necessarily to earn medals or laurels; pageants, awards, or accolades!

They strike different pleasing natural expressions and dispositions. Some are cute and coy; gentle and gracious; have affable smiles; some blush, shy; dreamy, thoughtful; stylish, elegant, and charming; some are more expressive and cheerful than others, exudes confidence and love for life. They are thoroughly enjoying the musical performance; the attention they are getting from the group of young men across the aisle even more. The two appear to be communicating without words- by juvenile curiosity, encrypted system of protocols, by languages of eye movements, exchanging flirting & furtive glances, telepathy; by means of invisible thoughts & dream waves.

The boys try to pretend that they are 'macho', aloof, and nonchalant that they are too modern and fully Westernized to a 'point of no return'. They try to present themselves smarter and wealthier than they look, or they don't care a hang, and so on. They like action & adventure movies, spy novels & intrigues; think watching cricket make them sportsmen and daredevils too. They bask in glory when India wins and think that they are at the pinnacle of the world; nothing else should matter at the moment. Within, they are spiteful and cynical of old traditions and outdated customs. The music they are hearing is not their genre; theirs are rap, hip-pop, rock & pop. Yet, they find this particular situation irresistible, in fact extremely stimulating. They want to do something spectacular and show that they are here too; are active participants in the 'Mela'. Nothing should take place in their community without their heavy involvement and active participation. They want recognition and notice; beam bigger than what they are. Right across from them, they spot particular individuals they want to impress, find good favor & connection; generally speaking, increase their 'star value' in the peer group. They just don't know how to accomplish all of that on this occasion. Neither do they possess exceptional skills, nor the wherewithal. But they feel the excitement; a spike in hormone surge, giving them some courage to overcome stage fright and an urge to perform some acts of show-off or adventure, 'out-of this-world!'

They consult among themselves; take out their smartphones [it costs several times more than all the textbooks they own!]; check out the latest craze in pop culture; what is hot and trending. They decide on a "fusion"; discuss some moves; quickly select a theme, a choreography, and a rendition. Without any more ado, they hand over their phones to friends and ask them to "keep them rolling", meaning to continuously shoot the video of the whole thing! They intend to stream and upload it onto Facebook to attract even more followers, earn 'likes'; carve out a niche and aura for themselves; already alerted friends to look out for a great surprise on their 'group pages'. One pulls out a peculiar-looking comb from his hip pocket and passes it around. Some comb their hair; others think their sparsely grown mustaches deserve more attention; assure themselves that they look good and handsome. Some keep the hair 'natural,' or unruly as at birth; some apply a thick slick concoction to the head, rub all the hair they got together; pull out a few here and there, looking like the prickly porcupine. Shortly, the costume part is all done; the rehearsal and preparation are completed; toss the books and bags to trusted friends; ask them to take care of them during the course of their uninvited 'cameo appearance'. They are now ready to 'roll & do their thing'!

Surprising everyone, they emerge from the crowd; stagger towards the arena one after the other, missing one step or the other like the drunkards before a fall. They proceed to perform a "Herky-Jerky & PuliKali" fusion; approach the drummer in jaunty steps and fashion. The drummer shows a welcome smile; seems to be saying "Kollam, Kemam thanne! [very nice,... go on..very nice indeed!] reassuring the performers with the trademark Indian head-oscillations. The dancers go berserk! They alter the routine to a more vigorous gyration of limbs, hip, and head; hands thrown around widely; legs moving in 'quick steps' and so on challenging the 'Bollywood dancers'. The crowd is on their feet; nudges forward closer to the action; do sorties of "Ku-Rava", the rhythmic howling/wailing [also common in the Middle East cultures] to mark their ecstasy. The ladies are mesmerized. No one in the village has ever experienced anything electric like this before! They thrust their eyes wide open; cover their mouth in disbelief; lean forward; make hushed comments; take some snaps and 'selfies' to share it at college with other friends. The place is in a craze, drowned in frenzy!

The ChendaMelam, PanchaVadhyam ensemble, and the Elephants; the indispensable performers in all the Kerala festivals

The boys are emboldened. But, the drummer has a sudden change of heart. He dislikes anybody stealing the show while he is performing. He is an opportunist, egomaniac, and a consummate showman! His status and image are challenged here, for the first time ever! He waits for his 'strike of revenge'. As they come too close to his face with a disparaging 'chicken dance', he starts banging his drum louder and faster than he has ever done before. The other drummers follow his lead; encircle the flash boyz. As the noise becomes excruciating and irresistible, the boys begin to run away one by one, helter-skelter, vanishing among the crowd, and in the diminishing light of the evening sun.

It is the same technique the American forces used to bring out Generalissimo Manuel Noriega, the former military dictator of Panama, by continuously blaring horribly loud and distasteful rock music towards the Vatican Embassy in Panama City where he had sought refuge. Among the secret services agencies around the world, this method had come to be known as the "Panama torture".

People started to poke fun at the 'dancing boyz'; ridicule them by making stinging and disparaging comments. The reactions from social media are nothing flattering at all, only scathing. The ladies think the boys were immature, cowards, and silly. They disapprove of it as ill-conceived and a botched show-off; form a low opinion of their counterparts. The boys are deeply feeling the insult and humiliation. They feel mortified; sternly, order their friends to 'cut' the recording, borrowing the phrase from the movie directors. Some had the camera on 'streaming' (continuously broadcasting to the Internet) mode- inadvertently! What goes into Cyberspace permeates and dissipates instantly. It's irretrievable! They fear that their friends and parents have seen this ignominious fiasco; feel humiliated, and ashamed of themselves. Holding the 'smart phones in their hands and technology at their fingertips', they thought they have become instantly smarter, attaining a level of maturity, knowledge, and capability to take on the bigger world. "When you think you know everything, you are shown, there is at least one more thing yet to be learned". They curse. "What a cruel world!"

Their ego suffers a tremendous beating; They feel depressed. Now, they cannot face their friends. The girls will giggle behind their backs and make fun of them. It's horrible! Just opposite to what they set out to achieve!

"This was all that silly, goofy Simon's idea...entirely his fault... he's stupid! What a moron! I'll throw him in the water, ...if I can find him; ...let him get lost!"

Some decide to skip the classes for a few days and let everything subside; it fades from everyone's memory. 'Time will heal most everything', as in the saying. They must refrain from mentioning "VallamKali" ever again. After this debacle, it can only bring back harrowing experience. The tough Electronics lab and the complicated Mathematic classes with 'Integration & Differentiation' starting next week are difficult to catch up- the two big monsters! Then, there are the nosy parents! What excuse are you going to give them?...yeh...guys?"

Damn! There's no place to hide in this world!

Having succeeded in chasing away the dancing boyz, the drummer is basking in the pinnacle of his triumph and glory; both his fame and popularity are surging simultaneously. But he doesn't hold back; looks for more! The crowd gathers behind him; shows various signs showing 'thumbs up'; some borrow certain patented gestures and poses of their favorite rap artists; take selfies with him. Gloatingly, he obliges; joins them in different funny poses. Suddenly he becomes a Cyber sensation! After all this hoopla, he takes a pause; 'stands at ease', listens to his musical challengers who have been struggling to get any traction or attention among listeners. He shows his approval of a rendition from the band by twisting his mustache pointing upwards like a buffalo's horn; if he doesn't like any, twists it stooping down like that of a tusker; scowls his eyebrows. He strikes several martial poses & gestures with the beat, like the possessed rock musicians, but restrains himself; refrains from smashing his drum on the stage, avoiding a loud explosion. He's one of a kind; reigns supreme here!

At this time, a tug boat brings farmworkers, predominantly women, from the rice fields far away. They are surprised at the crowd and all the noise heard from the distance. Now, momentarily, they have become the center of attraction! They possess a certain charm, vigor, and beauty. Infants jump out from the laps of their fathers like 'jack-in-the-box' spotting their mothers. They run towards them crying and sobbing. Fathers chase after them to protect them from falling into the canal. Mothers welcome them lovingly with open arms, lift them up and carry them with tenderness and affection. Suddenly, they become calm. They want the attention and affection only mothers can provide. They lean onto mothers' shoulders and feign sleep. "I wanna go home, and play!"; "I wanna sleep!". The infants laid out their wishes. They want to get away from this noisy and unfamiliar place. Fathers persuade them to stay a while longer and enjoy good 'family time' at the pristine meadows in the company of fellow villagers. Everyone feels relaxed!

The Oothu/ NadaSwaram (oboe) players have been vigorously blowing their horns for some time now. They feel unhappy and become jealous of the disproportionate attention the drummer is eliciting. They test different reeds hanging from a thread on their instruments and select one that is 'stagger /offset-tuned' (musette tuned, as in accordion). One chooses a reed with a tone and additional 'timbre'; the other with a reverberating tone and another 'tincture'. Listeners like the sum of the two, the wavy sound it produces, the oscillating 'tremolo' effect. At long last, the instrumentalists rest the initiative. Encouraged by the favorable reaction, they surround the show-man drummer and play their "Ooths". They swirl the "bell flare" (the mouth of the instrument where the sound comes out) like the snake charmers do, and force the drummer to do a cobra dance. Having no choice, he complies, albeit covering both his ears, suggesting their sound is too shrill and annoying. He alternates to a 'witches' dance' when the sound becomes so harsh for him, holding fingers resembling devil's horns. The drummer looks pale; lost all his glory he amassed just moments ago. People make fun of him; he chuckles; he winks; dismisses it as just a transient, harmless prank with no substance to it, and he doesn't care. Left with no option, reluctantly, he joins his rivals. At times, they play solo, abruptly merging into duets, as if they are willing to compromise the unfortunate rivalries in the immediate past. All join in the climactic '*finale*' in 'leaps' of resounding crescendo; exultant flare, flourish, and blaring fanfare!

The various percussion instruments of Indian drums and Tabla are very rich in harmonics because of their wooden resonance chambers, and shapes. They produce deep bass sounds; the playing by fingers is faster and more expressive. The drum-sets used in Western music sound a bit too 'tinny' because of their metal brim's rattlings. The Indian drums are hand-made, therefore never the same in sound or shape. It gives rich variety and listening pleasure. The extemporaneous improvisations are distinctive and enthralling hallmarks of Indian classical music renditions. This extravaganza generates auditory as well as visual impressions. It's effervescent! This particular feature is not commonly seen in Western philharmonic orchestras. They are performed under serious and solemn temperaments and acoustically 'appointed' environments. The conductor is the only one who is somewhat expressive, but he has to stand facing the orchestra, and back to the audience! In the modern consciousness of "show business", a lack of expression may be one of the reasons for the waning interest in Western classical music. On the other hand, if musicians try to do unnecessary things during the performance, they fear that they won't be able to play music correctly, as the composers originally intended.
That'll be considered an unpardonable disrespect of the maestro!

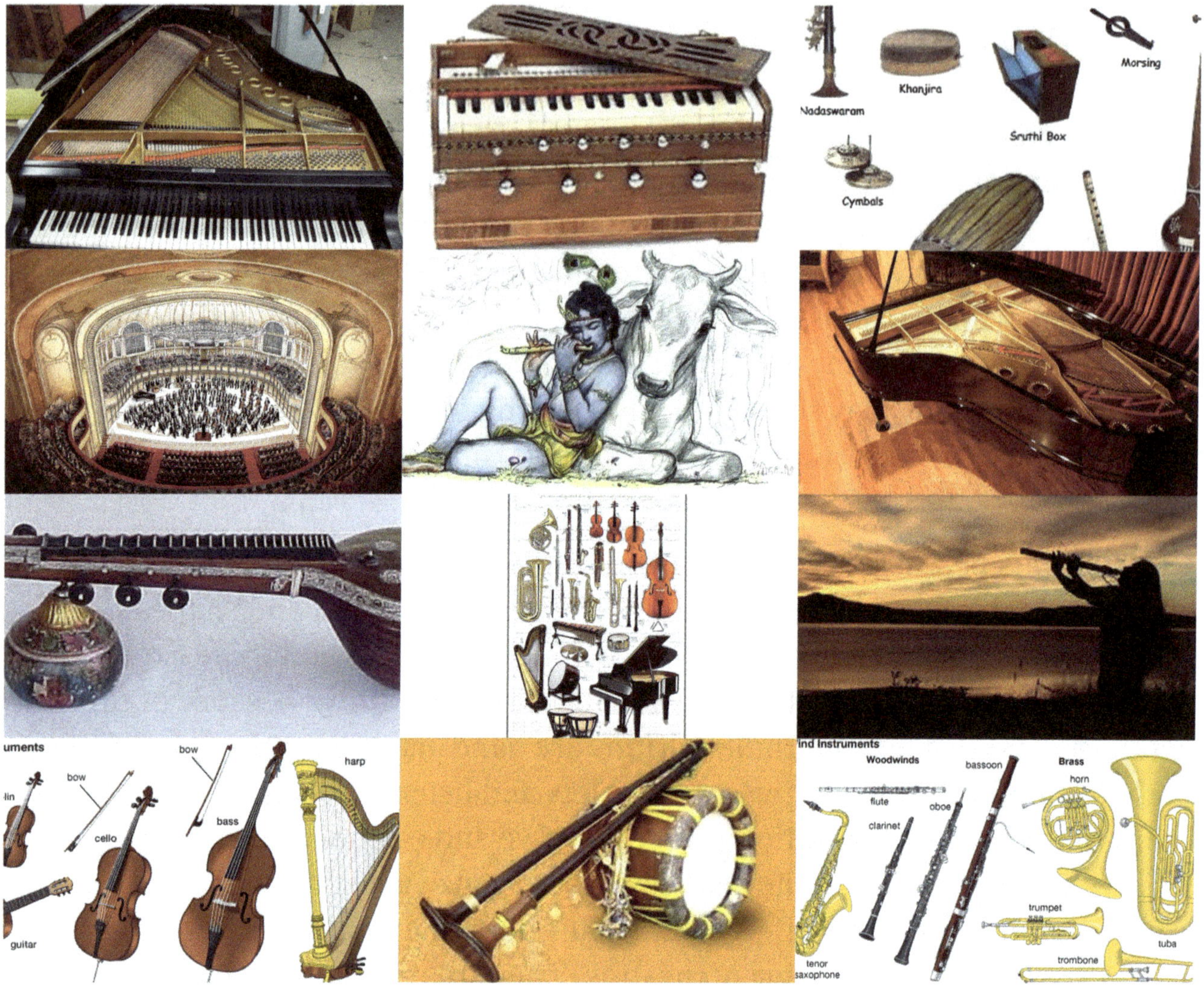

Music is made by instruments of many kinds- string, wood, brass, percussion; yet different in sounds, and manner of playing

The virtuoso and classical musicians are meticulous about the precision and the manner of playing the notes in perfect pitch, time values, and all its expressive forms of articulations and ornamentations. They are keenly conscious of their individual role in a philharmonic orchestra. For them, the symphonic sounds in its entirety and authenticity is of paramount importance. Western music is likened to be played in a controlled space, as in concert halls devoid of noise, rich in acoustic qualities; in hushed silence. Indian music, on the other hand, is usually played in open space, amidst the braying of donkeys, mooing of cattle, and chirping of birds; the wind blowing, people talking in the actual ambiance of life-like environments in harmony with life.

All creatures have their own place and significance in the world; they may be singing too and shall be also accommodated with music genres of many kinds! All are God's creations or manifestations according to Shri Buddha's tenets on beliefs!

Right now, here, the listeners love the sequence and the show; immerse themselves; the rendition penetrates deep into their musical /auditory senses; spontaneously tap their feet, slap their hands on laps to accompany rhythm or snap their fingers; let their arms soar freely in the air, swing their heads, sway their torsos, and jerk their bodies like plants in the breeze, in tune with the music. The musicians recognize the stimulating feedback, and attempt bolder episodes, complex phrases, and interludes.

The musicians and the listeners together experience a mystical, musical-transcendence!

At this climactic instant, the horrible "Che-nga-la" (gong) sound breaks out: interrupts the music. It is the sound of a heavy wooden hammer banging on a suspended wrought iron disc, primarily designed for schools, factories, and other noisy environments. It makes an oppressively painful sound in the low audio frequency. The powerful sound waves travel far and wide, as the materials in the surroundings are slow in resonating with it and absorbing its energy in its narrow spectrum of wavelengths. The sound cannot get easily dampened, lingers in the air for a while longer. Actually, it is designed to catch everybody's attention- in a truly annoying way! It has the desired effect here too!

The alert signal signifies the launching time. The music stops!
People get up; converge around the boat hangar; wait for the 'moment' to occur.

Strangely, the unpretentious Ashaari is visibly fidgety and tense, as nervous as Ludwig Van Beethoven was at the premiere of his 5th symphony. Since very early in life, Beethoven was hard on hearing (the cruelty of fate!) and couldn't sense beforehand how good his composition will be received by the discerning audience of the time, until he saw the enthusiastic encore cheering from the listeners. Here, in this case, it is all about how the boat floats in the water, its buoyancy, balance, symmetry, and many hydrostatic parameters. The launching pad was assembled with an agreeable slope. Skidding rollers are carefully placed at intervals on the launching platform while others insert 'cut' pieces of supple coconut stems devoid of leaves at strategic points as cushions and slipping sliders. They are interspersed with soft and polished wooden logs for smooth gliding and entering into the water flawlessly. Volunteers are stationed in the water to ensure that not a single drop should enter the boat. The onlookers watch in awe and let the more experienced volunteers handle the delicate work. They are too proud to be admonished or told-off for any improper act, or impertinent remarks. Here, the instructions and commands flow vertically downwards. The radical traditionalists are unforgiving; no one wants to hurt their sentiments, nor earn their ire either!

The ceremonious launching of the Chundan following a series of eventful and devotional rituals

An equal number of people are posted on either side at equal distances. Each person uses a soft cloth, instead of gloves; both hands on the Gunnel and their center of gravity leaning forward as in ‘tug-of-war’; ready to push. The cantor begins in a mellow voice the all too familiar 'launching song' in “Adagio” (in a slow tempo) as heard at the beginning of a military stroll, the emphatic and relaxing tones alternating in its characteristic cadence and crescendo. The crew responds each time with an obedient refrain or chorus, “Aai-Saam!” (meaning 'As you say!', Boss! Or Amen!), or the contextual exclamations given in the brackets. The chanter begins in the whole too familiar intonation: “Hey comrades, Listen yu, all! Ja, all ye faithful!(Aai-Saam!); Push this beast..(")... inch by inch!.(") slow 'n stea-dy! (") ; Use some might, Sirs!...(") ; ‘n guts, at will!("); All at- once! ...(") ; A stubborn load? Ye, there, cheating mates! ...(who? not Me!)....cut it out!('ll do!)... Do I spot them?...(Sure you do)....(shall I) 'send'm home?..(Oh, no, No, no!).. Ah,.. There I see!("); a sum (of) 'f something, ..(Kollaam -super!); Willing to move! a nudge in here..("); one there too.. (")...here, one; another there, too!(")...a little slide..(")... like a bride ..(").. a blushing bride...(oh, come on!) walk'n pride! A missed step!. A doubting bride?...(oh, dear, Not now!)... in little strides.. (Kollaam, that's better!)..left 'and right ...short 'n long...(towards?) 'A-waiting groom... (poor fe-llow!)...a merry old soul.. was he!..(yes, he was!); Ahh..way to go! ("); Not stop & go! ...(");..(Al)'most there! ....("); yeh, not yet there!..Ah! .oh,..oh! There you're!. The fin'st men of all! Yes, You're! (together) Arpo... Hurray!". With periodic midway corrections and re-positioning, also giving a chance for much-needed pauses (as in the 'commercial breaks' on Television) in action to recover physical strength and to regroup. Finally, the 'pushers' launch the boat gently into the water. Some volunteers are in the shallow waters cradling the boat in their arms; giving gentle support as holding a baby; facilitate a soft and smooth entry into the water, also preventing water from getting inside, the case of which is considered not at all a 'propitious' [auspicious] sign. Meticulously, they ease the boat smoothly into the water with a gentle follow-thru. An exultant cry of joy and excitement echoes. It's now the most recognizable village icon; hopes it will bring glory. Simultaneously, a thunderous cannon shot erupts, a single 'gun salute' to the Chundan, signifying its rank as # 1, followed by a series of scattered sounds and sparks from firecrackers hung around the hangar. The urchins jump off the trees in apparent shock, clutching ripe mangoes in their mouths; spilling juice onto their bare chests. Surprised housewives rush out of their houses caressing babies, for a peek to find out the cause of the sudden 'explosion' in the neighborhood. They watch with curiosity, momentarily forgetting what is cooking on the stoves. With all the fanfare and excitement, the boat is finally launched into the water; floats gently in the middle of the river!

Mukundan Malarkody, a Princeton professor on a visit to this homeland with his family was among the spectators. He was bemused by the fuss & frivolities surrounding the launch. He commented: "the boat could have been easily launched by a tug-boat!". Sure enough, his terse comment irked a proud local resident. He corrected the professor: "then, all these physical and emotional participation of the community would not have happened. It would have been entirely a stale mechanical event. By the way, the Chundan is not a 'thing', a venerable relic which should not be handled by mechanical devices, or cranes. Centuries ago, some wise men hatched out a plan and process with the intent of promoting goodwill, spirit, joy, cooperation, and harmony in village communities, also making people aware of the safety of the paddlers. We practice a policy "people first" and want to do things together! There is always significance and semantics in what we do here in our land. There's serious message beneath the frivolities; it has a much deeper substance than what meets the eyes, sir!"

"I do appreciate the great lesson and the sentiment you have so eloquently explained. In fact, in the time of the Vikings, perhaps even earlier, the launching of seafaring vessels were always accompanied by ghastly rituals of human sacrifice; later as people became more civilized, they replaced the custom with a sacrificial lamb, a 'milder aberration', still horrible! According to the beliefs and justifications, the sea travel were riddled with lots of perils involving storms, shipwrecks, mutinies, and human casualties. People wanted to appease the angry 'sea goddess' with appropriate rituals. In this country also, the goddess 'BhadraKali' is considered sadistic and blood thirsty, but revered. I understand we perform the launching rituals for the same or similar purposes, but in different style and civility. I thank you,... very much, Sir,... Absolutely Sir!"

Finally, seeing the Chundan gently floating over the gentle waves on the river itself is a thing of beauty to behold. People, freely share their excitement and express great sense of pride and affection for the Chundan. It is amazing that the carpenter ingeniously or intuitively applied laws & principles of hydrodynamics, hydraulics, Archimedes law of 'buoyancy & displacement'; Pascal's law of fluid pressure, and Newton's laws of motion. Apparently, in his traditional wisdom, he has been aware of a number of scientific theories that water is incompressible, has low viscosity or frictional characteristics; water quickly reverts to its permanent shape when disturbed; will exert pressure equally in all directions; the pressure acts perpendicular to the surface in contact, and so on. Are they all really simple matters of keen observation, or acuity (keenness in vision or thought) of human intuition? Or, perhaps a divine revelation!

Not at all! Viswakarman, the Hindu god of architecture, had been revered in the whole of India as the creator of the Science of Architecture and Engineering since ancient times. He built 'templates' and specifications for everything in his knowledge and purview. The Western world may wrest and give the title to one of their own, Leonardo da Vinci. The Ashaari and other 'skilled trade workers' belongs to the descendant community of Viswakarma and apparently came to know the science by virtue of the legacy and practice in the traditional trade from the early childhood. Early in life, he has witnessed, been taught, and asked to help in his father's work. What you learn in the childhood has the deepest degree of remembrance, utilization, and longevity! Not only they knew it, but practiced them as well!

The whole set of specifications is contained in the ancient "Sthapatya Veda". It entailed the information, measurement proportions, specifications, formulas, details on selection & treatment of wood, joints, and techniques for the building blocks of wooden boats. By the book, planks have to be precisely 25.3 m x 15 cm. But, people have never seen the carpenter looking up any of these specs or numbers, nor referring to diagrams in any book. One by one, the elders say "he is my age, I've never seen him in school". Everyone assumed that everything there is to know, he knew by heart! The skilled should know how to apply them without necessarily having to know the Sciences and principles behind them. Well, it's no wonder that people of this nation come to believe in myths and mythologies that is deep at heart and dearer than the Sciences!
Everything one is not able to do, must have to happen by magic alone! that's the general sentiment.

For the thoughtful carpenter, the boat is not merely a floating object in the water. It's about proof of performance, personal pride, reputation, and the glory it will bring to his beloved native place. It is like seeing a newborn child; emotions overwhelm; also a surge of anxieties and expectations. Involuntarily, tears drip from his eyes- the 'tears of joy' and the reward for the long hours of dedicated hard work coming to fruition! He's gratified, but humbled.

Now, the boat is a real thing; no longer imaginary. It has a shape and form, capable of moving fast in water. For better understanding of further reading, we need to know various boat/boating terminologies:

Bow: The forward or front part of a boat

Stern: The rear part of a boat

Gunwale/gunnel: The top edges of the sides of a boat (The term was borrowed from the fact that where guns were mounted along the "gun ridge" on sailing ships. This area was often reinforced). In race boats of Kerala, they are used as seatings instead of guns!

Hull : The body or shell of the vessel

Length *: the length is the measurement from tip to tip*

Beam: the measurement of width at its widest point; Amidships and perpendicular to the centerline of the vessel (**Athwart-ship**: direction across the boat, perpendicular to the centerline)

Helm: The driver-steering station/raised platform at the stern

Amidships: midway between the bow and the stern

Port: The left side of a boat, looking forward towards the bow.

Starboard: The right side of a boat, looking forward towards the bow

Waterline: This is the line indicated on the hull at which the boat sinks in the water when loaded with crew and all the things it is carrying

Draft: 'Draft' is the depth of water that the boat needs in order to float freely. The draft is measured as the distance from the waterline to the lowest point immersed in water.

Freeboard: is the 'clearance' of the boat from the waterline to the lowest point on deck

Children of the village- pesky, mischievous, seeking adventure & fun; add spice and fun to everyone's life

The villagers admire the masterpiece and the new treasure of the village. It is going to be a part of their lives. It'll be their collective identity bringing pride or derision depending on its performance in the races. They have great respect for the master carpenter; after all, he is the one who built the homes they have been living in. They withstood all the tests of time and extreme weather conditions. Amazingly, they stand strong and straight; firm and erect. The local comedian Sriman. Cinema-Kkaran commented: “these carpenters carry the lineage and pedigree of the master craftsmen who built Noah's Ark that saved humanity from the 'Great Deluge'. They came here after the fall of Babylon following the decline of the empire”. His comment fascinated people who heard it, spread the word quickly far and wide- a leaf out of the book of Joseph Goebbels, the minister of propaganda for the Nazi German government of the 'Third Reich'. Soon this factoid became ‘viral’ perpetuating it as a piece of newly mined factoid from the most ancient of history.

The ‘Jothisian’(Astrologer) assessed the Chundan’s “character & personality”; astrological signs and the coordinates of the planets on that day; marked on squares of a rectangular portable folding table; scribbled computations on a scroll of paper. He rolled it and tied it up in a queer ‘dragon fly’ knot with a blessed Orange ribbon and entrusted to the Kaaranavan, with the Vedic writings: “Sarveshwaraye, Namaha!...[“In the Almighty God, we submit!”]

With verbal instructions in suppressed solemn whispers, each one after making 3 paced steps towards the bow of the boat:

1. “3” is the magic number assigned to the boat
2. Always ask for lane assignment in 1, 3 or multiples of 3
3. It shall not be opened, before the final races in the 3rd year of anniversary of this launch date!
4. The boat shall be kept moored to a secure berth 3 feet away from the jetty for at least 3 days, before practice can begin!

Chapter 18

The training for the races

Trial and selection process for the paddlers go hand in hand. The arrangement is to have 64 pairs (a total of 128) of paddlers sitting in tandem on the port and starboard sides; each row supposedly representing a unique Kerala art form. Being selected as one, is a personal honor and social recognition. They are required to take a 'vow of abstinence' for the duration until the final races. Besides the paddlers, there are 5 helmsmen; a team of at least 8 Rhythm makers; a captain, and an assistant. Thus, there is a minimum of 141 team members on the race boat and a few officials/monitors at the start, finish, and various locations along the racecourse as part of the team. All these numbers have certain significance or reference. The paddlers sit in rows, pairs, and sections carrying out their varied and specialized paddling styles [as in football/soccer teams, the 11 players have different roles and expertise, as well as positions to play]. It becomes necessary to coordinate the actions to eliminate any chance of 'clashing of the paddles' that will upset the entire rhythm retarding the motion of the boat. That's where the absolute need and specific role of the rhythm makers become apparent.

Having selected the crew, the training & practice sessions can be scheduled. It's set for 6 weeks once or twice a week. The first session starts after a communal celebratory "Sadhya" (a sit-down banquet) at the 'culmination' of the day when the sun is on the meridian, speaking in astronomical terms, or simply at noon, in common terms. They are usually held at the temple premises, inside a temporarily erected "Pa-nthal", a hall with thatched canopy for shade and comfort. The ground is carpeted with mats woven out of the soft leaves of "Kai-tha", the tall grass, or the water cactus.

People form rows, sit face to face and back to back. There are no 'ushers'; people know standard practices and the customs very well. In Kerala, the mantra and 'modus operandi' is always to follow the unwritten protocols of traditions! They squat, resting on the soft mats, sit straight; legs crisscrossed into each other, in a 'yoga' like posture forming a linear esophagus pathway from mouth to stomach, as prescribed in the guidelines stated in Ayurveda. The food is served on the nature-friendly, 'right off the tree', biodegradable plantain leaves rich in phytonutrients, thoroughly washed and sized. It also adds a distinct, natural, leafy chlorophyll flavor. Keralites love to eat with cleanly washed bare hands. After all, the hands are given mainly for this purpose. It is the most natural, enjoyable, and intimate way of eating. In the mostly agrarian work, people handle mainly soil and plants; not rusty metals, grease, or lubricants like factory workers or car mechanics. They wash away the nature's elements of mud and soil with water and soap, hands rubbing together in a natural washing routine. The food is eaten only with the 'clean and sacred' right-hand; the left hand is leisurely resting on the lap for balance.

Though there are many routines, preparations, and protocols involved in the task, wearing any uniform, chef-hats or aprons by the cooks or servers is not one among them. It's perhaps due to the excessive hot conditions or by habit of wearing no separate attire for any work disciplines. They are habitually casual and use bare minimum of accessories or possessions. That's the way they like it. Habits die hard!

The traditional sumptuous Sadhya consists of 50 different serving items all prepared under the prescribed guidelines of Ayurveda containing specific nutrients for each vital organ and bodily function. The food is deliciously cooked with locally grown aromatic spices: chilly, onions, ginger, garlic, Cummins, mustard, turmeric (which can also be used as a yellow dye), curry leaves, cardamom, cloves, pepper, cinnamon, grated coconut, etc. They are all cut, sliced, grated, shredded, powdered, soaked, mixed, fried, baked, and smoked ingredients bringing out natural color, nutrients, and flavor. All finely cooked and seasoned in the most indispensable coconut oil. These non-chemical ingredients contain medicinal, and all nutritional categories needed for the body to maintain good health. The usual 'admixture' masala powder with curry flavor is added in prescribed proportions to savor a fine taste. Food is prepared under the close supervision of a gourmet chef; expert 'tasters', or food connoisseur, perform tasting, and recommend modifications if needed.

An all-important communal Sadhya, prior to the launching of the ChundanVallam

It is strictly a vegetarian dinner including all necessary nutrients. Carbohydrates from parboiled whole brown rice grains which have bran, fiber, and germ layers (refined grains lack them) for energy; proteins out of Legumes (beans, lentils, peas & peanuts, replete with rich nutrients) containing natural Phytochemicals distinctly different from vegetables & whole grains for tissue building; dark green vegetables containing folate & carotenoids to control cell growth and to protect DNA; also aiding organ and neural functions; fruits containing probiotic dietary fiber, vitamins, and minerals for digestion, for building responsive, and robust immune system; the indispensable curry flavors to please the palate, and stimulate the taste/saliva buds; helping to hasten digestion.

Courteous volunteers make up the serving crew; each has a designated function and knows the established serving course, order, and the routines quite well; coaching or instructions are simply superfluous. They stand in line, ready to carry food trays and containers. The first one, hand-carries a bundle of thoroughly washed plantain leaves, place it in front of each seat, facing the tapering side of the leaf to the left; followed by another server holding water jugs in both hands. He pours water into glasses placed on the left side. The next one brings salt and pickle dispenser; serves it on to the left of the leaf; pickle on the top edge, salt on the lower. The serving of thin, crisp & crunchy banana chips follows the order. Immediately, people at the Sadhya begin to munch on them until the items for the first course is fully served helping production of digestive fluid readily available in the stomach where everything lands,gets mixed, and digested. Every item has its own place and order for specific reasons. The basis of traditions evolved from a series of thoughtful planning in the ancient human civilization. The action plan was later developed into 'templates' so that common people can intuitively follow without any second thought, or hesitation!

The hot, crispy, fried 'Pa-pad' is served next; served in pairs. The staple, well-cooked white rice follows the order, served in two large scoops in the middle, occupying the most space and volume. Right out of the kitchen is brought the Lentils in thick and spicy gravy, the vapor rising and aroma of spices oozing from them. The server shuffles his holding hands to avoid burning; the steam contains 'latent heat'! That's why steam is more dangerous with burning than hot water. The lentil gravy is served spread on top of the first scoop of rice. Yet another server carries a large 'bucket dispenser', a silver-plated, pentagonal carousel containing assortments of curries-leafy vegetables of Indian spinach, the calciferous (containing calcium), "Avila", the vegetable medley in grated coconut mix, "Tho-ran", a hodgepodge of the dry vegetable dish containing seasoned beans, okra, shallot; jumbled yam, onion, garlic, potato, and carrot; then squash, and bitter gourd and/or cucumber slices in syrup with green chillies spewed around, (Pachati & Kitchati). This completes the first course. Eating can start at any time. Wait for no prompt or toast of any kind like "Guten Appetite !", "Bon appétit!", "Enjoy your dinner!", or any other curtsies or prompts. People have great respect and gratitude for the bounty in front of them. No pompous, pretentious dining etiquette, nor fancy formalities are pronounced; certainly none of the kinds shown on the TV drama series "Downtown Abbey" where snobbish conversations at the banquet gobbles up the bulk of the air-time. The affluent never think of the 2/3 of the world population who cannot afford a meal a day. They seem to be saying "it's not my concern; solely, God's!" forgetting the simple fact that they take too much from everybody, and everything.

To begin with, each takes a pa-pad in hand, crumbles it onto the rice heap of lentil gravy. The combined noise sounds like a series of firecrackers exploding one by one. The sound propagates from one end to the other, sounding like a long 'arpeggio' chord played up and down on a '88 key' grand piano, the 'sostenuto' pedal depressed when it reaches the end of each row. The crumbled pieces are mixed up with rice, salt as needed; each time making rice patties with an artful 'scoop & roll' maneuver or a few little 'single hand' juggling routines before throwing each one deep into the mouth and gulping the "U-rula" or the morsel one by one. Ayurveda reminds: "First chew well in the mouth before delivering it further down; spare the stomach an unnecessary burden!"

The all-important "Sambar" with similar ingredients as in "Avial" sans coconut ingredients, prepared in a gravy dish of tart tamarind syrup, seasoned with a pinch of acrid-tasting "Kaa-yam" or asafetida, is poured on the second heap of rice. Kaayam & turmeric have 'astringent' properties that can tighten gums and remove plaque from the teeth. The earlier mentioned method of eating is repeated with Sambar; the curries are tasted in between each bite and a 'lick' on the seriously hot pickles at random. The"Mo-ru" (spice-flavored buttermilk with sliced tender cucumber and shallot) is served to finish the first course. Some prefer the "Rasam' flavored with tart tamarind, instead. It moistens the morsel and makes it easier to pass through the constricted throat and narrow alimentary canal.

For the final two course dessert, yogurt, a choice of banana, sweet mango pulp, and a small scoop of rice are served one after the other. All the ingredients are mixed in proportions suited to the individual taste and eaten in a handful of portions. The rich and delicious "Pra-dha-man", the most delicious of all 'Paa-ya-sm' is served as the second, and final dessert. A mere thought of the now extinct "AmbalaPuzha Pal-Paaya-sm" is sufficient to salivate any dry mouth! People are not generally mindful of their public surroundings when eating while they 'enjoy the Sadhya'. The life sustaining food is a gift from the bounty of the Almighty and to be treated with utmost humility, and gratitude!

It is customary to perform 'ablution', a thorough washing of hands, face, and mouth, before and after every meal with water kept in different-sized containers and dispensers. Sometimes, as a gesture of courtesy and humility, people help the next one in line with pouring water for them for a tidy wash. Even for washing after eating, there are prescribed guidelines in Ayurveda, the Health Sciences.

When thinking about a sadhya, enjoying one, or reading about the lavish state banquets served to the G7 world leaders, which also happens to be the most important of all proceedings, in their annual summits, I was reminded of the brilliant MahaKavi (The great poet laureate) Kunjan Nambiar, and one of his poems written about it. He had such a unique and fascinating mastery of words, meaning, rhymes and style. He used homographs & heteronyms; synonyms & antonyms; dual, double, and multiple meanings quiet prolifically, and with ease. Kunjan was a prodigy on par with the greats like Mozart, Van Gough, Monet; Rabindranath Tagore, Sarojini Naidu, Bernard Shaw, and G. K. Chesterton. In my childhood, I heard a story from my maternal grandfather on our frequent outings and was very much intrigued, and marveled by it. I was also so-very privileged to share the personal libraries of my uncles on both sides; and fortunate enough to read and enjoy poems by world-famous poets, John Keats, Shelley, William Wordsworth, and Thomas Gray. But this one lingers in my mind the most, the others not well understood, and forgotten long since. I used to ask my grandfather to recite it every single time I spent school/college vacation days with him. Always, did he oblige me with his lilting (A cheerful or lively manner of speaking, in which the pitch of the voice varies pleasantly) recital in the typical cadence and prosody of Sanskrit 'sloakam':

"Pathram VistruthamAtta; ThumbaMalar Thotto-tino (ra)Annavum……". The plantain-leaf on which the food was served: big & large, spacious; greener than the green itself, glazed as a mirror;

The rice served on it, was so pure; shining white; rivaling the color of the whitest flower in the colorful royal garden, the beautiful "Thumba" (Edelweiss);…,

The delicious desert; the first of its course served with banana; the fully flavescent yellow fruit was the exquisite 'Poovan' (sugar sweet banana variety), the best and the most delicious of its kind on the planet!…yum,…yum…very yummy!

'Thumba' & 'Malar' are two of the homographs! If it is read transposing the last syllables of each word to the next line, (in one case, as individual words and in the next case, as connected words) the meaning amazingly becomes the antonyms. There were such phrases all along and on every line in the poem.

Accordingly, the other version of the same poem meant:

"The plantain leaf on which the food was served was stingily small, just about the size of the leaves of 'Thumba' flower; narrow spilling much of the food served on it, and it was as rough as 'Velcro'!

Well, the rice on it woefully under-cooked, worse than the "Malar", (popcorn, a favorite of American movie-goers), half-boiled; split open; non-uniform, white & brown pieces sticking together...; the banana was nothing better: far from ripe, half green, barely flavescent, tasting tart, gluey, sour on the tongue. Besides, it was 'Kaalan' (a green/blue 'Java' variety), the poorest tasting of the entire banana family; rubbery when chewed, just awful to taste!"

Kunjan spent a lot of time at the temple, practicing the 'Mizhavu', a temple musical instrument. The ardent devotees used to bring food as offerings to feed the idol-deities. In the pretext of wasting none of it, Kunjan consumed as much as he could; became a consummate connoisseur of food; making him large, obese, and over-weight. Naturally, he knew a lot about food as he ate both good and bad-tasting items without discrimination!

Judging from the enormous quantity of food he consumed voraciously, he had impeccable experience to be qualified as a food connoisseur & critic. Nonetheless, it is still commendable that he wrote this poem 'impromptu', in no time; with consummate ease, and fluency!
He wrote it ad-lib; extemporaneously; with virtuosic brilliance!

The story continued: the reigning King of Chembakassery invited the elite, preeminent guests and subjects of his kingdom for a grand Sadhya (Grand banquet) to commemorate an important temple feast. Kunjan Nambiar, the court's poet laureate and courtier (an attendant at a court), was one among them. The king was extremely pleased and wanted to immortalize the experience. He was a scholar and an accomplished poet himself. The king asked Kunjan, "How did you like the sadhya?". In the typical mannerism of a sycophant, Kunjan bowed to His majesty in obeisance and said enthusiastically, "Kemam, Ken-Kemam Thanne, my lord!" (exquisite,..sumptuous...splendid,... delicious indeed!); then hesitated; stammered,

stuttered in slurred mumble. “Well, write about it!”. The maharajah gave his command in his characteristic pompous, rumbling voice. Within minutes, Kunjan submitted to His majesty a roll of pure “Thumba- white” paper written in calligraphic handwriting with meticulous care, artistry, and finesse; tied it up with a colored ribbon, the yellow color of the king's favorite fruit, banana. It was a poem! The king recited it loud in his regal voice and in Vedic incantation, in front of the assembled subjects. He ordered his royal attendants to bring him two jewel-studded rings. They were fabulous and glittering; its reflections could be seen in the wide-open eyes of the invited guests, as well as the great amazement on their countenances. Unabashedly, everyone exclaimed: “Oh,..two rings, why two?”. The king sensed their jealousy and consternation; asked them to come up to the podium and recite the poem. Smartly, many came forward and recited it; again, and again; peevishly but obediently, went back to their seats. No one grasped the significance and the implied ‘doublespeak’. So the King took it upon himself. Once again, His majesty recited it twice, elucidating a single theme each time. At the end, raptous applause erupted! The band of sycophants lead the praise for the king in ‘royal anthem’ embellishing his boundless literary genius, acumen, wit, and wisdom, celebrating his attributes with all the “pomp and circumstances” fit only for the king and his royal court.
It was indeed as much a tribute to the phenomenally talented Kunjan Nambiar as well!

The Chundan, literally meaning the “beaked boat”, but in the parlance of the region the ‘snake boat’ is the foremost in the marine fleet of the Travancore Maharajah dynasty during their glorious reign. Nowadays, they are revered, and their memories are immortalized as the preeminent race boats. On the average, they are a 40 m long, sleek floating marvel. The apex of the stern stands almost 4 m high over the water. It possesses a massive, majestic, quaint, and charming look; commands respect and veneration. Its bow and stern are gilded with pure 21-carat gold in emblematic figures and calligraphic writings in the Vedic language of Sanskrit. The use of Gold is quite significant. It was a custom in those days that the subjects brought sovereign gold as presents to the Maharajah on auspicious, religious, and astrologically significant days. Without any knowledge of his subjects or members of his administration, the benevolent king had actually kept them in an underground vault in the premises of the temple of his patron deity “Shri. Padbhanabha”. Unlike the other greedy rulers and present-day ministers, he took none of it; instead wanted to preserve it as a ‘national treasure’ for the benefit of his subjects! By sheer serendipity alone, the vault was only very recently discovered, to the total amazement of the whole nation. The media was quick to call it the "Fort-Knox of Kerala"; perhaps with as much value! Hopefully, it will be there in its entirety as national wealth, and as intended, serving as a 'hedge against inflation' in the pertinent financial sense and terminology!

Well, the selected paddlers are invited for endurance and speed training. This is conducted by an experienced staff, the captain presiding. The yoga and ancient martial art routines are parts of the program. It gained added significance after the UN declared a day in June as the 'World Yoga Day'. People freely come to practice with the crew. Each paddler is assigned a position on the boat, the odd numbers on the '*port*' and the even number on the '*starboard*' sides like the naming scheme of houses on a street. They line up along the bank with their paddles in hand, in the same order as their seating arrangement. Each enters from the back, stern-end, walk along the "cat-walk' platform towards the bow, and takes up their respective seats. The first step is executed perfectly!

The rhythm-makers join in. The familiar Kerala oarsman's songs, the 'Vanchi-ppatt' describes vivacious, toiling damsels in the rice fields; the accounts of adventures & trysts of "Devan & Devika"; the deities and royalties, in a poetic mixture of mystery, humor, romance, and rhymes. It amuses the paddlers; clearly helps in relieving monotony and fatigue. This is an exciting time for the villagers. There is no dearth of volunteers to assist in the communal pastime. The volunteer craftsmen help to make paddles out of treated lumber from palm wood. The lumber is known for its sharp, fine, piercing fibers and splinters. They are carefully removed and stained using hand planes. The paddles are shaped, edged, polished, and greased. The paddles are very thin and light, but strong and easy to maneuver in the water. Science-savvy people nickname them 'graphite' paddles. [This and many such intricate work is usually carried out by the multi-talented and resourceful artisan community once belonged to a 'scheduled caste'. They have been always an integral, essential, and well-respected constituent In Kerala. Since decades ago, their status has been officially elevated ending any discrimination whatsoever, and fully integrated into the main stream of society. In spite of that, the world leaders continue to use their name frequently 'in vain' as synonymous with social outcasts. That's pathetic! This misnomer was propagated and perpetuated by the arrogant British colonists. It's high time to remove such 'racial epithet' from diplomatic dictionary. The onus is upon the Foreign Ministry of India!]

Nowadays, the paddles bear colorful icons of the sponsoring companies etched on them. The markings of commercialism have replaced the images of the deities and saints! People would rather follow the money, gladly of course! It's tangible, necessary, and has immediate purchasing power! Naturally, the prayers have to travel a long distance to reach the heavens. They may be altogether misplaced in the mail sorting centers that often bungle up operations;

delivered to the wrong address; 'hacked' or intercepted for ransom by the malicious imps. Who knows?. Everything has a cost; Even the heavens are not spared!

Formerly, the females couldn't participate in the races. This was purely out of reasons of respect and protection for the women from demanding and potentially hazards tasks rather than any of gender prejudices. They are considered the most important and valuable members of a family household. Speedily & dramatically, that has changed. Now, they can! They participate just as in any other international sporting competitions as equal, capable, competitive, spirited, and entertaining. Here now, a group of women possessing traditional artistic skills and talents sculpt the ornamentation and figureheads, polish the gilded decorative moldings at the iconic bow, while men do the same for the high arching stern. With the distinctive woman's soft touch and careful handling, everything becomes ornate and orderly; properly placed and organized- pleasing to the eyes. The resulting beauty attracts the attention of the fans. Another group decorates the boat with freshly plucked flowers, bouquets, and hand-woven garlands. The decorations contribute to the overall personality and appearance of the boat, that is also a relic revered by the community. Everyone likes flowers for their beauty, color, and fragrance. The passersby make favorable remarks and spread the word at no cost. They help form the spectators' view and impression of the race boat. Children watch with keen curiosity and wonder when they would be part of these activities. For the time being, they run only a few errands and not any meaningful work; actually not permitted to enter or handle the race boat. That doesn't prevent them from shouting out unsolicited tips and techniques to their parents from what they had learned from the nursery school. Parents don't like to be dictated by the underlings. They are annoyed, lose temper, become irritated, act impulsively; scold and ask them to go home and read some textbooks. Other parents give them a list of chores to do around the house, like feeding the chicken, collecting eggs from the hatcheries; dry out the grains in the sun and minding the cattle, and so on. Some children feel like they are 'told off'; treated like servants in public view and settings. Everybody has pride and self-respect; theirs suffer some damage; frowns; mutters to themselves and reluctantly leave the scene sad, dejected, and humiliated. A few of them stick around for a while, showing a valiant defiance, but soon, wisely, follow the example of their obedient friends. Their disappearance makes an immediate impact with quietness and dullness. Some youngsters comfort themselves that they will be included the next time around reaching adulthood, or voting age!

The training session starts early and methodically; paddlers must vow a period of abstinence!

The training sessions are conducted by paddling up and down along the local canals and the river under the prevailing current, tide, and wind conditions. People line up along the banks and cheer them on, voicing their appreciation and wholehearted support. Before the most coveted final race, the Nehru Trophy, there are several regional races organized by different communities and sponsors. This gives the teams opportunities to evaluate their strengths, weaknesses, and performance. They carry out different strategies before finalizing on the most suitable one. Besides, the events help to rekindle interest, support, and fan-base. The community responds in many encouraging ways. Well-to-do families and organizations host Sadhya or provide refreshments to the participants. After that, in the style of informal "jam session", they sit around and discuss suggestions from each specialized divisions of paddlers, Rhythm makers and the coaching staff. On their part, the fans offer their inputs and viewpoints. Well begun is half done; that's the common belief!

Chapter 19

Categories of participating Vallams, The race boats

On the ChundanVallam, the first five lead paddlers, the '*Fabulous Fives*' are seated astride, their feet dangling from cotton slings. They are ambidextrous, being the only paddlers who can paddle on both sides, facilitated by the acutely tapering and narrow forefront, the bow. Their paddles have relatively long handles and conical to suit their paddling style. Those at '*amidships*' (midway between the bow and stern) use paddles of standard size with edges of the blade rounded off, to facilitate a facile '*stabbing*' entry into the water. Their paddles have a slightly bigger surface area and can handle more water; that's where the '*center of gravity*' is largely concentrated and felt! Their action is rather abbreviated to accommodate the synchronized and rhythmic paddling of all the Hundred & Twenty-Eight paddlers with varying wiggle rooms. The boats are designed to accommodate all the specialized groups and taking into considerations of their functionalities. About 2/3rd of the way from the stern, there is an elevated platform. That's the location where the canon was originally placed in wartimes [as the boat is narrow, the "gunwales" were not suited for mounting heavy canons; the recoil from a firing canon could rock or even sink the boat. That, sure 'll be a total catastrophe!]. On the Chundan, as race boat, this is modified as a nice, harmless, and low profile soundboard. The chief Rhythm-maker stamps his cactus pole on it, setting the paddling tempo at each phases of the race following the command from the captain. The captain 'whistles' out his commands in 'telegraphic signal' modes.

The entry of paddlers is always from the stern end. There is a 15 cm wide plank fitted along the dead center from stern to bow. This serves as a walkway for the paddlers to reach their respective positions without having to hope from successive crossbeams rocking the boat or falling down. There are 64 crossbeams to seat all the 128 paddlers who are the only group that are seated. All the others (The helmsmen, Rhythm-makers, and captain) must stand during the entire race!

Several categories of Vallams with distinct shape and style participate in the races & festivals

On the average, the end-to-end total length of a typical Chundan is approximately 40 m. [Compare it to the Pharaoh's boat, the 'Khufu' ship of 43.4 m, although it was a ship, not a boat!; The longest NZ Maori 'Waka' is 40 m; it's a canoe. They are all wooden, crafted boats] It is the longest of any 'paddling wooden race boat' in the world! Besides, it's uniquely handcrafted; not manufactured. About 1.60 m of the bow end (fore-end / prow) spans in the form of a cantilever at least 80 cm over the water line, while floating. The 'plan view' of the boat from vertically above, and looking down shows a 'copper-age' (B.C.) arrowhead used in ancient combat, or simply as the shape of a sleekly formed nib on an old fountain pen. The overall shape of the Chundan looks like three arcs joined together, the "fuselage" has its center of the arc much further than that of the stern, or bow segments. In other words, the fuselage/hull is comparatively flatter than all the rest; the stern is the most curved, hence its geometric center much closer. The 'ornamental curl' on the mount of the stern is remindful of the stretched neck/throat of an enraged cobra engaged in combat; hence the nickname 'snake boat'; in that sense,"Cobra" boat would be more apt, and meaningful! The '*beam*', the maximum width from side to side, is approximately 1.20 m. The '*Gunwale*' or the rim is about 18 cm wide. There are '*cross planks*' or bench-seats just 5 cm below the level of the gunwale, at measured distances to seat the paddlers, in tandem. It also stabilizes the hull. The Chundan is broader at the midsection like most of us are, but shapely and elegant as gymnasts at both ends. The stern (aft-end), ironically called the '*Head*' (considering the head is in the front of all four-legged animals, also snakes & reptiles), carved as a '*totem pole*' held in an elevated position, the tip of which is over 4.00 m above the waterline. There are usually five helmsmen, nicknamed the '*Punchaman-mar*' reverentially remindful of the Five heroic Pandavan princes in the epic mythology '*Mahabharatha*' (The Great Magnificent India!). The five stand on their respective serrated platform steps in commanding positions. Their steering positions are situated above others for an unobstructed view. They use broad over-sized, 3.5 m '*rudder-oars*' of hard wood with long '*T*' shaped handles. The length of the handle varies for each of them. The # 1, who is at the rear most has the longest, as he is on the highest platform. The rest get proportionally shorter ones. The blade is marked, aesthetically and aerodynamically tapered to the sides; some with decorations or logos painted or carved on them. They are entrusted with the critical task of steering the boat along the course, which is not in perfect straight line along the racecourse [A naturally created waterway never runs straight for a long distance. Though the racecourse could be theoretically carved out as 6 lanes, it is not practical for other reasons. There are crowds of people on both sides of the banks that are watching. Therefore, the participating boats must be within comfortable view of the spectators!]. They do the steering by applying pressure with the blade in the water, using leverage against the reinforced sides of the stern. On critical moments in the race, they are called upon to use their big paddles with a distinctive, high arching paddling action known as the "*Kuthi-Eru*" or the '*jumbo splash*' from their respective standing positions. This action has great leverage and effect. I was reminded of an eyewitness incident, where in a spirited competition in the "Payipad Jalolsavam" (one of the oldest running boat races), my uncle slipped his feet and fell off in the water while doing the jumbo splash. The boat went off without him! He had to float in the water embarrassed, until all the boats have gone by him! My poor, brave uncle!

The last but not the least important of the crew are the rhythm makers. They are also the pace-setters for the crew of paddlers. Everyone is familiar with the role of the pace-setters in distance running events. But, here the tempo and dynamics changes frequently, as in classical music. A group of at least 8 men using supple '*water cactus* poles [the shredded leaves of which are woven into mats which are actually more flexible, durable than the modern yoga mats] banging on the soundboards set rhythm and rowing pace. It may be here mentioned that originally similar pole was used in identical fashion in conducting Philharmonic orchestras instead of the sleek '*batons*' used by conductors of today. The chorus sing the famous Kerala boatman's song "*Vanchipatt*"-with the refrain "*Thai Thai.. Tho, Thi Thai Tho..- Thi Thi Thara - Thi Thai.. Thom*", a relaxed 6 beats, with a heavy accent on the "*dip*" and a weak one on the "*take out*" phases of paddling. They face towards 8 distinct directions, enabling them to have an all-around view as well as all the other competing boats in the adjacent lanes. It helps them to choose the right time when they have to vary the tempo. A man with a rasping voice is most likely chosen as the *cantor to lead the rhythm*. For the sprint, they raise the tempo to (*vivace*/ fast and *presto*/very fast as in Western classical music '*Tempo* ' designations) "Aiyah-Da, *Poya-Da; AiyyaDa, PoyaDa*"(something like Hey, Mate! Firm, up!) a belligerent and spirited 4 beat chants. Then there is an even more hectic chant reserved for the finish: "*Ayyo – Poye!, Ayyo -Poye!...ai-ai-yo!*" ('Tic- Tac, Tic- Tac ' like the ticks of the venerable pendulum wall clock; literally cursing "Oh! Damn, All Gone!") Meaning doom, a lost cause, but in fact, sung in vigorous two-beat articulation with rushing crescendo and a heavy accent on *Ayyo*, suggesting an utmost urgency of "*code red*!"

Perhaps the most memorable and captivating boatman's song is the "Kuttanadan Punchayile...Kochupenne Kuyilale..." interspersed with a timely, rhyming rhythmic refrain sung by the paddlers '*Thai Thai.. Tho - Thi Thai Tho.. - Thi* ***Thi*** *Thara - Thi* ***Thai****. Thom..!'[the contextual meaning: listen, ye young, vivacious, lovely damsels toiling in our beautiful, sprawling, and reclaimed Kuttanadan Puncha-paddy fields! The shimmering waves in the canal and this boatman's song throng and thrill our hearts like no other. Hear! The clarion call rising from the banks that beckons us to the races on the pristine waters to join our comrades in singing the race anthem in exulting chorus;...hurry home now and get ready!; receive and congratulate us, when we finally arrive there triumphant and jubilant...!; Arrange Chenda Melam & Pancha Vadhyam to play in our victory parade. Let's celebrate the momentous win with fireworks and cheering fans, also with plentiful, colorful banners; Invite everyone, leave none! When we get there, let's enjoy our success together in great pride and honor as one happy community!]. A comprehensive 'song appreciation & translation' of the song are posted on this website.*

This is a splendidly composed poem with deep meaning, and context from the routine life of the local villagers. It is structured in the accurate metrical form of verse, rhythm, speed, pitch, and musical articulation, with distinct prosody and characteristic cadence. All these elements and properties make it a unique 'genre' by itself! It is highly fitting to the occasion and mood, as well as the simple, rustic lifestyle of the farming community of Kuttanad. To crown it all, it's equally well renditioned by the musical group.

The vanchippatt is a significant part of the regatta and the races. It evokes the memories of nostalgia, joy & spirit of festival season, and the seriousness of competition. Its history dates back to the 17 Hundreds, also to the era of one of the other benevolent and popular maharajahs of Travancore, Marthanda Varma, who was an accomplished literary scholar and an avid exponent of Hindu mythology. The fact-based story goes, that a destitute subject of his kingdom Shri. Ramapurathu Varrier, who was otherwise a gifted poet, wished for a royal audience of the king at a religious festival he came to attend. As the Maharajah was in great hurry to return to his capital after the conclusion of the pooja, His majesty, graciously, invited the poet to accompany his entourage in his own boat. It was a long, arduous journey over the vast and turbulent lakes and numerous intricate waterways. In such times, the Maharajah, habitually, indulged in his governance related matters, as well as literary & intellectual interests, and pursuits. Being briefed about this subject's special talents, the Maharajah promptly asked him "Well, I can listen to poetry and songs any time, of any length; let me hear yours!" Wasting no time, the poor poet began reciting his poem in great fluency and in pious Vedic incantations. Its rhythm and prosody jived harmoniously with the paddling action of the boat. Besides to the king himself, the tune was pleasing to the paddlers as well. It was an epic story-telling poem known as "Kuchelavruthm," which was a moving episode from yet another mythology "Bhagavatham". It was so fitting and poignant (to the point) as the case had several striking similarities of circumstances.

Kuchela was a very dear childhood friend of Lord Krishna from their days at the most revered Maharshi Sandipani's Ashram, a residential school for spiritual & theological Illuminati, and seminary in Hindu religion. After the end of studies, they departed. Both had contrasting fortunes & fates in their subsequent lives. Lord Krishna became a king, while Kuchela an abject destitute. With no means to survive and in their dire distress, Kuchela was coaxed by his wife into paying a visit to his old friend and king Lord Krishna and solicit much needed assistance. Kuchela agreed, but was quite reluctant to go without a gift. So, his wife went door to door to beg and borrow anything she could get. She managed to gather some grains of paddy which she laboriously ground with a stone; made it into a handful, looking like "Aval", a 'coarse cereal'. All the way to the palace, Kuchela kept his small pouch of gift of paltry material value, but great sentimental meaning, close to his chest and heart as mark of love for his wife and reverence & devotion to the lord. He walked a very long, laborious trek to reach the palace. Kuchela was accorded a royal welcome right from the gates of the palace by the waiting lord & king himself with great joy, affection, and respect reminiscent of their great camaraderie cultivated at the ashram. He was requested to stay at the palace as a guest of the king for some length of time. They enjoyed the company of each other and lost in the recollection of their days together discussing esoteric doctrines of philosophy and divinity. Then time came for Kuchela to depart. Kuchela was escorted back home in royal entourage in great comfort, and convenience. Only half way hence, he realized that he has completely forgotten to ask the king the favor he went there to ask for, in the first place. He was totally devastated. He became sad and distressed; wondered what he would tell his wife, who is eagerly waiting with great expectations. He knew it was not proper to ask the royal escort to take him back to the palace. Finally, Kuchela reached the outskirts of his hamlet. To his overwhelming amazement, he saw a new bungalow surrounded by beautiful gardens and gazebo standing in place of his shanty hut he left only a few days ago. A chariot was waiting at the front gate. He was ceremoniously escorted into the palatial house by a team of attendants and stewards. Immediately, he realized that his fate and fortunes have changed for good. Kuchela was at a loss to fathom the mystery, whether it were created by the majestic or the divine power of Lord Krishna, as the lord possessed both!

The boat arrived at the capital's docks earlier than usual. The epic recital lasted exactly until the very moment! The maharajah traveled the distance well entertained and in great merriment; readily understood the gist and essence of the theme. He was immensely pleased and greatly moved by the theatrical recital. Similar to the Kuchela's story, the king bestowed the poet with ample gifts of wealth and rewards. Since then, Kuchelavruthm became a popular theme of "KadhaKali", a pantomime enactment in the unique Kerala art form. Its popularity and familiarity swelled steadily. Its rhythm pattern synchronized suitably with the paddling actions. So much so, it soon became the 'de facto' choice of the genre of "Vanchipatt", that is synonymous with Kerala boatman's song of today!

Today, is the day of reckoning! It's the day of the race. This is the biggest regatta and continues into its third generation. It has survived lean times; periods of gloom & doom; cloudy and hot days; endured rain and thunder; experienced low and high tides, but always staged with undiminished enthusiasm and active participation from the indomitable people of the region. This has become an annual event with steadily increasing popularity, fanfare and media coverage. A growing number of tourists have been also showing a waxing interest in attendance and active participation in these phenomenal races.

This is also the culmination of the festival season in Keralam to commemorate the end of the cultivating cycle. It's similar to the "*October fest*" in Germany or the harvest festivals in many countries around the world. Understandably, the harvest is the most joyful occasion for farmers worldwide. They deserve the pride and praise for their noble occupation of growing grains and feeding the hungry in the world of ours. Their toils & struggles are our life-blood!

Here at the races, there are several categories of boats. All are made from timber, natural wood and materials (using no synthetic, vinyl, fiberglass nor carbon fiber). They are among the 100 + of such boats, distinguished and grouped by the distinct sizes, shapes, and appearance- Chundan, Oaadi, Vaippu, Churulan, Iruttukuthi, KochuVallam to name a few. Each has its own personality and glamor, similar to the newly unveiled car models. Evocative of my childhood imagination, the Chundan is majestic, reminding of an elephant, the bigness, the boat's movement similar to its gait with pomp, leisurely swinging its trunk, fanning its broad ears, occasionally looking sideways with arrogance and pretense; acknowledging the adoration it is getting (like Vladimir Putin), projecting a solemn disposition. There are 128 paddlers on Chundan sitting mostly in pairs. All boats and its accessories are made of wood or products fashioned from materials from the surrounding nature. The paddlers sit on fixed wooden seats.

The Oaadi is sleek and fast like a cheetah in its run. Its beam is narrower about 75 cm and has a '*hydro dynamically*' shaped hull; uniformly shaped end to end both the bow and stern converge sharply at the ends with flush platforms for 3 lead paddlers at the bow and 3 helmsmen at the stern. The *water line* is just below the brink by a few paltry centimeters. They are propelled by approximately 40+ paddlers. It's an exceptional skill that the paddlers keep the boat in such a critical balance without capsizing. The bow, the fore-end is vertical; its aerodynamic; the domineering ends smoothly curled back, as elegantly groomed burly mustache, staring at the paddlers. Both ends look the same, it can go reverse in equal speed in short order by changing directions of paddlers. It is rather light, picks up speed quickly, and streaks through the water, deflecting ripples and waves with its '*channeled*' bow. They were primarily built by the pirates for their clandestine activities. These boats were once used to prowl the big and narrow waterways in hiding and making little sound like the carnivorous 'big cats' do in the wild, or on the savannas. They plundered the unsuspecting and unarmed travelers; escaped capture by the versatile fleet of the Maharajah's Royal Navy at will. It seems to be challenging as in children's game: *"catch me,.. If you can!"*

Then, there are variations and wide imitations of Oaadi, classified as 'Vata-kkan Oaadi', 'Iru-ttu-kuthy' and so on. They are probably built by regional tribes, as disguises and designed for custom & multi purposes, or for additional capacity, carrying tools, and weapons. They were all speedier boats than the Maharajah's Chundan Vallams!

The Vaippu, strong and sturdy with a massive, carved '*figurehead*' of a certain ferocious mythological character, looking skywards; laying lazy, languishing, reminding of the male lion who shows up only to keep the '*spoils of hunting' by the lioness and her cubs*! The elevated stern is shaped similar, but much more diminutive, compared to that of a Chundan. Upon gaining full momentum, it sustains its speed well. A crew of 50 paddlers provides the motive manpower. They are the replicas of the original support boats meant to carry food & provisions for the fighting marine crew on the ChundanVallams. The term '*Vaippu' shall mean the place to prepare food. For all practical purposes, it's the 'pantry-boat' keeping the entire packed items of food for distribution.*

Like the flight of the bumblebee, Paddling is a high energy, and exhausting full-body exercise!

The races held in a fun and festive mood, but soon turns into stiff competitions & serious rivalries

It has to be remembered that some of these boats did find everyday use as commuter boats to transport farmworkers in the immediate past to the remote reclaimed-rice-fields in the middle of the lake. These fields are named 'Puncha' fields, where only one crop cycle is possible being below MSL. Whereas, most of the fields beyond the fringes of the lake in upper (South) Kuttanad, are "Viruppu" or flatland-fields, where 2 crop cycles are possible. The workers, as a group, paddled their boats to the fields early in the morning and back home in the evenings. On the average, the commute may have lasted over 2 hours one way. [Today, they are replaced by motorboats reducing commute time and increasing productivity of labor] Together, they have faced different incidents- of danger, thunderstorms, capsizes, sickness, emergencies, camaraderie, fights, joy, hate, love, and other memorable events. Essentially, these boats were once very much part of their lives and destinies!

The spectators enjoy the true feel, sense, passion, and the ambiance of the race Venue!

The 'Churu-lan' similar to the Oaadi, with more, accentuated, spiraled-back ends, the stern slightly elevated, looking at each other, ornate with brass moldings, reminding of the gentle fawn dashing across the woods. It belongs to the luxury transportation category. The hull is more or less round, making it faster to move through the water at some disadvantage of 'roll over'. Its beam is slightly larger, about 1.2 m comfortably accommodating 3 people in a row. The elders warn against using this boat in rough and stormy weather. They are well suited for punting with long, light bamboo poles.

On the other hand, the "*KochuVallam*" as the term indicates is the smallest of all; manned by one, two, or a handful of paddlers depending on their relative sizes; their rims dangerously flirting with the waterline, but the paddlers skillfully avoid sinking with steady balance and skills. They are sleek and shaped like a ‘sculling boat’. In daily life, they are mainly used for personal transportation; euphemistically nick-named the ‘manual water scooter’. Both ends are symmetrical, slightly fatter at the rear like us, the '*Homo sapiens*', slantingly carved heads facing upwards at about 10° slope. All boats are adorned with flower bands, blessed at local houses of worship for blessings and good fortunes.

The “palliyodams” are essentially the forerunner of ChundanVallams with exaggerated curves at bow and the stern. They are not really a raceboat, but used for ceremonious occasions to carry the idols of deities in processions around the river during temple feasts. Other times, it served the needs of travel and outings by the king and his entourage to ‘visit his subjects’. It was also employed to show around important sites within the kingdom to royal visitors in great hospitality and splendor. The Palliyodams are designed and built essentially for use on processional and festive occasions to impress the public. They are brought out in show of groups and pleasing formations.

Traditionally, the paddlers used to wear bandannas of cambric cloth with colored borders, identifying their villages, also a symbol of '*nonchalance*' and defiance. Recently, they swapped them for stylish hats supplied by the sponsors showing their artistically designed, embossed, and embroidered business logos. Formerly, the men's paddling crew were all '*bear chested*' (allowing natural heat dissipation under the hot and humid conditions, also happens to be the casual form of normal day wear). But lately, commercial interests entice them with jerseys made of glossy, air-cooled, sweat resistant fabric displaying captivating graphics and logos advertising their business. The coordinated colors and accents on them, form different shapes and figures in accordance with the movements in paddling. It enhances the look and glamor of the boats.

Thus, these classes of race-boats are relics and replicas from the ancient marine warfare fleet carrying an unbroken history from the early 14th century onwards. They have strong roots in Hindu mythology, deities, and dynasties. From the eventful times of frequent conflicts between the kingdoms & dynasties, the boats carry rich, checkered, colored stories, and glories with them without undergoing modification, except for the dimension of length of the chundan and its carrying capacity. They are built with the exact original specifications and treatise of Naval Architecture, again wresting another distinct place in human civilization. The boats are venerated by the people of Keralam for their role in shaping their destiny and history. They invoke nostalgic vibes from the dramatically narrated grandparent stories told regularly after nightly prayers or at bedtimes- of wars and adventures of the past; ecstasy in victory and agony in defeat; the fights between good and evils; conflicts for supremacy and dominance; strifes out of greed for gold, wealth & possessions of strategic land & lakes; for historical sculptures & statues; artworks & artifacts, to prove allegiance to gods & goddesses; to exact revenge and settle vendettas; to capture notorious robbers & bandits, and sometimes wars for no specific reasons. Usually, the stories are narrated in the literary prosody of "Sloakam" (art song), a Vedic recital style of poetic prose with rhyming intonations in archaic cadence creating periodic pause, suspense, drama, and wonder. This is a tremendously powerful tool for bonding of the generations and bridging the gaping gap of generational times. Children alternate between grandma and grandpa on alternate nights to hear a rich variety; different tone & feel. The impressionable children are kept in rapt attention throughout the story, each time telling them a new story. When they run out of unheard stories, quickly they make up one from their own unique life experiences to fill the time. Soon, the children will become dozy and drowsy, usually before the end of the story, fall asleep in heavenly bliss and happiness! The story-tellers can now go back and finish the rest of the house-hold chores, feeling happy and contentment.

Chapter 20

The grand arrival of the enthusiastic spectators

The spectators begin to arrive, by foot, on vehicles, in groups, with families, joining friends & neighbors, by a flotilla of public service boats, motorboats, speed boats, mechanized boats with inboard/outboard engines (Outboards have more power per Kg of weight than inboard engines do), luxury houseboats, wooden canoes; also the celebrities in entourage; tourist groups escorted by tour operators. Women are accompanied by responsible family members.

By anyways, or means! This is a sight full of passion and excitement; it must not be missed!

People wearing different colors, designs, and styles of dress converge to one location. It's a unique and wonderful sight to behold, much like the migratory birds swarming a resting area. The people who do not have their own means are generously offered rides. They arrive and take up positions along the embankments on both sides of the waterway of the racecourse, along the levee, quays, dikes, the ridges surrounding the paddy fields, on rooftops, on balconies of posh bungalows on the waterfront, on “Maa-tam” (the typical Kutta-nadan style 'standing hammocks' with coir woven seat / bed with a thatched canopy shade); tiny children piggyback on big brothers' backs; little children on fathers' shoulders; infants in mothers' arm cradles; some perch on tree branches, on single beam scaffolding of bamboo, which also serve as barricades planted for the event, and on floating platforms of tire tubes, driftwoods, coir/foam-ring buoys, inflatable floaters, and dingies, as well as many ingenious ways & means.

The camera pans, and the video shows images of throngs of people lined up along both sides of the course, stretching 1.50 km in length. The spectator boats are parked in a pattern, at an angle towards the finish line; in rows; the smaller ones in the front, the bigger ones in the rear, so that the view of none is blocked. Race boats are seen plying up and down the course in their warm-up exercise and preparations. Nowadays, they wear a distinctively colored set of uniforms supplied by the sponsors, helping them easily identified among the many others. Formerly, they were wearing a traditional all-white loin-cloth, “Banyan' an armless undershirt, and a headscarf or no shirts at all.

The invited guests and government officials are escorted onto the grand stage. Celebrities have bought and reserved all the high-priced seats with vantage points. They don't like to mingle with ordinary folks. They must be always aloof, seated in high places among the crowds or closest to the scene of action. The tourists are welcomed and given available cushioned seats near the finish lines. People watch from the roofs of large motorboats, not fearing the waves rocking the boats. It's 'free standing-space' in most other places! Everyone is searching for a vantage point, but not everyone can find one; each settles in the one they could get. Unlike in Football stadiums, there is no such area assigned to each team supporters. It's not a competition of just 2 teams. None wants to miss a glimpse!

This is a socialist country cherishing egalitarian principles. But, as everywhere else, there are exemptions for politicians as they enact laws with themselves, their beholden, and their vested interests always held deeply in their hearts. In racist and capitalist countries, there are 'privileged' classes that differentiate people by race, skin color, wealth, connections, contacts, positions, pedigree, education, and status in certain preferential social & hierarchical order. Even if they commit severe crimes, malfeasance, frauds, swindles, treason, or conspiracy, there are many loopholes that are unabashedly built into the system with glorified clauses and legal terms such as 'plea bargain', 'compassionate reasons 'considering charitable contribution', 'services to the country', etc. Besides, the judges accept feign excuses, before bestowing generous leniency to the accused.

Commoners, minorities, and the 'coloreds', often end up getting the harshest jail terms to the 'full extent of the Law' brandishing the maxim "this is a country of law & order!"

The justice is distinctly "Black" & "White" with widely contrasting consequences for the affected!

Spectators are arriving to see the regatta in great anticipation; in their own style, might, and means

The race is an event of 'superlatives'. This is the largest attended live paddling wooden boat races in the world; the biggest team sport of any kind; biggest, also in terms of the sheer size of the race boats; the largest number of spectators watching live and on Televisions, number of paddlers taking part, number of races held, distinct categories of boats, the intensity & passion, reverence, and the decibel level raised. In all, there is an excess of 100 boats and 5000 paddlers competing in the race today. The ambiance and the assemblage itself are so unique and simply awesome. It's an open water race, where the races are conducted in the natural conditions prevailing at that time. The races start from a 'stand still' position; paddles out of the water. Each year, the scope, and participation are expanded to additional groups- women, including foreign nationals, and types of boats participating. It rivals the 'gondola race' in Venice; Sydney Hobart yacht regatta which is the longest but few people get to see it in person, as it takes place somewhere out in the foggy remote sea-straits and out of sight. It seems to be intended mainly for the very rich & celebrities who have lots of spare time and ample means to follow the boats.

The VallamKali distinguishes itself also as a folk festival of all the people by the people, and for the people. Each boat is a 'communal property' of the village reflecting the same principle of 'community ownership' of land, and other resources by the indigenous people around the globe. The individual ownership and selfishness may have had its influence from the colonialists and the capitalists.

Currently, many interest groups and sponsored groups are trying to introduce Olympic events of "sculling' & "sweep" categories. The traditionalists fear that the move may diminish the surging interest in the traditional sports the communities can naturally relate to. Many of the young could be attracted to the lure and fame of the Olympic medals. In those proposed races, the boats are made mostly of synthetic materials and composite materials usually, carbon-fiber, fiberglass, graphene reinforced plastic and so on. They cost many times more, and remain beyond the means of local people. The accessories, bells & whistles for the boats are also too prohibitively expensive. All the items of competitions, its rules and the stakes are always heavily in favor of the rich countries and against the poor.

People say "let rich nations keep their gold medals; our wooden boats are worth more than those medals; we would gladly cherish our history deep within our minds and modest means!"

Chapter 21

The sights & sounds In and around the race course

Several motorboats are seen shuttling up and down the raceway. A boat passes by; the official seal, 'Two giant 'temple tuskers' facing each other in its natural stride; the long trunks held high in a cordial welcome sign, is prominently visible. It is the iconic insignia of the state government of Keralam. The sign is profoundly meaningful, signifies the Keralite heritage of love of nature, honor, hospitality, pride, and fairness. It's a Police Boat! It carries uniformed officers. The ranked ones are massively visible, the others are in obedient obeisance. The officers cast searching looks on each side moving across, looking stern and solemn, often readjust their gilded caps, perhaps showing off their rank and position. Mildly, they admonish and gesture away intruding, unauthorized boats & swimmers wielding their 'Laa-thies' (the solid, stiff, bone-breaker batons, a brutal colonial relic of the recent past) that they must clear the course at once. It's quite reassuring that they don't carry guns and tasers, unlike the police forces in certain countries where 'shoot to kill' upon mere suspicion, or a body movement is fully justified and generously condoned! This is a land that still cherishes the great Gautama Buddha's principle of "Ahimsa" and Gandhian principles of "Nonviolence"! The lives of all creations and ecosystems are cherished, respected, and considered invaluable. The crowd descends into an abrupt silence, reassure themselves that they have done nothing wrong . The police are generally lax and lenient; let people have their way and enjoy themselves. They are people too, and prefer to watch the race not being bothered; turn a blind eye to non-serious matters.

Many resources are deployed to ensure the smooth run of the races under safe and secure conditions. The authorities are worried about some risky positions people have taken, and the seemingly dangerous things they do. The easily excited and immature fans may be tempted to do foolhardy acts in front of a boisterous crowd; show off a little too much and endanger themselves as well as others. Law enforcement officers are fully cognizant of these types of impulsive behavior and resulting dangers they can pose.

The race boats ply up and down along a parallel practice course in great earnest, warm-ups, and final preparations

The Indian Navy squad is present. They conduct a small parade on water skies, holding national flags and making different formations. People applaud in appreciation and recognition in great pride. Another Navy troupe demonstrates diving and acrobatic routines. The third group demonstrates water rescue operations. An expert and qualified commentator explains these maneuvers. The navy rescue boat carries necessary equipment and trained rescue personnel with first-aid and emergency medical supplies. Three of their helicopters and an amphibian are standing by on specially built Helipads at both banks of the channel. Their patrol boats are vigilantly scouting the area with high-powered binoculars and communicate with their respective divisions; also operate mobile safety and security surveillance systems.

The scenes and activities create a unique ambiance; ambiance builds excitement; excitement builds expectations. There are plenty of things to see, learn, and enjoy. Some companies showcase their new and innovative products and wares; distribute brochures and promotional sample products.

The boat immediately behind the Police boat carries a number of people in traditional pure white Keralite dress of 'white Mundu' (waist-down, wrap-around), 'Juba', and sprawling scarf with thick borderline. Some wear 'Khadher' dress (a modest, roughly spun cotton fabric popularized by Mahatma Gandhi in the National Freedom Struggle against the British colonists, and thereafter by the members of the Indian National Congress party); the others are dressed in the so-called transparent linen 'foil' aka. Foil Ju-ba'-the 'F' omitted. Quite commonly, the Filipino folks also wear a similar fabric. It's comfortable under humid conditions. The people on the boat are definitely people of some importance, given that the police boat is in escort. It indicates the tenuous relationship of elected leaders to the public in a subtle manner. Some still wear the garlands they received at the welcome pavilion, in spite of its weight and discomfort; reminding people of their VIP status. Some others, ostentatiously, wear heavy gold metal chains around the neck instead, visible above the collar-less Juba, glittering under the bright sunshine.

A well known political satirist and comedian, who is often seen on the nighttime TV, hosting the 'Mockumentary' lampooning corrupt politicians, alias 'Ludicrous Lonappan' or Ludicrous, for short, explains: "those wearing Khadher, are congress party MLA (state legislators); those wearing the foil type are Communist party MLA. By the way, Kerala Congress MLA and many of the Syro-Malabar Christians like me, also wear the same, as they believe the 'Shroud of Turin' is made of the same linen". Some chuckled; others frowned in disapproval. Ludicrous is seldom politically correct. He is dangerously opinionated and provocative; probes the limit of jokes; consequently, has often been targeted by the political 'goons & goondas' (hired gangsters). People say he is a cat; has 7 lives! He goes on: "True! The very recent carbon testing confirmed that the cloth was spun in ancient India. The fact remains that the fabric was originally brought by Mar. Thomas Slihah (St. Thomas, the apostle, aka, the 'doubting' Thomas) when he sailed on a merchant ship from Tripoli, the then famous Mediterranean port, and arrived first at "Kodungalloor" the ancient port near Kochi. It was a very expensive fabric; my family was rich in those days and decided to have some exclusive style, and distinction for their elite status. Since that time, we have been wearing it. Simply, that's our fashion statement! We're the first to be baptized and became the earliest Christians. Well, now, those who wear gold chains have been recently either to Dubai or Singapore on visits at taxpayer's expenses. The gold they wear is bought out of the proceeds from the contract & business deals they graciously helped to consummate. The silly public and the overzealous media brand them as bribes, kickbacks, or ill-gotten wealth, depending on their vocabularies or level of indignation".

The women are seated, sheltered beneath the canopy, and obscured from public view. They are intentionally avoiding any tanning under the scorching sun. Fairness of the skin is a valuable personal asset in the culture, almost anywhere in the world for that matter. The men on the sun deck pompously look around, their faces masked by the over-sized sunglasses; wave to the crowd. But the crowd does not recognize them and therefore does not react. Some ego is bruised; nevertheless, the men hope that their faces will be prominently shown in the evening news in tight close-ups looking like sportsmen and in good standing with the 'adoring' supporters.

The election is looming, and it will be a free and powerful advertisement. Well, it only takes a couple of phone calls to the bureau chiefs in the media business to get a close-up picture and special attention! At one time in the local culture, the 'pot-belly' or euphemistically the 'omentum' was considered ostentatious signs of aristocracy & affluence, having enough and excess to eat and feast!
Now the sentiment is beginning to shift towards figure and fitness!

Chapter 22

The organization & preparation
The staged and orchestrated activities

The municipality has mobilized amenities for public comfort and relief; brought clean drinking water and first-aid kits. They requested additional transport services from the ministry of transport; sent newsletters instructing hoteliers and hospitality services to render courteous and friendly services; cautioned on price gouging. The police forces were instructed to be extra vigilant and surveillant on scammers, swindlers, and scoundrels; traffickers, fake currency note changers; predators, and 'Eve teasers'. They are asked to take additional steps for public safety & security. The mayor has generously approved overtime for essential staff. The sanitary workers did already clean and fumigate the venue in the morning. The Public Health Services bring in mobile health service units, volunteers, and medical staff; distributes illustrated pamphlets on health & family planning. Volunteers offer to check Pulse & BP for free. They show video programs on Hygiene, Diet, Exercise, Yoga, Meditation, and Diabetics, that is rampant in the community; mental health; children's & women's health. There are some religious affiliates offering counseling on family issues, although they left the family in their childhood, celibate, and have no intimate knowledge of the complexity of conjugal, and family life. The Social organizations provide lessons in civics, alcoholism, morals & ethics, abstinence; health services on family planning, preservation of life, showing anatomic illustrations along with the aid of slide presentation, and video footages. Specialized NGO services and charity organizations of several kinds are also well represented. Rent-free booths are allotted to them in the remote and obscure corners. Prime spots are reserved for rent-paying big businesses & large corporations.

There is another parade of a group of dozen boats with variously uniformed cadets of marine squads from the Indian Navy pass by. Pure white, Navy blue & white, are their favorite uniforms. They wear black hats with sunshade; A line of cadets facing one way, and the other just the opposite. They are seen 'standing at ease' on the deck. As soon as the 'Lieutenant' shouts out commands, the cadets go through their routine and salute the spectators and the paddlers in their view. People feel elated as they have not received such an honor from anybody. The closest thing is the welcome barks of their dog when they arrive home late at night, opening the heavy rattling front gate. The parade is intended as a PR measure for promoting national sentiments and attracting new recruits to the Indian Navy. Some overzealous fans jump into the water and approach daringly too close to the speeding boats, showing no awareness of their rapidly spinning propellers. They do this, perhaps to show their appreciation to the cadets. The police admonish and wave them away. The general frenzy is only growing, given that this is not an everyday event, and people are experiencing many new things that they are justified to display some exuberant expressions. But most spectators behave responsibly; they survey, study the sights, and things that are open to see; assess what all of these could possibly become the mainstay in everyday work and life. In the competitive world, they have to be on top of things to be successful.

The educated parents may ask them to describe their experiences and what new things they have witnessed for the benefit of parents and younger siblings who stayed at home. Besides, teachers are likely to ask to prepare a presentation in front of the class. Engaging teachers closely moderate and critique the presentations. They want the presenters to speak clearly with intonation and correct cadence without sacrificing the accuracy of the points discussed in the subject matter. They coach the presenter as to how to maintain correct posture, poise, and confidence throughout the presentations. They are allowed, rather encouraged to show slides & images to make it more interactive. Malayalam is commonly spoken rather too fast. When speaking other languages one has to be aware of the pronunciations and common styles, phrases; need to speak slower and coherently without mincing words or 'aspirated' sounds (expulsion of forceful and audible breath in speech) consonants (other than vowels; consonants represent sounds that are made when part or all of the vocal tract is closed). Depending on the contexts, words have to be spoken with accent, articulation; short or elongated, as the case may be. Western languages have fewer alphabets; English has only 26 ; Germanic 30 (including the 3 'umlauts' and special character; French 39, including 13 variants) while Malayalam has 56! Indo-Germanic languages (Malayalam among them) are spoken, same as they are written. English is not exactly so. So there's some confusion, hesitation, and deficiency in fluency also due to the scarcity of alphabets. Foreign English-speaking people get frustrated. Then there are 'migrated' words from other languages. To top it off, there are different accents prevalent in different English-speaking countries. It makes speaking English difficult and hesitant. Some words are constructed with vowels in them and provide no clue if it is pronounced short or long. Whereas, in Malayalam (അ, ആ; ഇ, ഈ; ഉ,ഊ; ഏ, ഏ, ഐ...) are a few examples of long, short and variant sounds; they are distinct; written as separate letters!

Well keeping all these factors in mind, this is the ideal time to observe and learn. Once the races start, attention will be focused on the boats with great anxiety and expectation of the results. Some are seen snapping pictures, adding captions, notes on their hand-held devices for references and reminders, also for preparing presentations. It's quite important to shoot some pictures and video footages; there are many, unusual settings, stunning landscapes, etc. but limited by the battery, and the memory capacity of the SD (Secure Digital memory card, a non-volatile memory card mainly intended for portable devices) card. I should have brought along some spare ones. The 4K video uses up a lot of 'juice'. As they say nowadays "a picture is worth thousand dollars!"; video footages, a lot more. There's big demand for them. Video editors, graphic artists, media companies, broadcasters, movie-makers, advertisers, sales people, video game developers, YouTube creators, websites, corporations, etc. use them for their edit and presentations. The growing demand and applications may justify the high cost of a professional DSLR (Digital Single-Lens Reflex Camera using memory card for storing/retrieving the gallery of images), professional video camera! Any presentation is bland and boring without multimedia mix. Demand is there for still and moving pictures, but marketing is another costly head-ache. Well, let me go around; make some reconnaissance, and video footages for my personal use first!

Chapter 23

A 'Cultural Heritage' Show!!

Various cultural, literary, and art performances are staged as part of the water festival. It is collectively labeled the 'heritage' festival. When there is a gathering of this size, there will be large groups of people with varied talents and interests, also many in fine arts and performing arts. Previously, most of the performances had their enactments only during temple festivals. “KadhaKali"(a unique Kerala style of 'pantomime' / costume / theatrical/ musical performance by means of just bodily & eye movements, hand gestures, without uttering a single word), “Ottam Thu-llal”, the traditional art forms, as well as Carnatic, and classical music performances are routine fixtures during the week-long “Ul-sa-vam”, the temple festivals attracting large crowd across all religions, and affiliations. They are public events held on “Mai-daanam' (open-air fields) on the temple premises. No ticket fees are collected nor donations solicited, financed only by voluntary donations and from the temple's cash offerings. It is free for anybody who can attend and enjoy it. It is like attending the US July 4th fireworks and the rock concerts after that. The nightly temple performances are commonly attended by the entire family and neighborhoods. My maternal uncle, Kuttappa-chayan was an accomplished student and proponent of classical music. He was a talented multi-instrumentalist in keyboard (Harmonium) & violin, besides also being a drama actor, and director. I was told by my grandfather and mother that he had his grand debut held at the courtyard in their sprawling house premises. Distinguished musicians of the time and a large crowd of music-lovers were invited. After a sumptuous Sadhya, accompanied on Mrudangam by his music teacher “Bhaa-ga-va-thar” (maestro) Sree. Govinda Pillai, the debutant played 'Kiritis' by Thyaga Raja, 'Ragas' of MuthuSwami Dikshithar, and 'Bhairavi' by Shyama Shastrikal. His 'repertoire' included rich combinations of compositions in Ragam, Tanam, Pallavi improvisations. The guests and the audience were treated to an enthralling performance. They applauded loudly and asked for encore performances with continuous clapping and cheers. The performers responded and wholeheartedly obliged. The debutant and the

Bhagavathar used to travel together to Thanjavoor (considered the birthplace of Carnatic music) and various temple festivals to listen and learn about the unique techniques and renditions of famous master musicians. Later, they analyzed the works and hosted seminars about their findings. In the absence of recording devices in those days, he could not resist these golden opportunities on offer, often neglecting his own law practices. In fact, my mother also took lessons. It lasted only up to her marriage. Her new family was apparently apathetic to music. Sadly, her musical instrument was promptly sent back, to the chagrin and mortification of her parents and siblings. My father was absolutely not the one to be blamed! In an extended family organization, such decisions were made by the eldest, most conservative, and most traditional matriarchs who themselves were married off in their early teens before becoming well-educated and trained in essential matters and social sensitivities.

For the sentimental consolation, and as a tribute to them, I strived hard to learn a musical instrument, although it was somewhat late in life. I was greatly gratified that my mother could hear and enjoy some of it when she came for visit. Besides, I was able to cultivate interest in my son and daughter alike to do the same. Thankfully, they didn't disappoint me though they had been preoccupied with a few other hobbies, and pastimes of their own, beside the most important studies and jobs.

Just like the Western classical music that flourished in the churches, Carnatic music finds its roots in temples. Many famous Western classical music composers were 'Kappel Meister or choir masters. I was led to believe that my father knew only two songs, occasionally sung in privacy, poorly renditioned "Vaikom Kaya-lil olam thullumbol ('വൈക്കം കായലിൽ ഓളം തുള്ളുമ്പോൾ...) ": 'as I see waves gathering and picking up in the lake, I fondly think of my beloved; I brave the waves and cross the lake in great bit of hurry to come home to my sweetheart!' Another was a poem by the eminent Malayalam poet "Chang am-Puzha". It tells the sorrows of a poor vassal. "Following the first rains in the early spring, in great joy, he planted a banana tree in the front yard of his thatched, shanty hut. Diligently and lovingly, he took care of it; periodically fertilized and strenuously weeded. Every day, with excitement, he watched it growing. It blossomed; buds sprouted, also the male & female flowers thick, full, and long; bees and butterflies flocked to it for its sweet honey. He smelled the fragrance in the air. In the following few weeks, it fruited; then gradually ripened. It became flavescent. Regularly, the poor tenant looked at his own plant in great amazement and waited yet another day to harvest it. By then, the landlord came by for his regular inspection of his landholding. He spotted the lush clusters of fully ripened yellow bunches of banana, together with its deciduous brown bracts, flowers still with sweet nectar within them, and carrying a heart-shaped brown bud at the tip, all together hanging enticingly on a green peduncle (the stalk bearing a flower or fruit) as a colorful decoration among the thick vegetation of green leaves and tall palm trees. Without a qualm, scruple, hesitation, or even uttering a single word, he harvested the bunch. The landlord took the whole bunch along with him, as the vassal and his family stood there sadly looking on with tears overflowing from their eyes!

Quite sorrowfully, an identical fate befell on my dad too, when he leased a block of paddy fields from his own rich and famous father for rice cultivation. It was poignant; probably was telling his deeply moving story or teaching me a meaningful life's lesson! Both songs reflected his sensitivity, character, and philosophy. I recall seeing him coming home late at stormy nights dazed, his 'Val lam-boat' capsized; drenched and shivering returning from the fields located far away by the lake, and catching fever the next day. He showed great sympathy for the poor and the underprivileged; felt much empathy; championed their causes and welfare throughout his life, oftentimes even forgetting the fact he had a large family to provide for.

My mother's love for music was undiminished. She, in her melodious voice, regularly sang to us Nursery Rhymes & tunes- Johannes Brahms lullaby, 'Minuets' by Johann Sebastian Bach and 'Cielito Lindo' the Mexican national song, also a Mariachi band favorite. Her father had generously donated prime roadside land to German missionaries who came to do charity work amongst the poor. They taught her and elder brother several music pieces during their frequent visits and sojourn at grandparent's house. The three Bees (Bach, Brahms & Beethoven) were all German composers. She performed these tunes in ceremonies at the convent-run boarding schools she attended, while her brother did the same at St. Berchman's college as a student. After several decades, when I heard the very first measure of the same Minuet [A stately dance for groups of couples in slow 3/4 time signature] on violin practiced by my daughter, I recognized it; felt great amazement that it has been hiding deep in my fuzzy mind since early childhood. I felt totally overwhelmed with emotions. In deep sentimental feelings, I recorded these songs on my accordion and posted them on 'YouTube' also in loving memory of them!

"All of them would have immensely enjoyed the jamboree and the events taking place here at the venue!" I mused in deep reflection and consuming feelings of love & nostalgia.

Today, there are a potpourri of events, folks music & dance programs, bands, and orchestras staged. There are various performances and artforms endemic and unique to Keralam. To be exact, there are 64 of them that are now enjoying a 'renaissance' of sorts. They were all held in abeyance and neglect for a long time. Meanwhile, the artists and performers had to seek other ways to practice their passion and earn a living. Now, in the context of the regatta, they garner wide interest, enthusiasm, recognition, and opportunities.

The enactments of various traditional artforms in renewed 'spirit of 'renaissance'

Several competitions are held in different performance categories and classes. They are held in the same spirit as the Olympic Games held in Ancient Greece, where participation was the principal motto. In Western parlance, it may be labeled as a 'show or extravaganza' but actually, that is not the case. They are created and practiced for aesthetic, intellectual, historical, and certain other meaningful purposes. It revives ancient, and forgotten arts, promotes a culture, cordiality, interaction; teaching poise, stage presence, discipline, etc. They improve mental faculties, coordination, besides gaining appreciation and providing entertainment to the public. The collateral benefit is to give the youth something to practice, keep them motivated, off the rowdy streets, and social media perversions; out of bad influence and inadvertent troubles. It helps to maintain the mind and body in a stimulating, energetic, and healthy state. The same spirit is manifested in several recognizable and beneficial ways in the community and daily life. In any need, danger, mishap, or emergency, neighbors respond spontaneously and come to the aid without any prompt, persuasion, pleas, or solicitation. Instinctively, they form task forces and perform relief activities. This is in contrast to Western cultures, where professional task forces or organizations are called in to manage the situations.

It's a stark contrast of empathetic socialism on one hand, all-consuming, self-centered, and indifferent capitalism on the other!

Ancient martial art-forms displaying agility, gravity-defying aerobatic & acrobatic maneuvers; showing remarkable poise, vigor, and concentration. **None** of it, is an Olympic event, but fencing, judo, karate, Taekwondo,'*horse jumping*' are!

On the other hand, on the negative side, it gives rise to vigilante groups who interfere with the 'law and order'. They thrust themselves into everyday affairs; take unauthorized actions in matters such as road accidents, petty crimes, property disputes, etc. like the Talibans, or "Muthawas" of the Middle East countries acting upon moral and social behavioral issues. Like the Chinese authorities, they do surveillance operations, monitor, admonish, and punish people suspected of antisocial & immoral activities based on mere suspicions and conjectures. They tie up and beat up offenders, meting out their own punishments and judgments then and there. In most cases, they turn out to be cases of mistaken identity, mere rumors, or wrong information or simply a case of innocent people caught in bad situations. Another manifestation of reigning anarchy!

In the villages, there is a tendency for 'kith, kin & clan' to live nearby. People are identified to their ancestral home and use surnames derived from it. The land they live in and cultivate are usually inherited properties from forefathers. It is also very common to adopt surnames from the cast, each belonging to or from a traditional occupation. This custom is prevalent throughout India. By virtue of its social setup, neighbors are well known to each other for generations; engage themselves in frequent conversations, social and religious events; maintain cordial relationships. New friendships are formed and maintained at schools, colleges, family relationships, business interactions, and at workplaces as well. It is often stated while you identify yourself as hailing from Keralam, people say: “I know a lot of Keralites; they are either cousins, relatives, school/college-mates, or neighbors; all connected and intimately known to each other in one way or another!”. Friendship is highly valued and strong in the community. It often transcends caste, creed, or other affiliations. I felt honored, also overwhelmed when my neighbor and childhood friend Sri. Unnikrishnan Nair visited me at home when he came for just a day’s visit from his military training from the faraway island of Andaman. He is the only child of his parents and had been away for a couple of years. I wondered if he shouldn’t have spent the entire time with his lonely parents! After exchanging all the common hospitality, pleasantries, and platitudes, I escorted him back to his own parent’s house and spent a brief courtesy time with them.

Referring to the whole of India, the first Prime Minister of India, Sri. Pundit Jawaharlal Nehru praised: there is “Unity among Diversity!”. I wondered if he said it, soon after visiting Kerala after inaugurating his namesake “Nehru Trophy VallamKali" taking place annually at this spot, on the backwaters at Alappuzha. There is an inherent civil and 'common cause' bond between people in any village. All island cultures have similar characteristics. I must hasten to add!

A friend in need is a friend indeed!; the corollary is: 'all neighbors are friends in need !'

Kerala may be counted as a large peninsula, if the breathtakingly beautiful Hi-ranges, the Sahyadris (strangely called 'Western Ghats') in the eastern border with Tamil Nadu is considered as an unusual 'isthmus'. Essentially, it is formed and situated like a fortress, or escarpment; a home of extraordinary biodiversity and a climate maker of great consequences.

Kerala, the exotic land of spices

Apart from the many large blocks of paddy fields, coconut groves, banana & arecanut (Betel nut) plantations, the state of Keralam is even more famous for its prolific spice produces of Tea, Coffee beans, pepper, cloves, cardamom, cinnamon, nutmeg, ginger, turmeric, tamarind, chillies, curry leaves, etc. Its taste, flavor, aroma, and uses reach far and wide in every kitchen, culinary preparation in almost every dish, food pantries, and gastronomy in every part of the world. The spices are also used for seasoning, food preservation, and certain industrial uses. They are dispensed in thoroughly ground and powdered form that bring out the aroma, and taste; readily available for mixing in food preparations. Imagine, for one, the bullseye with pepper & salt to go along with the hot morning coffee/tea from the start of the day and thereafter the many coffee breaks in between!

The serene settings of the thick plantations of thousands of green tea bushes on the rolling ridges and hill ranges that spread like a large green carpet fluttering in the gentle breeze; the beautiful maidens carrying backpack baskets delicately plucking tea leaves are the most beautiful sights in Munnar, Peeri-medu, Waynad, and other hill stations in the Hi-ranges of Kerala. The plant was originally brought here by the British who founded, and owned the estates to start large scale cultivation, and aggrandize international trade, and influence.

All the spices are grown as plants, herbs, bushes, vines, or creepers in the natural ecosystem of the tropical, high altitude environment and climate. India is the second-largest Tea, and the top 10 coffee producing countries in the world. Tea is arguably the most consumed herbal beverage supplying a good portion of water content the body needs. India is the top most producer of Coriander. Coriander powder is the main savory ingredient in the most indispensable 'masala powder' used in all spicy curries. The food connoisseurs and chefs critically smell the aroma; bite, chew, and taste its flavor, pungency, and piquancy before selecting the spices, and adding to the dishes. They must have sensitive taste buds and soft palate which sends the after-effect information of taste to the brains (probably!). Kerala spices often top their lists! The powdered spices show vibrant color; the aroma enticingly stimulating, and salivating.

As per Ayurveda, there is medicine in plants! Therefore, Spices may be considered as plant / herbal based medicine. Remember, the first known cure for HIV was an extract from the bark of a tree in Africa! Like wise, the spices have certain medicinal and nutritional benefits. They become household medicines; accordingly, garlic for cardiac health; cinnamon for lowering BP; ginger mitigating stomach pain; turmeric reducing inflammation, clove for soothing tooth-ache, and so on. Most of them contain antioxidants which naturalize the 'free radicals', thus protecting the body from cell damage. Mindbogglingly, a large portion of the local population is afflicted with acute form of diabetes!

The spices attracted cordial Arab traders as well as the wily colonial powers who pretended trade was their prime objective; civil & cultural interactions as the other stated goals. The monopolistic East India company established by the barons and royalty was a result of a 'Joint venture' which culminated in the brutal colonization of the country for over 4 centuries. Thus, the spices of Kerala, became a mute complicit and factor in several international conflicts and domestic wars, ultimately hosting colonization to India!

The spice trade has been a major foreign exchange earner and economic boon since the olden times!

The crowd, assembled here at the venue, roams around and uses the chance to see as much as possible. It's all open to the public. The local people are traditional. At the same time, they are intellectually advanced and like to find heuristic solutions to problems. The state boasts a perfect 100% literacy rate. Across the board, regardless of gender, they possess excellent academic qualifications; among them, several have achieved 'cum laude, magna cum laude, and summa cum Laude (the Latin honors). In spite of these coveted resources, the state of affairs is lamentably inferior to its potentials due to lack of visionary leadership, the scourge of corruption in politics, inept administration, frequent and self-destructive "Harthals" or politically inspired total shutdowns, interrupting daily lives and activities. The 'intelligentsia' pursue intellectually interesting careers; opts out of politics, yielding way to trouble-makers and revolutionaries to promptly fill the gaping void. People with a mere modicum of common sense, lacking sincerity, and who display deficiency in honesty become leaders of the political parties and heads of government. The public values honesty, but the parties dislike it. So the few good men who get elected do not get any position in the governing cabinet. They are relegated to the 'back benches', (a British parliamentary slang). Immediately, the politicians recognize the fact that 'honesty is **not** the best policy' in politics; the maxims are here inverted. Instantly, they convert! Consequently, the democracy has evolved into a "Kakistocracy", a form of government in which the worst persons come to power. Once they gain power, they automatically attain a status of invincibility and infallibility. It becomes difficult to dislodge them. They cling to power using all the vested privileges, media manipulation, populist reforms, and extreme measures. They don't heed the rules, laws, democratic and constitutional principles. The elected leadership is widely branded as ignorant 'backward-looking, stuck in the mud, heads buried in the sand', totally oblivious of the potential and opportunities the populace and the modern world offer. Ultimately, it deteriorates merely into a 'kleptocratic' body.

Apart from all the negatives, the activities, and events taking place here today are the testimonials of the vibrant spirit, exuberance, energy, and enthusiasm of people. The leaders are invited to preside, see and learn something. They come, see, and wave to the crowd; enjoy the festivities, and the adulation they receive; get an ego boost; go back empty-headed learning nothing from the experience! Smart leaders would have harnessed these positive qualities and abundance of talents in a number of ways to promote growth, prosperity, and advancement to the state. The highly educated chief minister of Andhra Pradesh seized his chance and position when he got one, to make his state a leading global leader in IT. A former Indian Finance minister, who was a graduate from a world-famous university also was seen on media using technology in actual budget discussions and in problem-solving exercises. Their initiatives stimulated interests and policies; provided incentives and opportunities to its citizens. Many availed themselves of these resources and motivating factors to greater advantage. Now, we find a number of people from these states at the highest echelons of corporate leaderships as CEO s , CFO s, chief technologists, and so on of global organizations in administration, finance, management, engineering, technology, and research. People hope and pray Kerala can elect such a capable leader at least once in lifetime and say: "good riddance" to the 'Miserables', the ignorant, undeserved, and unworthy!

The majority of its people in Keralam arc free thinkers and responsibly futuristic. They uphold values of fairness, equality, principles of equal opportunities, and sustainability. They are fiercely averse to exploitation of any kind, whether it is the case of human or natural resources. The corporations and capitalists strategically shun the state. The state is well known for its exotic spices since ancient times, and it is a top foreign exchange earner in the category. But the actual trading business is monopolized by a few industrial / business houses in Mumbai. The hard-working farmers in the state are continuously victimized having no say in setting the prices of the products and commodities they grow. The state leaders find the current situation well suited for their self-interests and enrichment; therefore, make no attempt to change the status quo. The citizens desperately hope that it will be brought back someday to the people, to whom it rightfully belongs. The state often elects a party to power other than the party at the Central Government, naturally inviting disfavor, hostilities, and political neglect. The ambitious, the qualified, well-connected, adventurous, as well as the desperate leave the state in search of job opportunities and careers. Some succeed; many become underemployed; some land in limbo; quite a few in dire predicament unable to find work; heavily in debt; others just stay hopeful and persevere until the tide turns in their favor.

Individuals who have succeeded and venture to return, heeding a tormenting inner nostalgic call from the homeland, quickly realize that things have changed a lot, mostly for the worse. They see strangers took over the land they purchased, by merely remitting a meager portion of the property tax in conspiracy with corrupt village revenue officers. The band of depraved bureaucrats, after receiving a bribe, effect this abrupt change in the ownership by a mere stroke of pen and stamping a Government seal on it. All the toils and struggles of the rightful owner, who worked for many years in distant lands to raise the means to own the property, were rendered futile in an instant. The owners are instantly deprived! This is the local equivalent of the “transfer of wealth” from the public to the top 'one percent super wealthy' through government legislation by the Republican administrations in the US. Some clerks intentionally jumble words in the owner's name in the deed of the property. One of my friends who is an accountant spotted the mistake; wanted to get it corrected then and there. He lamented that it took 10 years, a lot of grief, effort, legal costs, and intervention from the highest office to get it correctly recorded. Some bankers are caught trying to steal and transfer funds from dormant, or 'not regularly accessed' accounts. They are constantly dealing with other people's money and tempted to swipe some from the unsuspecting or deceased customers! An awful erosion of ‘law and order’ situation can be clearly felt. The most responsible group who are entrusted with fiduciary duties and to protect public interests as well as its guardians, aka. the lawyers, have become the most blatant and physically threatening; behave like they don't care about the law. People read about vicious assaults and violent altercations committed by some lawyers on media reporters, policemen, and ordinary citizens, with consummate impunity, within the sacred “Citadels of Law & Justice", the High court of Kerala!

Fabulous public transport network, educated population, but dysfunctional bureaucracy appalling law & order-disorder!

Similar incidents have spread to major judicial courts around the state. These lawyers are the ones who would become judges someday. What a horrible scenario to imagine! Civilians dread and tremble to tread into the compounds of courts, their public judicial institutions, to seek justice and truth. The culture originates in the Law school itself, where its students, since the ages, have been notoriously known for boorish behavior, unruly conduct, unworthy of their chosen profession. The indignant public asks: "Is there any code of conduct?; shouldn't they be debarred; lose license of practicing law?".

Perhaps, none exists, if there is, it is buried in the books; that's an awful shame, indeed!

Generally speaking, the world lacks great leaders, philosophers, and humanitarians who command moral and impeccable qualities such as Mahatma Gandhi, Da lai Lama, Mother Theresa, Rev. Martin Luther King, Nelson

Mandela, and the likes, who commanded decency, respect, and admiration, whom people were willing to heed.

Truthfully, there are many humanitarians, environmentalists, civil right leaders and social activists of many kinds. They are somehow silenced and rendered powerless by powerful lobbies canvassing favorable legislation for driving national wealth towards themselves; give much less to the treasury than they actually owe, though they utilize the lion share of the publicly funded infrastrucure resources. The government projects are funded by debt financing of loans from financial institutions who charge high interest rates, while the treasury offers them money sometimes free or at extremely low rates. A corporation is wealthy and powerful by itself. Still, they form industry-wide alliances on various fronts, to hijack, influence and force rules and regulation to promote their causes, exemption from liabilities, 'burdensome' environmental compliances. They increase their bonus & remunerations while keeping the wages of its workers low, citing a threat of 'rampant inflation' and consequent rise in unemployment, and so on. Also, obviously, they don't like uncertainties in the economy; the public don't either. For planning, forecasting, and profiteering, stability is vitally important. They want it to occur solely at the expense and burden on the poor citizens; totally exempting themselves! The interest groups use the print & broadcast media, as well as social media platforms to disseminate falsehood and misinformation. The data collection companies distort and manipulate algorithms to suite each intended purpose. They do it with blatant impunity!

In the election campaigns around the world, every politician says that "this one is the most 'consequential' of all!" Even in the advanced and developed countries, the voting turn-out is rather poor. But in Kerala, people enthusiastically exercise their franchise in every election scoring in the highest percentile in the world! Yet, they are so far, unsuccessful to elect a decent and capable administration!

Another riddle to crack, and a perplexity to add to the pile of 'unsolved mysteries'!

Chapter 24

The significant role played by the Malayalee Diaspora

The returnees also find out that the expensive items of electronic devices, appliances, etc. they gifted to the immediate family have not been properly taken care of. They find them rusted, and lay in ruins. The tools and accessories they brought from abroad and intended for personal use when they eventually return, are missing, or pawned to support habits of drinking, gambling, betting in card & chess games; or simply thrown away into the rivers in a fit of rage following some bizarre domestic quarrels. “It is obvious that our people don't know how to take care of valuables, other than jewelry!” one of my aggrieved relatives said. When the returnees approach the religious educational institutes for the admission of their children, the authorities mercilessly ignored the fact that some of the institutions were built on the land generously donated by their families. They demand astronomical 'capitation' fees (a fee of fixed amount/; here, it means enrollment fees. But, to be apt and meaningful, it should have been called 'decapitation' fees!). The bureaucrats like to use some newly found bombastic words to make the common man feel it’s a standard fee, something so vital, but that he/she doesn’t know right now what exactly it is for.

Another relative faced delay and dilemma for the funeral of his younger sibling. While the family was grieving, under duress, the parish priest demanded, on the spot, cumbersome documentation and proof that the deceased in fact belonged to his parish. The deceased has been the trustee and custodian of the congregation in the immediate past, moved away only recently! The land where the church and the school are situated was donated by the ancestors of the deceased. This well-known fact made the sorrowful experience much harder to bear. It’s not uncommon that the church waits for the funeral, or wedding time to exact a long brewing revenge for reasons for missing a church payment; being a communist (interpreted as atheist), committing blasphemy, or supporting abortion, totally disregarding the good deeds the deceased had done during the life-time. This is an appalling legacy from certain parishes in Italy. What an unscrupulous and reprehensible example of ‘forgiveness’!

The returnees sense jealousy, indifference, victimization, and alienation from their own people; experience 'emotional black mail'. Finding little in favor, they decide to settle elsewhere, cleverly citing asthmatic and allergic conditions afflicting his wife as a polite and plausible excuse. They are shaken to the core!

The favorite destination of the Kerala diaspora has always been and still is the Gulf countries. The state of Keralam owes so much to the nursing community. It still consists almost entirely of women as nurses, medical technicians, and other healthcare professionals. They single-handedly pioneered this movement and legal work-related migration. Following the legacies and examples of Florence Nightingale and mother Theresa, they went to far away countries and remote corners & islands of the world to serve and care for the patients. This trend owes much to the Christian institutions who built and ran many nursing schools on donated lands by the faithfuls. They provided teaching and training to the graduates. There is incredible generosity, sacrifice, goodwill, dedication, and purpose behind each institution. It is important to point out that the land and the necessary funding for such infrastructure were contributed by each and every individual of the Christian communities, in accordance with his/her financial means.

For more than three generations now, the nursing community has been faithfully pursuing this path of service. During my recent hospitalization, I heard it stated by the director of the health institute, who runs clinical practices in the US and Dubai, that “Keralite nurses are the best in the world; Filipino nurses have the best bedside manners; you could find them in every hospital around the world!”. I became an unintended beneficiary of special care and attention! Originally, the nursing professionals went to Aden when it was a British protectorate; gradually spreading to Kuwait, Dubai, and other rich gulf Kingdoms; recently to the USA, Canada, Germany, Ireland, Australasia- as Australia and New Zealand collectively called. They sponsor their spouses, siblings, and blood relatives to join them to feel at home. Eventually, they apply for migration and settle in foreign countries, expanding the Indian diaspora to the far corners of the globe. They persevere and prosper, earn respect for the native land; willingly and generously help others who are left behind, and always in need.

The NRI contributions: wealth, economy, education, capital, business, technology, know-how, Real estate value

They make vital contributions; remit huge amounts of foreign currencies to the treasury and banks as they do to their own families.

A contribution, sizable, remarkable, and valuable on multiple facets and fronts!

In Keralam without their support, many families cannot survive or maintain a decent status of living; wouldn't be able to afford essential things they need. The botched economy of the state would further suffer; will sink back into “Paathalam”, the abyss, mythologically 'where it surfaced from'. Unrecognizing all the services and contribution the nursing community render, some people, and religions disparage the women folks for the sins of the “the forbidden apple!”. It is high time to expunge such offensive 'misogynistic' language from the holy books; so also the doctrine of “sinner at birth”. Both stand against dignity, civility, and sanctity of humanity. Following the traditional beliefs and conduct of it, the religions only perpetuate these inequalities, offenses, and discriminatory teachings.

These statements shall not be labeled as blasphemy; It's only a thoughtful, fervent, and urgent appeal in good faith to condemn evil acts; mend what is broken; make amends as and when needed!

It seems that only unexpected cataclysmic events are capable of bringing about structural changes in society, as the necessary will and intentions are sorely lacking. It is a human tragedy for our destiny, when everyone thinks that he/she can get ahead on their own without the participation of others!

A silent and diminutive, but significant group of individuals who are recognized only by a very few of the public, are the dedicated people who serve humanity in many ways, and capacities. They are the selfless men and women who own no personal possessions or belongings in the material world, who have joined various religious orders, convents, communities, and charitable organizations. Their contributions, service, and activities are largely visible and felt throughout Keralam. They run schools, colleges & vocational institutes, trade schools, and Nursing schools; provide healthcare services; organize and manage social service operations by building, and setting up hospitals, orphanages, homes for the destitute, nursing homes for the elderly who need long-term care. In other cases, they visit homes of lonely senior citizens and take care of their needs with kindness, compassion, and providing companionship. The institutions administered by them are always considered the best of its classes. Beyond the state of Keralam, they extend their services to all across other states in India, and has spread to far corners of the world including North America, Europe, Africa, Asia, Australasia & Island nations. This is over and beyond their spiritual, counseling, and civic services to the communities they become an integral part of. For several decades, Kerala dioceses rank as the top in such religious "calling & services" in the world. My own aunt, who was the youngest in the family, now 96 years old was a principal of a school. After her retirement, she joined a religious institution who has been running a nursing home in the USA. She has now become one of its residents. Three of my cousins also joined different convents and serve in responsible capacities. A younger cousin of mine has joined the Salesian order dedicated to education and social work. The same is true for my wife's family as well. Similarly, almost every Christian family in Keralam has at least a member in one of these voluntary, religious, or humanitarian congregations. Their charitable services are truly invaluable to the humanity that are downtrodden, abandoned, and marginalized. The only reward they reap is self fulfillment and an occasional public recognition!

The movement of people and worker migration that has been taking place in Keralam gave rise to having 4 international airports in this relatively small state, proportionately far higher in preponderance to the rest of India. In spite of these international interactions, some people fiercely resist changes and want to desperately cling to their unfounded myths, customs, and traditions. The officials who are employed as 'facilitators for the causes of public services' actually become 'obstructionists' instead. It's like the secretary of EPA (Environmental Protection Agency) behaving like the secretary of EDA (Environmental Destruction Agency) during his tenure in the 1981 to 1989 administration. The political leaders literally feud and disrespect each other on personal vendetta shedding all public respectability and sensibility. They discuss no concrete programs addressing the immediate needs and welfare of the citizens; seem determined to perpetuate the cultural, business, and economic isolation, and backwardness of the state. Despite the prevailing radical ideology, it is most impressive that there was never a Cultural Revolution, although there were several major socioeconomic upheavals. It proves that the citizens are capable of seeing the limits of their ideological extremism. They are profoundly aware of the human cost in the Chinese "Cultural Revolution" [where they seem to be obsessed with creating a monolithic society] that was one of the worst human tragedy in the whole history of mankind, estimated to have cost lives of up to 45 million people through rigorous indoctrination, ethnic cleansing, violence, and starvation. The genocides and occupations of foreign countries and even the entire continents such as North America, South America, Australasia by the White colonialists must also stand out and rank as among the worst episodes of World history while they criticize others ignoring their own grave transgressions. All of them owe humanity irreparable debt!

Keralam boasts its dubious distinction as the first to have a 'democratically elected' communist government in the world and may very well remain as its last vestige. The distinction of electing the first national communist government rests with Chile. With that, it suffered deep social divisions and severe episodes of economic collapses until ultimately a democratic system prevailed. The Chinese retain the traditional name and perception as communists, sans (without) the principles of the original doctrine and practice of the ideology!

Chapter 25

The psyche of Keralites

Water, Vallam(boats), paddy fields; coconut groves; rubber, coffee & tea plantations; educational institutions; places of worship of many religions, fiestas & festivals are the constant sights in Kerala. Most people live in villages and work in the fields, or plantations. Modernization & Urbanization are rapidly on the upswing.

Anywhere, people can be fairly accurately identified and distinguished by how they present themselves, as they carry numerous ostensible signs on them; not just a 'mole (A skin lesion that is typically raised and discolored) at the corner of the left eye; dimples in the cheeks & chin' [borrowing a few of the old passport identification phraseologies]. A tropical hot climate allows acceptably decent, minimally required dressing style, making body a convenient 'billboard' for small items of decorations and symbolic announcements. Unlike the tattoo marks, they are easily erasable by simple means and methods! Some are obligatory to wear, such as school uniforms, business, and professional attire at work places. Others are mostly done according to personal choice, or purpose. The markings on the forehead are quite common among both genders, but quite different in their depictions. Women have them for beauty, fashion, and piety depending on its figure, and form; men for status, statement of themes, identifications; also showing piety; children are given them to look cute, lovable; to treat them as adorable. They like to look like the grownups! Usually, women and girls wear flowers for hair decorations for its fragrance, beauty; to look attractive, or cryptic way of saying, they are 'soft as a petal' [treat me nicely and respectfully!] lovable, and natural.

Many men sport mustaches, parted at the 'philtrum'; on upper-lip, lower-lip, under-the-chin; thick, bushy, sparse, overgrown; combed, trimmed, groomed, waxed, shined; of different shapes: Spiral, Horseshoe, Tusker, Pencil, Handlebar, Walrus, Goatee / Sheik / Super / Bulganin / Ho-Chi-Min / Hitler's broken 'Half-toothbrush' mustache, [the latter, used to be a favorite choice of Malayalam Literary

Munshies & poets] and styling; the ends made into various common shapes of curl, stoop, horn, taper, straight line, question, exclamation marks, etc. People want to express their personality, character, cast, creed in several ways, and means. Some want to be funny, others serious, rebellious, daunting, defiant, or simply attractive; want to look like a thinker, philosopher, celibate, sentimental, or deeply spiritual.

The dress, hairstyle, speaking accent; greeting & communication styles, mannerism, as well as the ornaments & flowers persons wear speak about themselves without the need for lengthy introductions. The speaking accent, phrases, choice of words can identify the native region of a person. The bodily markings a person put on oneself may tell a tale of their cultural and religious belongings, or simply as a tradition. The actual substance, shape and/or color of the forehead markings may tell the finest divisions of cast, one's traditional skills & qualities, social ranks & legacies, or certain cultural identities. An elongated 'heart-shaped' pendant say the lady is married, in the same fashion as the people in Western countries wear wedding-rings. Hindu married women put on red 'sindhoor' on their forehead, while the unmarried may use Vermilion (red/orange) colored Kumkumum, or other color (Both are actually the same and identical. The married puts it up on the partition of the hair; while the unmarried in the 'nose-center' of forehead between the eyes!). Both symbolize a particular sentiment, though the average people draw different impressions. The original intent was to provide extra vigor & radiance to a female person energizing the 'Third Eye' which contributes to extra-sensory powers of focus, intuition, and positive vibes. It goes without saying that plastic, metallic, or solid objects cover it and nullify the benefit. People may wear objects such as ornaments, rudraksha, rosary; neck / wristband, or religious icons suggesting their obvious affiliations. During 'Lent', the Christians might wear a suggestive icon for fasting. Muslims may carry prayer Dikr beads. The Hindus who plan a pilgrimage commonly wear rudrakha, and so on. The dressing style/fashion has also a few distinguishing characteristics. Some wear dhoti in a clockwise wrap; certain others in the reverse. Politicians tend to wear certain fabric, dress, and style common to their party affiliates. A cap, a certain type of head-wear, or the stripes on the shoulder scarf, are other distinguishing items. Sometimes, there are indications of the events a person has taken part that day such as a pooja, ceremony, wedding, etc. There are symbols for special days, festival time, dates of important personal or community events. Besides, there are distinct and deliberate markings for group identification. The school uniforms belong this category, as it is very important to easily identify children going to schools, in public transport networks, park, and public places. It is much prominently visible than the name badges common at business enterprises. They are needed for the principles of equality, unity. It's paramount for children's safety!

The practice might have come into vogue by customs, fashion, habit & life-style rituals, teachings, personal discipline, or for particular reasons. All are considered normal and harmless; but, all do have purpose, significance, and utilization, such as remembrance, wish, need, signs of respect, reverence, authority, rank, or a special occasion. In its universal meaning and intent, they are all sets of mute tools intended for personal pronouncements! Among all the prevailing local and national customs & codes, the office culture, uniform & attire requirements are seamlessly accommodated. These manifestations now represent a considerate, multicultural, secular, all-inclusive, and reverberant population. The intention and purpose are clearly rooted in the teachings and writings from the ancient philosophy, and the social order prevailed in the pre-colonial days. In those days, the Maharajah's administration of various dignitaries, knighted citizens, staff divisions of cabinet ministers, bureaucrats, military personnels, physicians, educationalists, entourage, etc. belonged to separate casts based on education, knowledge, and traditional skills they possessed. They were directly related to the class systems prevailed in society. Each class was entitled to certain authority and privileges; ranking, levels of respect, and status. In public gatherings and festivals, the distinct markings and appearances on them gave the subjects their social identifications, so that they will be recognized and treated accordingly. Similar "tell-tale" expressions may be clearly visible in all regions of the Indian Subcontinent.

Generally speaking, people of Keralam hold a stoic, serious, somber, stern as well as stolid disposition. They have a holistic, socialistic, and sharing philosophy, and approach. They tend to be laconic and less expressive, but emphasize actions to mere words and routine pleasantries. Although people engage themselves in new and modern pursuits and occupations, every one is self-conscious of their agrarian roots and the traditional beliefs; show no pretensions of urbane transformation. They are thoughtful, educated, intelligent, and shrewd, yet suffer from lack of a general 'entrepreneurial' spirit. Strangely, the performance in this field is woefully underwhelming, retrogressive, and depressing. The general socio-political environment is not conducive to any ambitious undertakings, or such activities. Furthermore, there is an appalling dearth of infrastructure and lack of capable institutions to support such a push forward. People seem to be reluctant to use proper concepts, solutions, and technologies in daily lives, and situations. English is often termed the 'link-language' within India as a large population can communicate fairly well in the language. Indisputably, English is still the preeminent language of education, knowledge, technology, trade, business, banking among professionals, Internet, communication, and media. The proficiency in English is adversely impacted by the insistence of the local languages as the medium of education. It limits employment opportunities outside the state despite the fact, within the state itself the opportunities are lacking. Furthermore, pride, an egoistic personality, temperament, and lack of application, stymie progress and advancement in business and corporate careers. By any measurement or matrices, huge potential does exist to be one of the most technologically advanced places in the world, like the Baltic country of Estonia or the Asian model of Singapore. Instead, at this rate, it will soon earn the dubious distinction as the 'most backward' state in capital investments in the whole of India. Counter-productive policies, rampant corruption, selfish interests, and frequent harthals dissuade entrepreneurs. The Keralites are known to have an insatiable 'wander lust' which is closely linked to employment prospects. They are staunchly family oriented and are mindful of the well-being of the collective group of family members; make enormous sacrifices to fulfill those burdensome obligations. As a cynic put it: "the giving keeps on giving; the recipients keep on receiving, in a constant stream of one-way traffic!" The charity is not often permeated likewise, replicated, or reciprocated across the society. When people suffer, others sympathize and volunteer to come forward with assistance in so many useful ways. On the other hand, when people prosper, others become jealous and want to bring them down". My cousin who maintains dual residency in two countries said in a thoughtfully eloquent way. Occasionally, the Central government entities introduce enticing programs, incentives and 'SOAPs' to FIIs (Foreign Institutional Investors) & NRI (Non-Resident Indians) to covet investments and know-how. But due to its inefficiency, insincerity, intricate bureaucracy coupled with hostile and atrocious attitude, people lose their faith and their hard-earned money. The programs gradually lose credibility and draw disdain from the participants.

Another seemingly hapless situation that must be rectified immediately!

Chapter 26

A Mecca of Opportunities

Here, at the venue of the regatta, the opportunistic advertisers shall make their heavy presence felt. It provides them a great opportunity for visibility, positive PR announcements, and brand building. They set up stalls and distribute cheap goods of plastic bags, hats, mugs, and other souvenirs similar to the Las Vegas trade exhibitions or the Hanover Messe. These activities are identified by the business entities as "marketing & product-unveiling campaign". The young, vivacious day-hires catch the attention of passersby with attractive smiles. They promptly give out kits of toys, pencils, and coloring books to take home to the children, and a Geisha style thin, foldable 'bamboo-slat' fan for wife, all the items bearing business logo. Thus, they cater to all sections of the valuable consumer groups. At another booth, brochures and free samples are distributed to all and sundry who care to stop by. A 'live' TV feed is leased from the Broadcast van to attract and keep the crowd at the stalls constantly looking and listening. 'For-profit' private colleges promote their institutions; boast their track record in successful job placements; justify the exorbitant fees they charge; give out application forms. Cleverly, they hide the vital facts that the majority of its faculty & staff are fresh graduates from the immediately previous years or are 'adjuncts' lacking experience in the fields they are offering courses in. Their teaching is solely 'text books based' and doing only bare minimum laboratory experiments. Parents often fear the fragility and sustainability of these colleges. 'God forbid', if they would end up with the same fate as the Corinthian, Trump colleges, and ITT Technical institutes in the US, leaving their hard-working students 'Not-graduated', failing to get a recognized degree, and their hard-earned money along with years of studies wasted. Private employers scout for promising talents and take job applications on the spot. Exceptional prospects land job interviews after a brief consultation among HR executives and a few phone calls to their respective main offices.

Right now, a lot of wheeling and dealings are carried out. Where ever there's a crowd, there are people who find opportunities to make some money. Some are here just for that! Others are just vendors hawking their products and trying to make a true and honest living. A few harbor devious intentions because they don't have a 'fair-value' proposition; believe in their persuasive and narrative storytelling talents. They want to be instantly rich and famous. If you lend your ears, they become a total nuisance 'as ear bugs'; will keep munching on your ear-lobes.

A host of other important activities & product exhibitions take place at the Venue

There are various print displays of magazine & book covers with sensational headlines and seductive poses intended to wake up some interest and curiosity. Some people don't care much for them today, as their backpack is heavy and full of free reading materials, job applications, and business cards. They want to be as free as possible without carrying heavy load of possessions. Now, they want to enjoy the races; jostle around freely as they wish. Perhaps, they simply make a note of the title and decided to buy it later at leisure if anything at all piques interest. They are smart and mature; always evaluate multiple options and advantages before committing or investing in one. The ambitious always look for privileged information, tips; notice new fads and trends for assurances and authentication. They are adept at finding them in any situation. They must be seeing plenty of it here today!

There are many vendors and peddlers hawking their own homemade products, widgets, and edibles. The coffee and Lemonade stands sell local fried favorites and savories "Va-dai, Mu-ru-kku (Pretzels), Unni-ya-ppam, banana flakes; sweets coated with 'Sar-kkara'- a sweet 'jaggery' extracted from locally grown sugarcane crops in addition to coffee & tea; often complaining 'Starbucks' stole their business ideas and promptly patented. There is bitterness throughout India that Texas patented its famous 'Basmathi rice' brand in spite of its irrefutable origin in the Punjab regions of India and Pakistan.

The brand of rice is to India as '*Cognac* ' is to the French wine!. But it doesn't work that way!

In this wonderful world of ours, laws, rules, and media can 'manufacture' facts; make fallacies the actual missing truths, as illusions creating 'reality and belief' in magic. Laws do not have to be always complying with truth and facts of nature, or basic social principles; it simply needs a 'legal seal & signature'. In history and also in modern times, autocrats and tyrants made the laws first, before claiming that they had the legal backing as they were committing atrocities against their own people, democracy, and constitution. Like the chess players, they foresee multiple moves with a specific outcome! The vested interest-groups and the lobbyists dominate the process of writing the laws. The legislators simply accede to their outlandish wishes without even reading them; take a huge sum of rewards as "campaign contributions" for their generous considerations and accommodations.

Their votes and consciousness are 'on *sale* ' in the legislative assemblies and government departments!

Well, here also, such opportunities abound, if one is willing to seize it, stake a claim, and want a piece of the action.

This is the new 'world order'!; Play it, and prosper with it; or fall to the bottom, if you don't!

Nevertheless, the new doctrine is certainly much more palatable and tolerable than the notorious "Eine neue Ordnung" of the Nazi doctrine, or other similar decrees and proclamations in the history of the world. The Nazi doctrine explicitly stated the superiority and privileges of Aryan races and that the geographic, economic, and political needs of Germany justified its invasion and seizure of other lands.

What a selfish and radical notion! Eventually, it led to its downfall, submission, and obliteration.

Some politicians try to revive those extreme ideas. The vigilant world must thwart any such ambition, or resurgence!

Today's Germany is remarkably different, democratic, socialistic, stable, altruistic, kind, and compassionate, deserving of a lot of praise and respect in the world. Recently, VW, Daimler Benz & Deutsche Bank had severely dented that glowing reputation. Their greed and fraud got the better of them and resulted in grave consequences. Except on Wall Street where no matter what, the perpetrators thrive and flourish. They win exemption from any wrongdoing or punishment, under the protection & patronage of governments and legislative bodies. They are honored as "rare breed of visionaries & entrepreneurs". All the brunt and burden is quietly transferred to the hapless masses who sit there whaling.

In the same fashion and measure, the colonial countries received enthusiastic patronage from royalty and church; therefore deserve censure and utmost blame. The former believed that the conquered land was theirs to keep; the vanquished deserved nothing, in spite of the fact that the tribes inhabited and held their 'sacred' lands continuously for many generations spanning thousands of years. In the mutually benefiting arrangement, the church bestowed their spiritual blessings on the 'exploration & development' undertaken by the 'faithful braves' for such takeover, simply saying that the indigenous natives were 'non-believers' and had no right to the lands that God created! (The tribes also believe in God, fervently indeed; but a different God, or a 'pantheon' of gods! None matched the one the invaders worshiped. Therefore, they were eternally condemned and damned). To date, it has been a symbiotic relationship that besides the wealth & territory, the colonialists wanted a spiritual ruling, and support from the 'Servants of God' to absolve them from the mortal sins committed against the 'Ten commandments' and a constant feeling of guilt and offense. On their part, the church wanted to spread its faith and spiritual domination. Naturally, the two facts and forces blend harmoniously together without any feelings of hesitation, reservation, regrets, guilt, or shame! Shouldn't they be condemned to 'perdition' (eternal damnation and hell) for the genocides and atrocities committed, based on their own teachings and preachings?

Just wondering!

Colonization & evangelism were the two voluminous episodes of the same World history of conquest and exploitation practiced by the invaders. Both did always interpret the laws as suited to their goals and intentions; preached to the believers and the vanquished, summarily saying: "That's the Will of the Almighty God!"
In the process, several flourishing and ancient civilizations were destroyed and buried in their own ruins and remnants without a trace so that they cannot be unearthed, understood, nor revived! One must wonder what would have been the fate of humanity if they were allowed to continue and flourish further?

History repeats itself! Similar things are repeated as subsequent causes and effects of the previous one. Every time, it leads to an entirely new order. New entities, regions, countries, or religions wrest the power and proceedings in unexpected ways and means, or as usual, by waging bloody upheavals, and armed conflicts. In the scope and timeline of History, they are actually all dynamic, periodic, and continuously occurring. Amongst all these turbulent sagas, struggles, and changes, people still live their lives somehow. Certainly, some perish; others survive and succeed; still, others doomed, or nothing really changes for them. The world turn its attention only to the survivors, successful and prosperous; the perished and the vanquished are quickly forgotten. In life, people under the same circumstances and conditions reap different rewards and face dissimilar outcomes. The spirit of survival is a human trait, a valuable asset, and a strength of character, or a turn of individual fate and destiny.

Even in this festive and high-spirited occasion, my mind wandered waywardly. It occurred while contemplating from this spot on the origin and significance of this regatta which pointed to local conflicts and warfare waged right here for some curious reasons significant in those times. None has been quite evident!

It is hard for me to imagine that the very spot was once a war zone of frequent bitter battles and bloodshed only a couple of centuries ago. Now, on the very scene, its inhabitants assemble here as a civilized community of people enjoying a drastically different experience- one of goodwill, cordiality, and mutual respect. It's a loud testament that humanity can transform in ways useful for its own common good, progress, and peaceful co-existence.
Let's fervently hope and pray that such changes last; always lead towards peaceful solutions, and happiness to all humanity without an exception!

Chapter 27

AROUND THE VENUE

The commercial video production companies are busy at work. They plant their actors and celebrity spokespersons strategically amidst the crowd to do extensive video shoots. They are paid to promote products they know nothing about and seldom use. The actors perform fake 'lip-syncing' reciting from the script they have 'prerecorded' at the Studio. It is similar to the familiar, but hilariously awkward 'playback singing' in the Indian cinemas & films. In the final 'commercial/advertisement' video, they are presented on the TV screens with body curves they really don't have; hair that is not theirs; the faces that are so finely airbrushed into smooth polished, and painted skin, as well as the cute dimple that is entirely missing when you see them face to face. There is reality, virtual reality, and augmented reality but, not 'actuality'! There are plenty of fakes in visual media, thanks to makeup artists, 'digital image manipulation', and studio technology. The producers underestimate the intelligence and perceptions of the adoring public. They know that the working people, come home after the days' hard work, needs just some 'low-exertion' entertainment, and not inclined to do any critical thinking! Oh, look at them, the poor souls!

Post-production requires a great deal of imagination, skills in graphic/image manipulation, as well as editing technology to make everything look apparently realistic. That is done in 'photo-shop & 'visual augmentation'. The directors of these commercials and TV serials check the 'storyboard' (a record and checklists describing scenes, screenplay, sound bites, dialogue, media elements, images, etc.) to ensure that they have shot enough angles, footage, scenes, and sequences for subsequent post-production purposes, and fulfillment. They interpolate these visuals as video backgrounds and scenes by employing sophisticated editing process. Several transition effects & techniques such as cuts, wipes, dissolve, key frames & 'chroma key' (key-off a uniform color background, then filling or substituting with other themes of stills and/or motion video clips) are duly applied. Many visual effects such as 'rotoscoping' (superimposition of artwork or cartoon character in the 'live' video scene), 'compositing' (the process of combining different media elements such as still images, movie clips, etc. shot separately into a final product to give the impression that they were all shot 'live' at the same time. They are actually finished in a 'virtual reality' mode). These effects render the scenes look realistic and authentic. As a pertinent observation, it is noted that "3D Virtual reality" is extensively and seamlessly used in Hollywood films, Video games, TV commercials, Studio sets, game shows, etc.

The elaborate Television set up for the international satellite telecast of the Vallam Kali

The same techniques are also applied for simulations, training videos, and operations training. What was done previously by hand-drawn pictures called "flip book" is now done electronically, by making "clones" and applying 'predictive' movements of motion. The Graphic and Animation studios take this technology to a very high level (to name a few: Disney, Dream Works, Pixar Studios in the USA; Crest, Pentamedia in India). They use specialized, sophisticated, and dedicated 'super-speed-processing' workstations as tools as well as complex and powerful 'rendering' program software for it. The 'raw footage' is imported into the 'editing server' at the studio, where skilled editors collaboratively finalize the work for airing in the same nightly transmission. This is the glamorous-looking virtual world of the Television & Movie industry. The educational and persuasive power as well as the impressionable effects are impressively efficacious. They are meant to create sudden, surprise impacts, and get the biggest '*bang for the advertising bucks*'!

In today's Industrial engineering, as well as design & manufacturing practices, animation, and multimedia are indispensable tools. As a matter of fact, they are the first and foremost steps in conceptual product design & development. Similar multimedia tools are swiftly adapted in education, presentation, research, robotics, medical, and analytical studies to get realistic and visual perspectives, and insights. Specialized industries use several other signal sources, besides electrical; such as, sonar, light /Laser, thermal, radio frequency, echo, radar, and magnetic imaging. Various sensors convert other forms of energy into electrical signals which render themselves suitable for digital processing with the help of computer and software products. They have opened an entirely new and effective paradigm (pattern or model) of doing things in better, faster, efficient, and extremely accurate ways.

Computer workstations fitted with multiple microprocessors capable of processing information at several Giga bits/sec; Hi-speed R/W (Read Write) SSD memory devices, and compatible Graphic cards using specialized software applications, make hi-resolution pictures and motion realistic, and seamless. They render the 'Real reality' & 'Virtual reality' practically indiscernible!

Chapter 28

Topics, conjectures, science & culture

Discussions & interaction between people

The crowd is swelling!

People take up positions wherever one is available. They squat, climb, lean, or stand. Some don't find any; jumps into the water at the shallow end; decide to watch it from there. A few have brought inflated tire tubes to use as floaters or life jackets. The water is pleasantly lukewarm thanks to the tropical sun shining down continuously. Some spectators make frantic use of their binoculars, camcorders, and smartphones. Bunch of people wear headphones; talk too loud when addressed, or asked something- a natural reaction and effect. They text, tweet, post, podcast and live stream directly onto their Social media pages. There are plenty of 'Wi-Fi hot spots' provided by business sponsors. One does not know for sure whether it is a courtesy service, or provided with ulterior motives for 'sniffing, spoofing, snooping and spamming, or planting cookies'. All are notoriously known hackers' tool-set! Reputed corporations like Google, Facebook, Huwei, and data analytic companies harvest peoples data paying nothing for other people's 'identity-property', bundle them into "intelligent data" packages, and sell it to any entity that is interested, and who can pay for it, whether it is Government agencies, industries, clandestine operations, drug & human trafficking, crime operations, or simply anybody who has money and the willingness to pay. Money is the same no matter how you get it! People and corporations are always eager to take advantage of free stuff in addition to government subsidies, 'soaps' and discounts! But, must be aware, vigilant, and discriminating while using free apps, and connected smart devices as they secretly collect personal, private, financial, and even intimate information of the browsing 'visitors'. They tag you, and know where you are at any given time. Sounds horrible, strangulating!

Some spectators are able to catch the full view of the races from their current positions, others only partial. Partisan fans are habitually restive. They, like the birds on the perch, are never satisfied with their current position;hop around a lot. Actually, if one is willing to wait long enough, some new places may open up, like parking spaces in crowded parking lots. In any case, no one can miss the Chundan Vallam because of its high profile of domineering head. They tower over everything, but needs to be in a good spot to see all the other boats. Many don't worry or care. They crowd around numerous Flat TV screens set up around the venue. In fact, they offer the best view from different camera angles, powerful zoom lenses, and all the "virtual" markings, as well as a stream of up-to-date race data. Most of the tourists occupy the highly prized gallery seats. They read-up on information materials, statistics, betting odds; listen to the tour guides. They ask around to identify each boat; find the names too difficult to pronounce and remember; repeat the same few times in vocalization to memorize them all; ultimately decide to go by the identifying numbers. The guests compare each with the photos; annotate on them. The illustrated guide shows the location of its home-village on the map. Many tourists will visit the winning one and watch the victory celebrations in the actual rustic settings and style. Most have done their homework and are well-informed. They have some notions & expectations; mingle with the crowd, strike brief conversation to get the real feel, and mood of the moment. They display some genuine excitement and camaraderie. One American gentleman points to a paddler on the women's race boat, says: " Look, this is Stacy, I know her from college; The college boat club of the University of Pennsylvania. I'll be damned!; We practiced on the Schuylkill river. I remember her well. She was the first in the "Cox less four"; was the captain. I, too, was in the college rowing team-the 'quad scull'". He said rather excitedly. "Oh, I'll be really damned! It's a small world after all... Oh, My...!". He has the biceps, quadriceps, deltoids, and long arms to prove his boast! Nobody can doubt him. "Does any team Chundan takes foreign men as paddlers?". He inquired. One suspicious looking guy comes forward; offers his help; asks for his entire contact details and that of Miss. Stacy; exhibits all the looks and trappings of a hacker! Readily, the unsuspecting visitor obliges. After all, tourists come all the way to enjoy and get the best out of the money they spent!

One among them looked up the guide, turned to his wife and started explaining, then pointing finger "... This one is the beak Chundan! The other...". A Malayalee gentleman in the hind row corrected "*literally, it's Lip Chundan*". The tour guide interjected "*well, it is locally known as Snake boat*". "It is a misnomer; the snake wags its tail in moving through the water, the Chundan does not; it's steady and held high; besides, the stern is fatter! The bow and the stern are held way above the water; we must stop using this silly term; it is totally false, incorrect", another opined in firm voice. The debate takes its course, but fades away quickly as just a trivial matter!

Nearby, another discussion is going on about the peculiar shape of the boat and its propulsion methods. Most regattas and Olympic events are in sailing, surfing, rowing, sculling & canoing. The influential countries use their clout and leverage to introduce events in which they excel, have an edge, and stand a good chance to win medals. Everything matters in the final medal tally giving the countries some bragging rights for a short period while its memory persists. Occasionally, the media replay and re-live its memory to make people feel good. They believe, it strengthens their standing further on the world stage. The commercial world ensures the participants /competitors can make good money out of the glory of winning medals, exploiting their charisma, and star-power as a magnet for attention and the 'eyeballs'! It is an industry with a lot of money, endorsements, promotional events, celebrity-prestige, and power. In the poor countries, the sports people face poverty, destitution, deprivation, and abandonment soon after their brief stint and unsuccessful careers come to an unceremonious end. The public and media rue the fact of winning no medals in international competitions without assessing the reasons. It takes more than able bodies. It needs infrastructure, organizations, funding, science, technology and many more. Some clever doping techniques don't hurt the cause. Recently, cannabis and marijuana have been elevated to the status of 'medicinal' products; they are natural, green and 'legislatively' made a legal product. At the same time, what was grown in Afghanistan and extracted out of it, is absolutely illegal, addictive opium (which was widely used as a cheap painkiller in remote and poor countries), a deleterious substance and harmful 'opioid'! The conquering troops destroyed all the crops and fields; came back and founded profitable business cultivating "legal marijuana".

World tourists are also familiar with kayaking, gondolas and venetian rowing. In rowing, each person uses only one oar- both hands are on that oar; in sculling, each person uses two oars - one oar in each hand. The rowers sit on sliding seats which slide back and forth like a piston in an engine with corresponding motions in the propulsive & release phases of rowing. They use torso and legs in the actions. The seafaring fishermen of the Malabar coast do rowing as well, although sitting on stationary seats. They generate power mainly from the torso and arm action. Paddlers paddle the Kerala race boats, dragon boats, kayaks, canoes, and rafts. Their seats are stationary. Also, paddlers face and propel the boats forward while rowers do the same backwards. Paddling is widely seen in the Amazon River, Lake Tanganyika, Mesopotamian/Iraq marshes and many other waterways. Anything resembling Vallam Kali may be the dragon boat race in Hong Kong harbor or in the Chinese Pearl River Delta. History has it, that the boat races originated in China 2500 years ago. A Math & Science teacher, a pedantic pedagogue, in the adjacent seat, could not resist. He began by saying in his teaching parlance "the boat is without exception constructed out of aged hard wood, devoid of any synthetic material, fiberglass, metal or carbon. They may have a soft keel or none at all. Originally, it was mostly carved from a single piece; the stern and bow as discrete parts joined to the hull. In other cases, it may be construed as made up of several sections, each being an arc of giant circles, the centers of which are at different planes or elevations. It will be an interesting study for mathematicians and computer techies. They could write modeling programs; then apply modifiers of various hull shapes, hydrostatic parameters such as drag, weight, buoyancy, displacement, hydraulic /flow characteristics, etc. to come up with a theoretical optimum design". A lone contrarian retorted: "that's exactly what the Vedic architect Viswakarma invented; he did it without the aid of computers..., thousands of years ago!" The professor didn't garner any further attention as people didn't understand what he was talking about. Therefore, he sat back in his seat; kept mum thereafter!

An elderly gentleman, apparently a philosopher/thinker, senses a favorable opportunity to express his own thoughts. No one at home or in his community would normally listen to him, as his thoughts are considered too idealistic and esoteric. Villagers simply cannot relate to his viewpoints. He says: "The Westerners always want to break down things into bits and pieces, examine and analyze it, research, and postulate. After intense discussions of various scenarios & possibilities, they generate inferences; draw conclusions which are then treated as absolute facts, while it is only a conjecture at the time, or '*just about right or right enough*'; true only within certain constraints, and not proven beyond doubt under all operating conditions. Some examples are: Theory of Relativity for one; another is existence of "*God's particles*" which gives mass to any matter (Recently, the theory was positively proved in tests conducted at the Large Hadron Collider in Switzerland, the particle accelerator & atom smasher). They use the findings, apply it and make timely modifications along the way, as the concept evolves. Citing the business model of Microsoft, it started marketing the flawed OS (Operating System), added versions & revisions along the way, collecting 'repetitive' revenue streams at every release. The consumers will have to pay for the same product several times. They did pay dearly indeed; keep on paying every year, making the founders and the company the richest on earth! On the other hand, we orientals tend to keep everything in a shroud of ambiguity and dismay until an ultimate proof and validation are obtained. The unending wait cost time & resources; seldom find a chance to launch anything. If by perseverance, should we invent anything, they will be casually termed as 'an act of God', attributing all the credit to the Almighty, making the inventor baffled, and dispirited. We're afraid to be proven wrong; feel devastated by criticisms and ridicule; retreat to our protective shells like the turtles! Besides, we like to associate and attach a good deal of divinity, mystery to everything we deal with. We simply say: *"God only knows and leave things hanging,... unresolved!"*

Some listeners are amused; others bemused. Both groups prodded him to continue his monologue. He did continue, gladly, albeit with some repetition: "people used to believe that the wisdom is obtained by practicing "*Tha-pass*" or mysticism, a practice of contemplative prayer, recitals, and spirituality aimed at achieving direct intuitive experience of the divine. We don't particularly like lab works; build something and test it. By nature and tradition, we are theoreticians, not builders. We are hesitant, circumspect, lethargic, shy of action, and want others to prove something and hand it over to us to imitate, or talk about it profusely in Scientific, Mathematical, and philosophical terms!

A Maharshi (sage) attains enlightenment, wisdom, knowledge, mystical powers after performing "ThaPass"

'Thapass' begins with a form of deep meditation, with an intense focus on just one thing, one thing alone, ignoring worldly woes, mundane pleasures, needs, other matters, and even the surroundings. After reaching a state of spiritual transcendence, the '*god of wisdom*' bestow supernatural qualities and enlightenment to transform this person into a "*Maharshi*" (mystic, sage, savant, or yogi). Just think of *Valmiki*! This is much similar to the Christian belief that the Holy Spirit is the source of all blessings. They, and also in Islam, are called the 'blessed' person, or a prophet'. People venerate the Maharshi; flock to him for advice, spiritual guidance, and judgment. On the other hand, in the modern Western world, especially since the beginning of the Industrial Revolution, people like to study, learn, research, tinker, experiment, then deduce, apply, commercialize, monetize, improve, and make products & machines to simplify daily chores. They create demand, sell the products, and make a profit and move on to the next big thing in the upward cycle of invention and prosperity. Both are limitless entities that cannot be quantified! The process makes them rich. The habit becomes a culture that spreads.

In our world, even though we thoroughly weigh out pros and cons, commonly, the deep-rooted primal instincts of traditions, and customs prevail in the end. The cognitive thinking process whittles down somewhere along the wayside. The traditional wisdom which is tried and tested over the long, past years helps common people make decisions without a lot of processing, deliberations, or hesitation. It is a dependable tool for anybody to make decisions before acquiring full knowledge, experience, wisdom, proficiency, or proper understanding. Most of the education is considered to be just for passing exams and getting a good job. It is basically text-based and deemed sufficient for the intended purpose" of getting into a job and progress from there.

........"Ladies and Gentlemen, fellow countrymen, all the race fans, Your kind attention please... !".
It's the PA announcer speaking again, abruptly breaking the lively discussion.

"The boats are assembling at the starting deck at this time to start the races. Favorable settings and conditions now prevail for an exciting afternoon of competitions. There is a gentle, soothing breeze, 2.3 k mph North- Westerly in the direction of the race; the tailwind is less than the allowable limit of 2 m/sec or 3.9 k mph to affect record books! Temperature is comparatively cool at 28^{0} C; humidity 87 %; visibility 16 Km, quite far! The drifting white clouds in the skies occasionally masking the blazing sun, providing some shade; no chance of precipitation! The mild traveling waves sparkle under the bright sun!". As always and everywhere, the weatherman gives his upbeat forecast, assuring fans that the weather will fully cooperate this time on a good day of the races, discounting the seasonal thunderstorms that may show up somewhat early.

People curiously look around, see new things; blinking neon and LED lights; news crawls and advertising signs; technology of many kind & applications; see new styles and fashions; people of different skin colors in attendance, foreign languages are heard-English in different accents, German, Arabic, etc. a festival spirit of fun and passion; an air of curiosity and anticipation; find themselves among uncharted situations and novel experiences. The ambiance is effervescent! Minds become excited and expectant of many things to immediately follow. They witness the constant dynamics of change happening around them and worry if they would be able to cope with them at all. The intellectually curious folks jot down certain questions in their mind to pursue discussions with their own children studying in colleges; snap photos and 'selfies' heralding their own 'esteemed' presence in the arena to others. They will be portrayed in local media and social platforms as a "cool" guy in the thick of things, moving with the times!

Now, the course is fully cleared for the race; no obstacles or intruding boats is seen. People and boats are ordered not to move unnecessarily and make waves or splash water. There is some gentle wave that changes direction constantly; nothing of consequence. A brief spell of silence fall at the site. Suddenly, a somber sense of seriousness reigns around the premises. Spectators waiting at the finishing point look expectantly towards the starting point, about 1.4 Km awy. Some take out their large zoom binoculars; some have motorized variable zoom ratios helping them see things far; the objects appear much closer, clearer, and magnified. Firstly, they spot the rising splashes of water (a phenomenon due to the curvature of earth) from the rudder-oars that help the boat to overcome the force of the stubborn inertia. This fact of Physics was first theorized, and explained by Sir. Issac Newton. Initially, the helmsmen make this rearguard action to shake up the heavy rear-end of the boat that sends out a feeling of motion, and encouraging vibes to the paddlers sitting in front of them. Now, all the paddles are in the water and making a combined effort. There is a uniform motive force from one end to the other. It exceeds the force of inertia; feels the boat is loosening up somewhat; gradually, it attains steady motion, and speed. More and more of the spectators on the banks get to see the boats as they pass by them; feel the ambiance; see the scenes they wish to look at. The Television viewers get a continuous view that is presented on the screen, but unable to see many other things that are happening in connection with the races. The emotions and impressions of the two groups of viewers are different. The spectators get various other stimulations from the reactions and opinions of the people watching nearby. It's a long race; heats follow at periodic intervals; one after the other in succession, offering continuous spectacle for the viewers.

Chapter 29

The Pavilion & Reception Center

A passel of ministers, high-ranking government officials wearing official badges, invited guests, VIP & dignitaries begin to arrive with their entourage at the pavilion built on the quay. The pavilion is decorated with many iconic symbols and items from the daily life and nourishment of the people of Keralam. Each represents a particular significance and assumes a certain place in their life. It consists of several items:

- Twin bunches of "Koomb", the pale ivory-colored apical buds of coconut flowers (palm Cabbage) extracted from the tip of young palm trees (delicately plucked so as not to harm the trees). It signifies growth and vibrancy of the youth,
- Fully grown bunches of smoothly shaped golden coconuts; a symbol of wealth and dependability,
-
- A Peduncle stalk of dense banana blossom; its bud, bearing tiers of thick white female flowers and yellow male flowers full of honeycombs enclosed in their dark brown *bracts at the end of the long Rachis, symbolizing fertility, and the gift of nature,*
- *Another one of a similar kind, but flavescent,* (turning yellow) *indicating maturity, and full growth;*

The banana plant belonging to the 'genus' Musa had its origin in the tropical West African countries, but now grown in many other tropical, and subtropical countries. The easily and ubiquitously growing plants, are also extremely useful in many ways, second only to the 'perfectly' resourceful coconut trees. Besides giving soft, delicious, potassium & Magnesium rich fruits, each of its parts is biodegradable and useful. All of its body, the stem, leaves, stalk, and 'corm' (the bulb-like underground growth storing its food) *are used as organic manure/ compost for coconut trees. Potassium is a mineral electrolyte that aids the minute electrical signals to flow in our body systems that are essential for the heartbeats and the neuron flow in the brains. The flower bud and the stem are used for cooking. They are said to be good for cleansing the intestines. Most other parts are also edible after proper cooking. The plant carries both male & female flowers. As children, we used to pluck off the flowers and suck out the drops of nectar from them, beating the hungry beetles and bees to it, not knowing it affected pollination adversely; also that the female flowers become young bananas. Parents, often forbid such misconducts without giving any reasons. Also, its petals/sepals/calyx are the largest in any plant. Each plant produces several tiny 'sucker' or a secondary 'shoot' of its own, which can be transplanted. The leaves of the plant are larger than those of any other plant. They are chloroform-rich and used as disposable food packing for short duration like 'lunch packing' for farmers working in the remote fields, as well as eating directly off them as in 'Sadhyas', the local form of 'squat down banquets.* [In this society, the BPA containing items of plastic are avoided as much as possible]. *Parts of the stem and branches are used to make synthetic yarn for textiles. Its stem/trunk is supple; holds a lot of water content. It's buoyant. Children use it as a float while learning to swim without any supervision, with the disadvantage of not learning the proper and effective way of using the hands in swimming! Consequently, the local swimmers lack speed and efficiency.*

The traditional greeting rituals of hospitality & reverence at the magnificent reception pavilion

- A pair of "NilaVilakku", oil lamps on ceremoniously sculptured, shining, multi-level brass pedestals; the wicks of spun cambric fabric immersed in light coconut oil glimmer faintly in the eight directions around, indicating the 'radiance of life', also a symbol of worship; remembrance of the presence of God
- "NiraPara" a bushel with a full heap of brown paddy grains, some spilling over to mean 'abundance and prosperity' to humanity
- Sandalwood carvings of mythical figures, for godly remembrance, cleansing the air in the assembly of so many people, and spreading the pleasing aroma
- The burning incense wicks oozing mild & mellow fragrance to signify a ceremonious occasion taking place at the premises

The guests are welcomed with “NaMaSthe,” a folded, courteous greeting gesture. Unlike handshake, no transmission of bacteria or virus takes place during the traditional greeting! Beautifully groomed ladies dressed in 'Kasavu Saris” (bright white woven cotton cloth with golden borders) wearing glittering gold chains & stylish earrings, assorted jingling glass bangles, and adorned with fragrant jasmine flower-tops (cautiously avoiding Talcum powder containing asbestos particles which are found to be carcinogenic--ask J&J!), offer garlands or flower bouquets to the celebrities following the age-old customs. The recipients feel elated; their hubris swell. Their social standing and importance are once again reaffirmed. The Malayalee beauties feel honored for being selected for this honorable duty. They come to believe that they possess exquisite charm, congeniality, poise, and personality; dream of good prospects in the upcoming beauty pageants or the audition in film roles. Truly, they dream that ultimately this will help them in their relentless pursuits of the irrepressible desire for stardom, wealth, and fame.

The members of the “reception committee” are very astute. They know all the age-old traditions and sensitivities. Though ridiculed sometimes and derided other times, the traditions cannot go wrong. They are vetted, tried, and tested. In fact, they have their own written ‘tenets’ to follow and practice! People don’t know that for a fact. It’s not born out of anybody's thought or deed on the ‘spur of the moment’. They follow and practice every step with meticulous compliance and utmost respect.

There are no institutes or academies to teach and groom people in this art of hospitality. It seems like certain individuals become one, by doing and getting noticed. In many Western countries, each skilled occupation requires a certification. For example, in Germany, one has to be qualified and licensed from a culinary school to be accepted in the occupation as a ‘baker’, or restaurateur; certification from a sports academy to be called ‘Football coach’, and so on. In this society, however, such documentation is not necessary. Whoever volunteers, or boldly comes forward and starts doing something is generally accepted as an expert. That’s not the case in education and other well-known professional categories of work. Now, as new specialized fields emerge constantly, people with somewhat similar qualifications are retrained in that particular discipline. There is interest and eagerness to adopt Western and modern business, commercial, and corporate practices.

Keralam is acclaimed for its educational institutions, professional colleges, trade schools. It has the largest number of per capita educational institutions than anywhere else in the world. There are many trade schools that teach and train in basic types of common skills, such as carpentry, electric wiring, painting, metal fabrication, etc.; specialized schools for Lab technology, cosmetology, Aviation, Culinary Arts, as well as colleges offer courses, degrees; postgraduate degrees in Nursing, Hotel Management, Travel & Tourism, Business Management, etc. There are several professional colleges in Medical, Engineering, Science & Technology, Business & Finance, Agricultural, Ayurvedic Sciences; Fine & Performing Arts, Liberal Arts situated in almost every city and town. In the state of Keralam, an educational or training institute can be found in the surroundings of any 10 km dia. area. They are mostly Private or community enterprises. The Christian, NSS, and other communities, as well as private organizations are the pioneers and still at the forefront in this endeavor. Lately, there are many others who are coming forward to serve the people of all denominations and classes. It's heartening to know that the generation who went ahead still cares for the next one! The progressive community organizations never wait for the Government to act. It's just an unavoidable dead weight and a huge burden for the people to carry and endure their misdeeds. This is in stark contrast to the well-functioning and farsighted vision applied by the Western Governments and administrations.

Here, if you need to build and operate something, one has to just go ahead, organize one's own community; raise funds by asking donations going door to door in the neighborhood; find sincere volunteers, and do necessary things on one's own accord. Keralam had been particularly fortunate to have had a continuous reign of good kings and benevolent royal dynasties in the long past who cared for the well-being of their subjects. Since the advent of the white colonist's occupation and after that, with the rule of the democratically elected governments, people became accustomed to the new reality of 'community reliance' instead of 'dependence' on the elected government. As to be expected, these institutions are severely underfunded to such an extent that they could not afford 'state of the art' technology, equipment, measuring, diagnostic instruments, telescopes, electron microscopes, fabricating machinery, expensive biomedical equipment, tools, devices, etc. This limits abilities to conduct extensive lab, research activities, and experiments. This deficiency can be clearly seen and felt! Soliciting assistance or joint ventures with corporations, International organizations such as UNESCO, World Bank, etc. must be urgently considered!

For at least three or four generations now, a very high percentage of the state's population has been living, working, in developed countries; engaging in and contributing to multinational corporations, and international organizations in many fields, and top ranks. Yet, it's mind-boggling and disappointing to see that people are still unable to elect a good government of educated, and talented people from a pool of exceptional individuals who can 'make positive difference'. Occasionally, an individual with great ideas dare to attempt something extraordinary. Quickly, that person is pulled down into the quagmire of bureaucracy, inefficiency, and corruption. This serves as a true lesson; dissuade potential entrepreneurs. The prevailing norm and perceptions are: "the practice of law, and running a government is not for the meek and honest, nor the intellectuals! People can change individually; collectively, they can't!"
That' a contradiction and a conundrum! any solution,... anyone?...please!

Chapter 30

THE TELEVISION BROADCAST OPERATIONS

An informative & in-depth technical Analysis

A 'Remote Television Broadcast Van' referred to here only as OB Van (Outside Broadcast) belonging to "Door Darshan" (All India Television Networks) is parked adjacent to the Pavilion at the finish line. The van is a '16 wheeler' , a 'Straßenkreuzer' (Ger: a road blocker/ jammer) carrying sophisticated 'HDTV' High Definition video production equipment. Yet another Trailer van, as big, carries the Uplink, an over-sized parabolic dish & transmitter which is needed to upload signals directly to one of the InSat (a consortium of Indian-owned communication satellite systems). The area is cordoned off, security guards stand by on duty. The camera crew had previously scouted the venue and identified positions and shooting angles. A few drones are hovering over the racecourse with remote-controlled Pan/ Tilt / Zoom/Focus (PTZ) video cameras. They are more resourceful than the sky cam familiar to cricket fans. One camera is permanently set at wide-angle, but can shoot closeup shots by remote control; yet another, slowly rotating at 30 scanning segments in its full 360° circles, covers the all-around area; serving the dual purposes of event coverage and security surveillance. There are also leading and trailing boats with roof-mounted cameras. At various predetermined locations in the water, 'submarine cameras' are installed to show the underwater paddling actions. They are triggered when boats pass by certain predetermined locations.

"*All angles shall be covered!". The Television producer exclaimed!*

All the remote cameras are now sending signals to the OB Van via wireless or microwave links. Formerly, thick multi-pin cables had to be laid out between each of the cameras and the OB Van. The crowd used to 'choke' the signal by inadvertently stepping on it or by bending at turns. The 'kinks/sharp bends' in the cable were yet another problem. Additionally, there are cameras mounted on articulating cranes & jibs; on tripods with fluid heads; on motion-controlled rails; suspended from, and mounted on to erccted poles for the live telecast. Some 'Go-pro' dash cam cameras can also be sighted on the tips of the ChundanVallams looking down on the entire length of the boat. The audio is cautiously eliminated to prevent any 'undesirable phraseology' (words that require bleeps) from being aired! Most importantly, there are cameras positioned at the start and the finish lines. A 'virtual' line, similar to the one seen in Olympic swim events, is precisely aligned by Laser beams that quickly identify any infringing boat, and that are promptly repositioned. Similarly, there is a laser and hi-resolution camera at the finish line. The camera captures the crucial final moments of action at least 40 seconds before the finish and 30 secs after crossing the line, allowing the entire part of the ChundanVallam seen passing the finish line. This is an “ISO” camera- a broadcast jargon, whose signal is continuously recorded on one of the digital recorders with slow-motion playback capabilities. It’s capable to show the slow-motion in forward and reverse, at variable speed, frame by frame, as well as 'freeze' frame or a single frame of the picture [as in ‘lbw’ review in cricket; offside/penalty check in Football]. This helps to accurately determine the final result, actions, emotions, and progressions leading up to the finish and beyond a reasonable period of time. Above all, it helps the viewers to recapture and re-live the exact ‘fleeting moment’ of the ‘finish,’ and the excitement multiple times in 'Slo- motion' allowing their own critical analysis. This technique defies the saying: *'Time and tide wait for no man!'*

Boom-Mics are set up along the racecourse to capture the ambiance, crowd noise, and the Doppler effect (a sound sensation as experienced in the classical example of a passing train) while the boats speed past, to capture that enigmatic ‘audio sense’ of motion and speed. In short, it’s the effect of the ‘sound following the motion’.

The **DoorDarsan** Broadcast Van

The commentators, statisticians, and reporters pore over their respective scripts and data for one final time and memorize the keywords. Actually, they read off a teleprompter in the form of commentary. A large display monitor in front of them continuously updates all the dynamic data & information on each topic for accurate and timely reporting. They adjust their head-worn IFB sets (an all-in-one, integrated announcer microphone cum communication set; the mic to speak into & 'listen only' headset to receive instructions & cues from floor managers/producers/directors). Each functional category has its own assigned communication channel for targeted, point-to-point, and efficient communication.

The announcers work in tandem partnership; one of them is designated as 'play by play' commentator, telling what he sees out there at the venue of action; his colleague usually an analyst, spewing out statistics, data, critical action analysis, explanations, etc., giving "color" to the listeners. There is also a 'narrator' to describe any incidental matters. They hawk their throats and clear vocal cavities in typical Keralite style; practice breathing exercises like runners do before the race. They want to bring out their best voices and make this moment as exciting as the scene the fans are witnessing. He/she transforms oneself into a hyper-excited spectator, yet taking no side. They have done their homework and rehearsed in rigorous preparation routines. *It's show-time now!*

Every one of the Television broadcast crew is in their respective assigned positions, well in advance. The production coordinator is tense; yells out final instructions to camera operators; the specific shots he wants- its composition, leveling, framing, zoom type & speed, point-of-view, etc. over the head-sets. "Imagine that the cameras are actually the eyes; the sound systems the ears of the viewers!" he reminds repeatedly in 'refrain'. The video engineer uses the

'Signal Analyzers' to check numerous video signal parameters-brightness, color, hue, Waveform, Vector, Gamut, Audio, Picture, Timing synchronization, etc. The 4x4 grid matrix (row & column) 'multi-view' monitor in front of him helps to see all the camera previews and other signal sources on a single, very large, 180 cm Ultra High Resolution 4K QLED monitor. All pictures are scaled or squeezed as in smartphone, 'pinch & squeeze' onto the big monitor. His task is to ensure that all camera signals can be 'switched', 'dissolved', 'wiped' seamlessly without picture break-ups, color 'gamut' difference, etc. He enters each relevant parameter, activates the trigger relays and switching protocols for the 'auto-magic' insertion of the "streaming race progress data" to occur at certain predefined distance intervals.

The audio engineer conducts final Microphone tests. He carefully labels and identifies sources or Mics and adjusts the fader levels on the prolific audio mixing console consisting of knobs, switches, levers, faders, and LED display readouts. He compensates for the ambient noise and performs audio equalization of incoming signals. The PA (Public Address) system engineer adjusts his amplifier volume so that the commentary can be clearly heard over the noise, without delay, echo, feedback, or excessive distortion affecting diction or clarity of speech. He has the facility to send sound to a selected zone/location or certain targeted audiences, such as starting platform, all, or any sides of the racecourse. Besides, the commentary is broadcast on a HD (hi-defenition) FM radio channel to listen from anywhere, while driving, etc. Remember, no fan can actually see each moment of the race from one place. The sound helps them to imagine the progress and status of the race.

The video production engineer sets up his complex Vision Mixer and rehearses video switching sequences, transition effects, such as cut, dissolve, wipe, 3-D warp effects, PIP (Picture In Picture), 3 Way & 4-way split-screen to show different views of the same action in the perspective of a viewer, giving as much visual information as possible. He has formed a crew of qualified assistants or technicians to share this enormous 'real time' responsibility. The commentators will be alerted to watch the monitor and adapt their commentaries to the images the viewers are currently seeing. The broadcast usually opens with 'intro' and ends with an 'outro' of media clips/graphics, highlights, all pre-programmed beforehand. Each monitor identifies by superimposed 'labels' text; cameras by position; images from Digital Video recorders; Playback sources; Titler Graphics; Slo-mo, Key Effects, Preview, Program buses (rows of many buttons for selection & switching). This job requires special skills; multitasking, coordination of seeing, selecting, switching, compositing, and presenting scenes in the 'point of view' of TV viewers [like a keyboard musician find the right keys on the instrument]. In practice, the director calls out "the shots' ; the switching actions are performed by his/her assistant. These actions take place in real-time in a high-paced live telecast environment. 'Live broadcast leaves no room for error'! Many of the effects can be called back from nonvolatile Micron Technology's new breed of '3D Memory' module locations by names, '*mnemonics*' (A cryptic code consisting of abbreviated instructions), or simply by number IDs given to each effect, simplifying the operations. The video switcher console has evolved into an intuitive workstation from a multitude of push switches, knobs, faders, levers, and other mechanical components of the earlier era. Now, they use 'soft keys' and touch screen commands instead. It is all net-worked, allowing access control across devices and operators.

The technical director asks Titling/Graphics/Animation group to cue up their Lower Thirds (still-title or name inserts). They are shown centered at the bottom of the screen on a specific location known in broadcast parlance as the '*Lower Third*'. The *crawls* for *PSA* (**P**ublic **S**ervice **A**nnouncements, public notifications, and other secondary text-based information) appear at the bottom, moving across from Right to the left of the screen, contrary to the natural reading order from left to right (just the opposite for Urdu, Persian/ Farsi, Arabic, Aramaic, Azeri, Hebrew...). *Rolls* for the 'end of show' credits, which will be seen moving from bottom to top of the screen at the closing of the show, again in the opposite direction of normal reading. Besides, there are graphic images of boats, signs & logos, drawings, and other visuals stored in *SSD*, (solid-state memory devices). They can be instantly accessed and recalled as image files in any format (.jpg, .bmp, .tiff, .png, .gif,.flv, .mpeg .pdf and so on). The Slow Motion and Instant Replay crew are confident that they can play back in slow motion any part of the race, event, incidence, or highlights when prompted. They are capable of accessing video segments instantly, randomly, or in any sequence, thanks to the advanced digital recording technology. It allows them to set up '*loop play*' to continuously repeat a chosen segment without any perceivable break when rewinding back to the beginning of the segment, unlike the old tape machines where one has to stop, rewind, search for the exact spot to playback the segment. Technology has advanced so much in a short period of time in all the associated fields. During my career, it started from CRT, to LED; Printed circuit boards, to transistors, Integrated chips, System on a chip, memory modules, mass miniaturization; analog, terrestrial TV to digital HDTV; cable TV to Satellite TV; Internet streaming, Tel-conferencing; Interactive & multimedia, streaming TV; analog to digital post-production editing; cloud technology and many others I haven't learned about, or used. Simply amazing and stupendous! I'm gratified that I was fortunate to have participated in the mentioned technologies from the advent of each of them with hands-on experience. It's equally fulfilling to have trained many talented students into and imaginative professionals in various disciplines.

The satellite engineer has set up the 'up-link' on the assigned transponder 9, on InSat 4. The broadcast engineer alerts the (MCR) Master control Room at the Broadcast Center at Thiru-vanantha-Puram, the state's capital. Methodically, they check each of the 'composite' signals, video, audio, data, and communication to assure broadcast compliance; confirm that the signal feed from the Remote Van is in true *HD* (High Definition) and is robust. The satellite uplink is verified to be now fully functional. The Broadcast engineer schedules the transmission to automatically start precisely at the stroke of 13:00 hrs. The broadcast signal is fully digital 'end to end'. It samples and digitizes signals, filters out redundant or repetitively occurring picture data (by a mathematical function called DCT, for Discrete Cosine Transformation, mentioned for the benefit of Mathematicians & broadcasters!) into a compressed 'packet', *(Mpeg2)* HDTV transmission format.

Aerial view and general scenes; the warm-up preparations before the serious races

The processed signal is then wrapped and packaged with 'metadata'-source and destination addresses, as in Internet or IP protocol. This technique reduces the Bandwidth requirement by a mere fraction. Now a total of 6 TV channels can be transmitted in HD (High Definition) in the space of 1 formerly used analog TV channel. This frees up a large amount of 'frequency spectrum'. These newly available additional channels are the ones which are auctioned off to Mobile Telecommunication entities which have become a subject of bitter bidding wars, ugly corruption & political scandals; also necessitating a rate increase that is passed on to the powerless consumers!

The public have to be constantly reminded that each stage of technical progress is actually contributed by many entities in various fields. They may start from mining, material sciences, conceptual & prototype building, electronics, chip/memory design, manufacturing, testing, software developments, marketing, application, incorporation, so on, and so forth. In practice, they occur in many phases and different time periods. All are intellectual, capital, manpower, technology, and supply-chain intensive. The world is on a trajectory of exponential growth and advancement helped by the sum of all previous knowledge, discoveries, and inventions working together with accelerating speed, and synergistic forces. Some activities may be hazardous and unfriendly to environment, health, and human existence. Waste, obsolescence, and desertification are some other major issues. These are only a few of the costs of progress, convenience, comfort, and '*instant gratification*' demanded by human population. Many of the ramifications will be felt only gradually, without warning, or at the very 'end of time' leaving no opportunity to implement remedial measures. At this hectic pace, and point in time, there seems to be no room for retreat, or even moderation.

Chapter 31

THE PREVIEW & THE HYPE

Prior to the 'live' transmission of the regatta, is the pre-game show. It starts with showing images from cameras, and the race schedule from Video content/data servers on the numerous HDTV flat screens set up at the pavilion and other crowd centers. There are several hi-speed Wi-Fi hotspots to enable spectators along the racecourse to see the transmission on their own tablets or smartphones. In any crowd situation, a person's view is limited. The crowd feels that they are missing certain things happening alongside; becomes restive. They are perhaps oblivious of the fact that the race can be viewed significantly better on these screens than seen by the naked eye, thanks to the power of camera lenses with huge zoom range as well as a rich variety of camera angles & perspectives obtained. Besides, the images will be up to the very instant in actual real time. People who are watching at home on the broadcast TV screens receive the signal with a couple of seconds delayed. This phenomenon called '*latency*' is caused by the additional signal path it takes to reach the home TV. The broadcast satellites are in the geosynchronous orbit at 36,000 Km high over the equator. The back and forth distance itself is (2 X 36,000)= 72,000 Km and takes a finite time though short, in spite of the fact that the microwaves at GHz frequencies travel almost equal to the speed of light (approximately 3.00×10^5 km/s). Digital Signal Processing equipment and signal amplifiers in the transmission paths exacerbate the condition a bit further. In addition, there are some noticeable and annoying 'lip-sync' artifacts due to the unequal lag between the audio and video channels in the composite broadcast signal. The annoyance of '*lip-syncing*' becomes perceptible when the lip movements fail to synchronize or coincide with the video when words are spoken, and when 'talking heads' are shown in tight close-up.

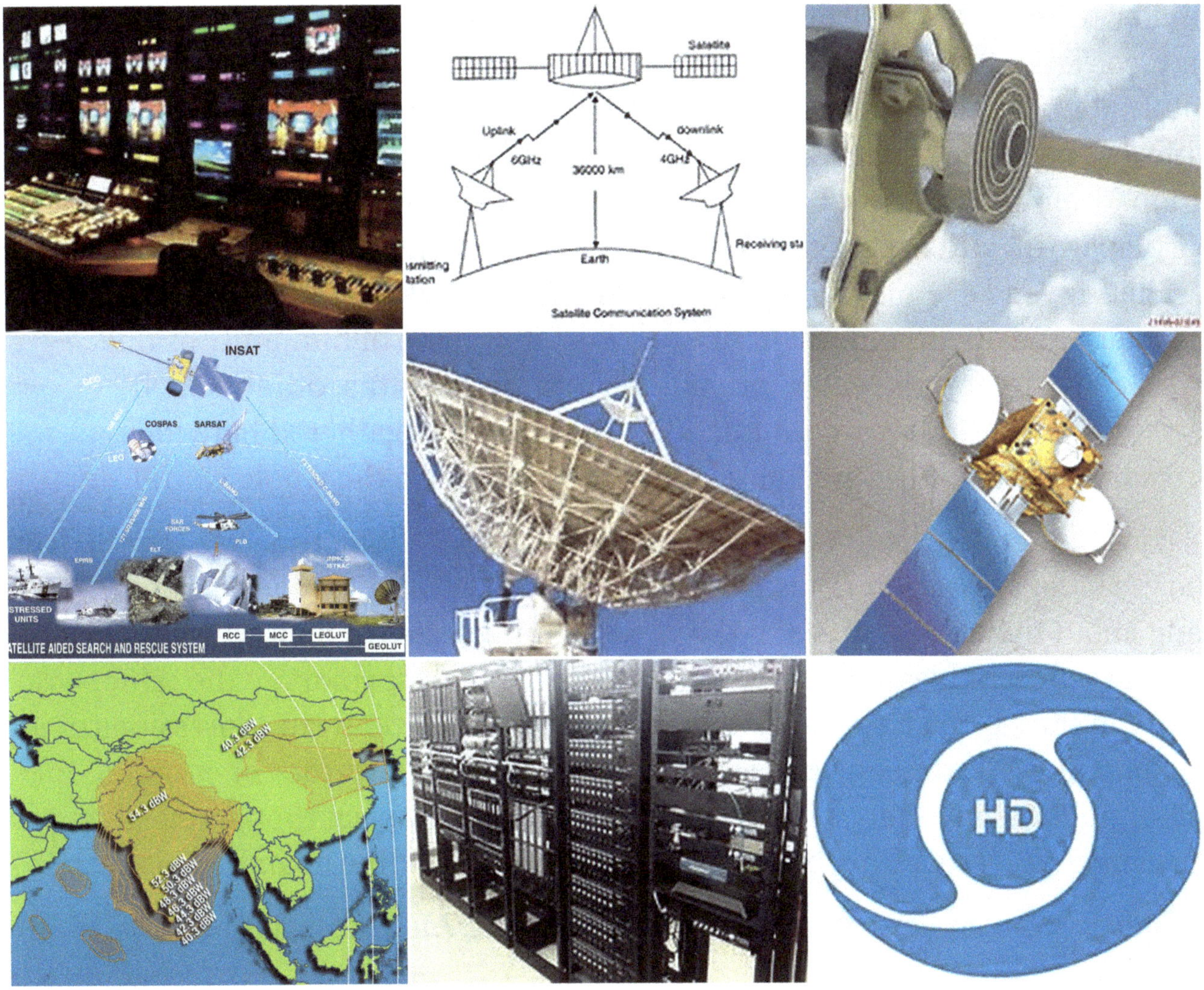

Satellite transmission facilities of Television signal transmission systems

The commentator comes on screen, greets everyone and everybody; describes the weather once again, the wonderful view, and the ambiance in flowery phrases; recite selected poems from popular literature on the beauty of the land, how wonderful its inhabitants are, and so on. He muses, in a melodramatic tone, "I am born and brought up here. The sound I have been constantly hearing was the roar of the Arabian Sea, which is less than 5 Km from this spot. But today, I hear a different sound, sound of people, their excitement, exuberance, and exclamations. Certainly, as you are, I witness an unusual atmosphere, see a different world altogether. I see wonder, colors, excitement, on the faces, the sceneries, the decorations, the different uniforms the rowers of each of the boats are wearing, the gathering crowd of the excited fans, and the spectators assembled all around! It's quite overwhelming to see the enjoyment and expectations on those countenances. I've never been in such an electrifying atmosphere, I dare say perhaps none else exists in the world. This is not just a small world; it's a wide world- a joyous world indeed! I see people of different ages, various status, civilians, military personnel, Navy boats, tourists, International Television crew, academicians, Air force planes, people wearing different national dresses; tourists not only up in the spectator stands but also in competition in visible positions performing and enjoying the things they came to participate in, and experience. What a wonderful sight to behold! There are new, fascinating and exciting things in front of you, behind you, on the waters, up in the skies, and all around. Right now, we feel the whole of humanity as one, irrespective of any perceived, prejudices; privileged status, or fabricated divisions prevailing. We are humanity! The humanity gathered here today proclaims the same sentiments so loudly and visibly to the whole world!

With this backdrop, I'm reminded of a song heard a long time ago, a song from the great Jazz musician Louis Armstrong, titled "*What a wonderful world!*". He sang about the trees of green; red roses blooming; skies of blue and clouds of white; the bright blessed days & the dark sacred nights; colors of the rainbow so pretty in the sky, also on the faces of people going by; showing expressions of joy and happiness.... the song ends with the deeply moving words "*then I think to myself what a won-der-ful wo...rld!*" I have the same exact thought, feeling, and amazement reverberating between the walls of mind & the chambers of heart!"

"What do you say, my honorable fellow-citizens..?.Is that what you are feeling too?" He asks.

"*Yes indeed*!" "*Shari- Thanne*" (Amen!, as we hear in churches), a resounding reply echoes in affirmative, followed by cheers and rhythmic applause.

"Well, now! May I introduce to you, the chairman of the Boat racing organization, RACE?. He is also the District Collector, Hon. Ananthan Abhi-Kamyan IAS, MBA. (Indian Institute of Management) Allahabad".

"Hello, All!. My fellow citizens, officials, judges, dignitaries, celebrities, respected tourists, competitors, and the most important of all,... you! The fans & spectators who are assembled here, also the esteemed viewers who are watching this event 'Live' on various Television & satellite channels at home and abroad! What a magnificent atmosphere you have presented here today!

Welcome to this year's Nehru Trophy races! All the signs and indications suggest that we are going to have a wonderful time, exciting races, and memorable experiences, as I see the boats line up at the starting docks, rearing to go. These race boats own a distinctly honorable place unrivaled in its length, size, weight, and other important measurements of any other paddling boats in the whole wide world! They carry a rich history dating back to the early 17th century, as they replicate the same designs of the original ones. These 'navy boats' propelled by the subjects of the Maharajah have helped him to win many battles, so that we are proudly united today as one community, one nation. The rowers remind us what our forefathers did generations ago so that we, the descendants are living today in peace, harmony, and contentment. I pray these traditions and festivities of Vallam Kali will live through eternity in the same sentiments and traditions I have elucidated a moment ago.

Now, without any further ado, may I announce the prize-money pledged by our successful, distinguished, and very generous sponsors and benefactors?. It is as follows:

The **1st** prize for the winner:

Rs. Ten Lakhs (Rs. One Million) by AmBadi-Kemi! (Hello, Dolly!) Jewelries, Changanacherry,
The **2nd** prize Winner:

Rs. Five Lakhs (Rs. 500,000) by Vaayoo-Sanchar (Air travel) Airlines, Kochi,

The **3rd** prize Winner:

Rs. 2.0 Lakhs (Rs. 200,000) by Moolyam! (Value!) Electric Scooters Ltd, Kottayam,

The **4th** prize Winner:

Rs. One Lakh (Rs. 100,000) by Krishi-Karunya (Liberal Lending Farmers' bank) The Farmers Co-operative Bank, Alappuzha,

For the **5th** prize:

Rs. Rs. 50,000 by Ko-njhu (Prawns & seafoods) fisheries (Pvt) Ltd. Kollam,

For the **6th** prize:

Rs. Rs. 40,000 jointly by Kuttanad Karshaka Sangham (Farmers Union); Nalikera Vikasana Kendram (The Coconut Research Center); Thotta-palli & Nellu Krishi Vikasana Kendram, (Paddy cultivation Research Board) Mon-Kombu,

All the participating Vallams will receive equal monetary incentives from the funds created by the Kerala Tourism board & Chamber of commerce!

And now...without any further fuss...

I,.. Declare....the races O-oo-pen!"

Cheers arise from the crowd gathered along both sides of the course, creating a periodic and enigmatic crescendo and decrescendo. They chant the race song or the anthem "*Oh!..alll the boats that race over the shimmering waves of the majestic Vembanad lake*" ("Eee Vembanattil ..Olathinu-melel- Kuthikkum Vallangal…!"). It is done in the famous style of the 'crowd *waves*' as seen in the Football World cup matches in which spectators rise and sit section by section and row by row. Here, the school children from the many schools in the greater region who are stationed at specific distance intervals lead the anthem joined by the fans. The sound travels like a slow-moving locomotive departing from the Pavilion end to the starting point at 1400 m away, gradually decaying into stillness.

The scene is now set in a magnificent way. It brings the viewers into an '*immersive*' presence, to make everyone feel that they are there at the venue among the thick of things. The Announcer uses a computer tablet as his '*teleprompter,* and talks from it in a skillful manner so that viewers are led to believe that he had memorized all of it, and is actually speaking off his mind. He has an 'on-air preview monitor' to see what the viewers are actually seeing. The commentator outlines the scenario and the prelude to the races, the video often cutting away from him, showing live video of the venue; the countenances of the VIPs, dignitaries, celebrities, sports & film stars, pretty faces, and children in wonderment among the fans. The camera pans to show the profile of the crowd- the odd, the witty, the adventurous, the passionate, the flirtatious & the coquettish; the indulgent, infatuated, inebriated, jaunty, and the tipsy. The camera cuts to a close-up view of faces expressing a gamut of emotions of excitement, expectation, nervousness, anxiety, tension, along with joy.

Now, it's the turn of the color commentator, or the analyst. He begins eloquently in his baritone voice, while the video shows stock-footages, the description of the boats, its physical, and speed characteristics, race statistics, track records, the village, and the community each of them represents, highlighting their locations on the map on the touch-sensitive interactive TV screen. It starts in the typical style of the 'interactive' election reporting display screen used by US Television networks, accessing the preprogrammed demographic data, seemingly hidden within it, of respective states and counties.

There, the 'projected' election results are announced even before the polling stations in the West coast states (who are some 3 hours behind) close! They use a combination of 'hard to predict' technology and a risky 'speculative guess' (in place of 'educated guess') in the mad race to be the first media network to announce the results, which matters much in the TV ratings and the advertising world. One might argue it undermines the basic principles of democracy and the franchise rights of millions of voters! This is the only 'so-called-democracy' on earth where the candidate who gets less popular votes can still win the election and become President. The much boasted "greatest democracy" is, in fact, not a democracy at all; at best, only a 'crooked' or 'Topsy-turvy' system that is often ridiculed elsewhere. Not surprisingly, several times, the pollsters & forecasters wind up with 'eggs on their faces' for their botched "projection' results. But they still carry on the same practice undeterred and unembarrassed.
The same is not even remotely possible in the case of the Vallam Kali!

Perhaps because it's a large team-sport of strength, stamina, synchronization, and many other intangible factors.

Having completed the introduction, the commentator begins to introduce the captains of each of the boats, one by one, in a full-size picture and video gallery. The first one sports a feisty extra-long mustache spanning over to each side, his curly hair, unkempt, looking like an uncoiled slinky-spring toy familiar to toddlers as 'Jack-in -the-box'. He poses with a formidable posture. To his credit, he has won a couple of races in the long past. The next, massive in size, leaning back, possibly to counter-balance his mid-body protrusion, with his shiny long silver hair folded backward, exudes true confidence. The third one is quite familiar to the crowd, a veteran of many races; winner of none; witty and charming, blustery, burly with dark thick eyebrows but a barren skull. Then, the commentator announces in gradual, exceedingly dramatic crescendo "the last but not the least...;" pauses, the crowd waits, the figure fades up on the screen, a giant frame straight out of the mythology books, thick long beard neatly trimmed, falling beyond the navel, resting on his heavy belly, his long hair propped high in several stepped up, bunched, matted layers, like the village damsels carrying several water jugs stacked on their heads.
The crowd goes berserk!

Each Chundan is a communal property, treasure, relic; pride and joy of its respective villages

The children recall the cartoon series they watch in the morning while mechanically swallowing breakfast. They yell out in awe and amazement in their innocent voices, "dhey, look! Parasuram,..Parasuraman; Parasuram in flesh & blood". The brash captain interrupts the announcer; grabs his Mic (as wrestlers often do in the pre-game show, before their 'Rough & Tumble'); introduce himself in a rasping voice reminding of Stenka Rasin, the Russian Robin Hood/Pirate who once prowled up and down the mighty Volga River and the Caspian Sea, challenging the marine forces of the Czar, and terrorizing the rich people living along the banks. He brags about his training in "*Kalari-Payatt*" the ancient Keralite martial art; the beautiful boat he steers, its brave crew, and the historic hamlet his boat represents. He makes a bold prediction that everything is ripe for him to make history today! Furthermore, he is the captain of the debuting, brand-new boat "**Maha Keralam**" (The Great, Keralam). It is carved out of Mahogany, which has an exotic, beautiful, light brown distinctive wood tint and saturation. After his vaunting bravado, he brushes aside the spectators assembled at the jetty; jumps onto his boat in 'pole vault' style, pivoting on his long, flat Rudder-paddle; lands right on the top tier platform at the stern; takes up a position as the top helmsman, already looking triumphant. He does the trade-mark full circle cartwheel action of “Kuthi-Earu,” the stabbing maneuver with the oar with a roar, kicking up and throwing water high up in the air with brute force; drenches fans watching from closeby, as the proverbial ‘vindictive elephant’ did to the tailor who pricked him on the trunk with a needle. They are dismayed, amused, but now have to take off their shirts and look for some sunshine to dry them out. The video shows the boat paddles away in great style and elegance, people whistle to show their utmost admiration. A lot is talked about this Chundan, as it is expected to 'make waves'. This debuting boat is regarded as the ‘dark horse’ in its very first race, according to some pundits.

Then, there is the ceremonious Captains' pledge, honoring fellow participants and pledging to uphold good sportsmanship and camaraderie throughout the races. It is followed by a drill and exercise routine with all the paddlers raising and moving their paddles in tandem, shouting words of valor and might. This routine is similar to the New Zealand Rugby team or the Japanese Samurai wrestlers performing the traditional rituals before the start of their competitions. It is possibly to psych themselves up, and raise the adrenaline levels for 'enhanced performance' without using any anabolic, which is mainly ‘testosterone’ hormones, or some other banned stimulants.

The TV screen shows, that the boats have all assembled at the starting dock. They got their lane assignments earlier in the day. It is a nationally televised event and by satellite channels. It has to be punctual to the second. The premium satellite channels carry it 'live'. The commentaries are carried on multiple audio 'sub channels' in Malayalam, Hindi, English, Arabic, Persian, Chinese, Tagalu, Polynesian dialect, and so on. Depending upon the 'carriers, providers, and packages' that is subscribed to. The program can be viewed virtually anywhere in the world. To reach across the globe, more than one satellite is required. A minimum of three satellites at the right positions at 120° apart can cover the entire globe. There are also several free 'streaming' channels available on the Internet. One has to be extremely careful and circumspect about using a ‘free service’. In any case, the popularity of Vallam Kali is steadily rising. The Polynesian Island countries have recently expressed interest in participating; so also the Amazonian tribes. Both were the original “Navigators’ in the world; used to trade by barter and visit neighboring islands on rowing boats. They mapped out routes based on positions of the stars, each identified by a particular stone placed in the appropriate configuration, to determine locations and destinations. Their goals were to promote cordial, diplomatic, and peaceful relationships with neighboring islands. Propitiously, in the process, they came to discover that they were mostly distant relatives after all; perhaps in similar expeditions became ship wrecked and stranded; left to build their own island communities. It is believed that the coconut came from one of the main islands in the Pacific region. No one is exactly sure if it was brought here or came on its own by floating in the sea! It suffices to know that the unassuming, but the 'most useful' coconut is our common bond and sentimental link of enduring friendship, and goodwill!

On the other hand, the wily European colonists came in disguise, professing good intentions. The British, French, Dutch, Belgian, Spanish, Portuguese, and others used their seafaring and navigation expertise, fighting might, '*Butcher's Lust*' to conquer foreign, far away-lying lands, countries, and continents to plunder their wealth, and invaluable national treasures claiming them as extraordinary acts of adventure, and exploration. When confronted in their own lands and turf, we were told "they were all done (tactfully avoiding the word "committed") in the past by the people of that ambitious generation, and therefore, today, it bears no relevance whatsoever. No culpability shall remain; besides, there's the law of "Statute of Limitations"! (perhaps ignoring the fact that under international law, **crimes against humanity, war crimes, and genocide** have no 'statute of limitations'!). It was pointed out that guilt and sins will pass down through blood and genes, as stated in their own holy books. Disappointingly, there are still many forms of the very same horrible attitude and wretched behavior practiced throughout the colonially established countries. The same privileged justice system for one race, and a ruthless one for the others are still carried out just the way it prevailed throughout the colonial times. As a result, the privileged win easy acquittals, or leniency, while the colored & minorities are meted out unlawful convictions, jail terms, and extreme punishments to the "full extent of the '***unjust***' law". Truly, there are now many who are feeling genuine guilt, horror, and shame about the past deeds of their forefathers; want to make amends 'without admitting any blame' - a hypocritical legal clause enacted for the benefit, comfort, and absolution of the elite, and the well-connected.

It's high time to admit the truth, and the full extent of responsibility; negotiate reconciliation & reparations!

Well, here now, the organizers are mindful of the importance of their tasks. They strive to deliver the best quality, coverage, and capabilities to match the expectations of the international viewers. It is to be conducted in its authentic, traditional fashion and spirit. It has to attempt to highlight the pristine ecosystem of the land; the accustomed hospitality; genuine warmth, and the people's love of nature; the socialistic and altruistic system that's inherent in the lifestyle; the uncomplicated way life is conducted; the ancient philosophy, the mythology that is comparable to that of the Greek, Egyptian & Babylonian; all preceding that of the Roman's. Then, there is arts & music, Sciences from the Vedas which were all swept away into abeyance and later to its total oblivion by the long and brutal colonist occupation and its self-serving regime. Over time, the measures presently undertaken will pay rich dividends in many ways; most importantly, leading to a recognition and appreciation of an ancient culture, as well as the need for an eventual 'truth-telling' of the world history.

This annual regatta and race here today, as well as the trophy were named after the first Prime Minister of Independent India, Pundit Jawahar Lal Nehru. It has come to be known as the Nehru trophy boat race. Nehru was one of the foremost freedom fighters of India in the footsteps of the great Mahathma Gandhi, a distinguished orator, a first-class author, and a charismatic statesman. The non-violent struggle for independence they lead in India against the most violent world power of that era did indeed contribute favorably to other commonwealth nations successively gaining independence, and becoming their own proud republics. Therefore, the event of today bears important national, international, as well as moral, and historic significance.

Thus, the humble regatta festivals and competitions of the Kerala 'Vallam Kali' embody and symbolize several subject matters all at once!

Chapter 32

The Malayalee diaspora gather to watch the race on satellite Television

The Gulf states are home to the largest Kerala diaspora. The hospitality workers of the three artificial islands that make up the palm islands in Dubai, the largest '*man-made*' island in the world -*Palm Jemeirah, Deira, & Palm Jebel Ali* assemble in their community lounge; so also their compatriots at "*Burj Al Arab'*. Other expatriate communities gather around TV screens in Potluck dinner style, bringing international cuisines -the Arabic favorites of Pita pocket bread right out of the blazing clay oven, *ShishKebab* on skewers, *Chicken Shawarma*, stuffed aubergine, and many other menu items. As gracious hosts, the Malayalees themselves cook Rice, *Vadai, Palappam* & Fish curry; North Indians bring Daal, Chapathy, *Alu-mutter* & *Tandoori chicken*; Pakistanis the *Queema, Korma, Biriyani & Pula*o; Bangladeshis their chicken *Pakora, Stir-fried cabbage* with garlic & Bengali spices; Filipinos offering their delicacies of *Fish Tinola* in sauce, flavored with onions, tomato, tamarind; fried & stuffed *Empanada*, boiled duck embryo, *Egg Balut*; finally hot and sweet Arabic lemon tea to go around multiple times to make it all glide smoothly down the esophagus into the stomach pouch. For the first time ever, the tracts in the bodies will be required to manufacture new types of digestive enzymes to break down the mix of newly received food molecules into their smaller building blocks, in order to facilitate their absorption by the body. Possibly, a few cases of constipation might ensue the next day!

Warm words of greetings and pleasantries resound around the crowded space, "*Assalamu Alaikkum*!, *Ahlan Wa Sahlan*!; *Bhai Saab*!; *En-thuttu Mashe*..!; *Marhaba*!; *Keef Haalak?.. Namasthe, Ji!; Baath Kaise?; Hey man!,..; Good evening, sir*!; *Nice to see you!*; *Namaskaram Saar!*; *Salaam,.. Bhai Saab*!; Thamam, Kullum Thamam!...*Are Yaar!*; *Dhanya-waad!*; Guten Tag, Herr Meister!; Bonjour!; Are Bapre..!; Shukran Saab!; En-thu Vi-sesham?; *Su-kham,.. SukhamThanne*; *Maa SaLa Mah!* The multilingual and customary greetings and platitudes transcending cultures, languages, nationalistic spirits echo against the hard and hot concrete walls. It fills the air in cordiality, congeniality, curtsies, and sincerity signifying a true international camaraderie that's rarely seen, but truly exists here between common people. It's here to behold and experience!

Some Malayalee fans proudly associate themselves with a particular Chundan; boast how uniquely qualified it is to win, given its track records and bet that waged on each of them. One person claims his High school had won a regatta competition three successive times in a rare record 'hat-trick'. Proudly, he displays the engraved medals in a sequential order for everyone to see, in his carefully organized showcase.

Right now, the all-important parade or procession of the boats is taking place projecting the ceremonial pomp and splendor of the races. It's a processional regatta held in colorful pageantry of all participating race boats, preceding the races before the 'whole in earnest' contests begin. The spectators and fans get a final chance to preview and compare the boats within each of its peer group. They could assess many things; form impressions & opinions; determine betting odds. It's a great photo-opportunity with the boats paddling by in spectacular formations and order, as in a military parade consisting of all the divisions of the armed forces wearing their respective colorful uniforms.

The chundan Vallam is the preeminent race boat, the most prestigious and readily recognized. Its status can never be challenged in any way, shape, or form!

To lead the parade are the 6 Chundans, one in each lane, all winners in the immediately previous years, then the Oaa-dis, followed by the Women's teams of Oaa-dis, Vaippu; Iruttu-Kuthi and Chu-ru-lan; close behind is the KochuVallam (small canoe race boats) then, the men's teams of Vaippu and Chu-ru-lan. The same order and sequence are repeated until every participating Vallam rows past the entire stretch of the course in front of the cheering spectators.

This is the much anticipated ceremonious regatta in the traditional style of ostentatious display, and ceremonious grandeur. It is carried out in leisurely manner, good jest, and cordial spirit. It serves as a warm-up exercise as well as a chance to 'get a feel' of the course, the crowd noise, wind, water, and current. The rhythm makers sing their songs of praise of their chosen patron deity, saint, great men, and women whose legacy they cherish and honor to this day and hour. They wish good tidings and prosperity for all the people in the land, protection from natural disasters, and other mundane miseries. They don't neglect to boast about their valor; promise to chase all the demons out and into the deep seas, or wherever they originally came from, giving firm reassurance of peace and security to the public. The crowd gets energized; the self-esteem swells; get goose bumps!

The boats begin to line up for the ceremonious parade and procession of all participating Vallams in the regatta

In modern times, no contest or sports event can take place without the involvement of betting. Things long considered as evils are no longer viewed as such. The money, its power, and the prevalence make many things acceptable, though some are the least honorables. Man invents things for the good of humanity; unscrupulous people often hijack, and divert them for evil purposes of crime, fraud, gambling, human trafficking, hacking, blackmail, greed, terrorism, and destruction. The lawmakers condone many on the hushed pretext of providing occupation to the participants, especially when the citizens cannot be provided with ordinary jobs. The governments get tax money, sometimes unaccounted. They can pilfer some of it and the rest be allocated for the "*good of the public*". The bookmakers recruit race-savvy people to get tips; post them at strategic locations along the course to report on the race progress to update the odds, They use their own private '*shadowy backstreet*' communication channels, and networks. The syndicate employ techies to develop computer algorithms and fuzzy logic to predict the outcome, using the up-to-date data and archived statistics. They claim that they are simply doing the same things as Google, FB, MS, Amazon, and other such global conglomerates do with data mining & *AI*. The betting syndicates and online betting sites are largely interested, and heavily invested here. Even a 100th of a second, matters much, when the stakes are this high, as recently gleaned from the infamous '*flash boys*' traders of the Wall Street. As in any business, the more the revenue, the better the *payout* for the top operators. People who lose the bets blame their misfortune and regret their misjudgment. On the web of betting circles, there are many unethical practices as in any other cases of high stake manipulation. They resort to fake news, inside information, persuasion, intimidation, coercion, violence, sabotage, cheating, fraud, nonpayment, and many others only the insiders, and the operators alone would know about. The Far-Eastern syndicates are very much interested in this particular races as it's a rare event, and it brings together a large number of fans. They have gathered all data, results, and track records since its inception. Somehow, they are better positioned in this sport than other rivals. They prefer to stay '*invisible*'. Here, they deal only in 'buck', and keep the tax men at bay.

The Monaco & Las Vegas casino operators envy them!

Chapter 33

The Race Starts!

'Jet drive' propelled speed boats thrust themselves into view racing like hydroplanes [in 'planing' mode when power is increased to the maximum so that the hull glides on top of the water], one on each lane, carrying white flags. They hover over the surface of water and move in skating speed, generating trails of fast moving traveling waves. The boats are seen periodically making different formations, arrow-head, parallel, and line-ups resembling MIG, Sukhoi 'multirole' fighter planes in the 'Republic day parade' emitting colored jet plumes of the tricolor national flag. They seem to have learned a lot from the flights of the migratory birds. The spectacle catches people's attention and awe. It signals all clear sign and call it to the attention of the imminent start of the race. The spectator boats and Vallams sway in the mighty turbulence. Infants in them are amused by the rocking oscillations. They giggle in mirth in heavenly bliss; possibly thinking that they are at home in their swinging cradles; finally fall asleep helping parents to enjoy the races without any more preoccupations. Parents believe the images, sounds, and sights will stimulate their developing minds, remain and grow with them. When they mature, it may trigger great imagination which will be useful and wholesome for their welfare!

Right now, the finalists are all lined-up at the starting point!

This is the season-ending grand finale of VallamKali season, and also the most anticipated. The starting decks are the temporarily made floating platform on 'dinghies', making up 6 lanes. Traditional methods & commands are used for starting the race. A slender 'flat board' is seen hoisted out of water by manpower to align the boat, serving as a start line and then held in exact position until the race is started. High technology is currently in its testing phase only. In the preliminary heats, leading up to the finals, one race half way, the next one starts providing the spectators a continuous view and excitement. Sometimes, the fans are left to wonder how their boat in the previous heat fared; if it has qualified, or not. For that, they need to check the Television screens, or access the custom 'Vallam Kali app' downloaded for the purpose. All such heats of each category of boats has been conducted one after the other. Then, the quarterfinals and semi-finals follow between the qualified boats. Six boats qualify for the finals from each category. The main, the ultimate event, is the finals of the ChundanVallams. Saddam would have called it: "Mother of all regattas!"

The Elite Six, the finalists are drawn side by side in each of the 4.0 m wide lanes. They look like some prehistoric, dinosaur dragons raising their hideous heads on the horizon, when viewed from the distant finish line. Currently, they are "on the mark" and held there for few moments. "Ready", a Pause in between, then "Go" with a loud shot from an electronically triggered starting horn facing backwards, towards the race boats. Actually, the traditional canon shot heard all the way up to the finish line triggers all the electronic timers. It, starts the race! The finishing point is approximately 1400 m away requiring roughly, 4 minutes 30 seconds to complete without being current aided. For comparison, in athletics for the short sprint events and long jumps, the corresponding term is 'wind aided'!

The moment of truth has arrived for the fans. Expectingly, all eyes are turned towards the starting horizon far away. They strain their eyes and scour the area for any glimpse of the towering boats, or water splashes. The colorless water provides a poor contrast in background to be able to spot an object or figure at that much distance. Furthermore, there is always moisture and light fog over such a large body of water which obscure clear view, and fine determination. So, they turn to the TV screen nearby for the first moment of discovery; lively call out what they see, mainly the names of the chundans and the lanes they are on. They have to patiently wait for a few more seconds to feel the excitement of seeing the race directly, and plainly in front of them with own eyes. In the meantime, they weigh the merits and demerits of each lane. A few conclude the ones who had the same lane in the immediate previous heat shall prevail in the end; others are not sure. Some cite this as a disadvantage, as they are more tired and hadn't had sufficient time to recover. Simply, the other teams in the competition are better rested. Though it may sound logical, the majority refute all such unscientific assumptions. Some think that better experience is the ultimate key factor in winning as often proven in most sports. Some state their opinion vehemently; others in somewhat ambivalence; the rest like to wait for a few more minutes to find out, and be sure.

Yes, now indeed,... they catch the first glimpse, and become instantly excited!

Chapter 34

The running commentary

The running commentary is narrated in its typical style, using incomplete sentences of broken words, adjusting to the thrill and pace of the races. The commentators go back and forth at frenetic pace and frequency; vary pitch and style in order to capture listeners' attention and imagination. The commentators are well versed in literature, vocabulary, perception, and observation. If they cannot find the right word, they make up '*compound words*' which convey the meaning, although making it long and hard to pronounce. Many languages of the world do the same. Indo-Germanic languages tend to utilize similar word formations quite commonly. The German language prefers to use extra-long compound words rather than synthetic words or '*coinage*' new words common in English literature.

Sriman. DhanaVan Kuttappan is in a motorboat moving in line with the leading Vallam. He updates the commentators with current positions, speaking directly into their '*intercom*' which is heard through the headpiece they are wearing. This is analogous to the mobile announcer on a motorbike following the "*Tour d' France' cycling competition*. These techniques lend accuracy, credibility, excitement, and great 'feel' to the running commentary.

"There goes the flare and an accompanying sound! By the way, the flare is from a Hydrogen gun/*torch, seen in welding*. When the gas interacts with air it becomes flammable and rises fast, vertically upwards without spreading fire around. Hydrogen burns with a pale blue flame that is almost invisible during daylight time-hours. It is made into an orange flame by adding impurities, or sodium dust. Then, it becomes visible all the way to the finish point. The very loud "bang" heard is from the H_2 collected in a specially designed, elongated balloon held at some safe height and distance in front of the boats."

The race is now on!

Also, the running commentary; the commentators go back and forth during the entire race; a narrator also contributing at times. They speak very fast, often mincing words; use incomplete sentences. But the people complete the missing elements in their minds, making sense and understanding; no worries there!

Sri. Thommikkunju MalikaMukalil, the leading '*Play by play*' commentator, is accompanied by

Shri. Kesavan Ashan:

"The paddlers stab their paddles into the water, that's for the first stroke

Make a laborious sweep to overcome the stubborn force of inertia, the biggest the paddlers have to overcome

They have learned it in school- from Newton's first law of motion!

A Boat in the middle lane experiences a '*yaw*' (side to side movement)

The one in the 1st lane, a little '*roll*'

As the rowers at the starboard and port sides apply unequal forces

And wild body movements, shifting their center of gravity while exerting excessive force

It may take some brief moments to feel the right balance and fluency

They have to build up a constant and sustained rhythm past the 'acceleration phase'

The Helmsmen want to ensure

That the heavy Chundan doesn't sway from its optimum course and direction

Any tiny extra deviation from the path may cost too dearly at the result tally

The first Helmsman, the strongest and the most skillful,

Make a high swashbuckling stab into the water with his large Rudder cum paddle

Leaving the task of steering to the 4 other lieutenants

Gives a fillip to the boat, breaks up the lethargy somewhat

Rowers feel a shake and a jerk as well as the resulting fluid motion

As if the boat is untied and let loose to go

Like a cow which is released from the leash, or the dog from its restraint

The Rhythm makers know their tasks too well

Try to influence matters right from the outset"

Paul Jose Praa-vady, the second color commentator is a Post-graduate in Physics from the University of Kerala

He takes over the commentary:

"The boat must pick up steady speed, maintain it,

Avoid frequent episodes of acceleration and deceleration

Newton's second law of motion says F=m*a

It means, in each instance of acceleration, additional force is required

Which in turn is additional energy expended; and unnecessary; it's not efficient!

The paddlers will tire themselves out without getting a proportional or favorable outcome

The acceleration is a *vector* force too, or simply speaking, 'it is directional'

Therefore, it has to be in line, along the line of motion!

The helmsmen should bear this important fact in mind and

Steer the boat in a straight line,

Avoid jerks and deviations or drifting in zigzag direction

Although a '*pitch*' or a '*yaw*' may be expected at the start

In this long race, it is to be expected that a few other issues may pop up

In the frantic effort to overcome the dead weight of 'inertia'.

The boat must accelerate at the sprint phase of the race

That happens in the final moments of the race

Must do it without feeling much exhaustion

The stated sequence is considered the most efficient

It was tested and confirmed in practice. It is a matter of common knowledge now, and putting it in practice!

Yes, of course! it's widely known and therefore the accepted strategy by all the race boats!

Everyone knows it; wants to apply this principle

It's all in its execution, You know!

Over to Srimathi. Parvathy Pankaja-kshi!"

Srimathi. Parvathy is a popular drama actress, articulate and talkative; witty, occasionally whimsical, and scolding; talks with good diction and proper enunciation; mindful of the metrics and cadence; intonation and voice modulation. Listeners describe her voice as "*kuyil naadam*", (birdsong) that of a singing bird in the woods. A female voice adds a different tone; alters the mood, and sensation. Her presence alters the ambiance, as well as the course of proceedings. She is well known for her contrasting perspective in all matters affecting life. Women, especially mothers, are the closest to life, and bringing up future generations. Their perspective is important; it must be heard, respected, and heeded, although they tend to be distinct, forceful, and occasionally repetitive.

"Now all the boats are moving along well in a normal, sustainable pace

The cantor casts a look aside; sees his lead slightly dwindling

Increases the pace, stamps his '*rhythm pole*' harder, alters the tempo; from 'andante' to 'allegro' (walking spped -lively tempo-brisk)

Prompting increased effort from the paddlers

The crowd knows how important the start is

Some infatuated fans want to follow their boats

Run along the banks

Immediately find others blocking their paths

Decide to stay where they are at, right now

Frantically wave their head and shoulder scarves

One loses his grip on his scarf; the wind carries it away

He doesn't relent, uses his heavy arms devoid of towels

Still keeping a folded fist

People on the left and right of him

Evade his jabs, as in boxing matches.

Now, all the boats are moving at normal speed

The paddlers seem happy with the response from their boats,

The singers sing songs of praise of their boats and the crew

Assuring that they are doing pretty well!

The *Vanchipatt* sings more than a lot of stories

The words, tones, and rhythm carry substance and instructions

The paddlers sharpen their ears

Grasp the meaning; follow the tempo set by the singers

Relish the feedback they are getting

From the spectators and their own rhythm makers

They are consciously aware of the cardinal sin of 'looking aside'

Know for sure their own family members are somewhere in the crowd

Besides the large and noisy "race watch-party" in every village, watching the televised broadcast together on large screens

Stubbornly, resist the temptation for a searching look

Instead, concentrate on what they are supposed to do

Keep a constant rhythm

Continue with their task 'at hand'!

As a crew of smart robots '*programmed*' to operate by the pace of the rowing chant

They know they could gloat and boast in front of everyone, if they win

They maintain stern discipline; the yoga routine they practiced during training exercise does help here a lot!

The '*fabulous Fives*' at the bow think they can do a little bit better

They have strong torso, biceps, and long reach

Want to '*up the ante!*'

Lean further forward; get ready for huge 'grabbing' sweeps

The chief helmsman approves the move bypassing the captain

He ventures his own trademark cartwheel sweep to announce his intention

Effecting a regenerative feedback with his bold execution!

Srimathi. Am-bili Varada-Rajan, is a linguist, fluent in Tamil, Hindi, Bengali, Arabic & English. She had assembled a team of commentators in foreign languages. Their audio is transmitted in the "*sub-channels*" of the main Satellite signal. They provide the running commentaries in all these languages:

"The first rower counts out. One, Two, Up; -- One, Two, Down. This tells the other four front-rowers behind him that after counts One & Two, take the paddle out and hold it vertical in the 'Up' position; at the count 'Down' slash the paddle on the '*opposite side*' and continue paddling normally. When properly executed, the first, Third, and the Fifth will be paddling from the same side. The second & fourth paddle from the opposite side, all in unison". They execute the maneuver seamlessly in delectable style and ease with just a momentary 'pause,' but without missing another stroke.

The fabulous Fives need much body flexibility and strength. In the sweep phase, they keep their body streamlined like the hurdlers clearing the high hurdles in a 110 m race, bent at the hip, hands stretched in good balance while preparing for a smooth glide over the hurdles. Another way of explaining is; as in Freestyle swimming, it involves 3 distinct actions, 'grab, pull, follow-thru-push off'.

The push is an accelerated action, as the water grabbed on the paddles progressively slip aside fast.

All the international betting syndicates from Mumbai, Macao, and Manchester are keenly interested and heavily invested in the races. Not to be outdone, is the online betting sites. They won't miss any opportunity in competitive events like this, of unsure outcomes; demonstrate a penchant for reckoning the best possibility and most likely probabilities; they apply statistics, historical data, and specially developed algorithms for each of the specific sports to call. They use permutations and combinations of advanced Mathematics to help in determining the odds. Some bettors go to the extreme extent of betting the 'trifecta' or triple, a form of high stake betting in which the punter selects the first three place-winners in a race in the correct order. Invariably, this situation makes rare winners, but certainly an overwhelming number of losers. Gamblers must have guts too, but they trust their home-work better. They must be prepared to hold their breaths until the final results are declared, and shed all the tears afterward. To achieve their goals, some may employ various tactics and strategies: such as bribes, threats, provide performance-enhancing drugs, coerce performers for deliberate under performance, and judges for their 'judgmental decisions', etc., etc.

Nowadays, any event is open to betting, directly or indirectly; even cock-fights, bullock cart races, and Cobra-mongoose encounters. A fight between an Indian king cobra and mongoose intrigues many people, mostly the villagers. The two creatures don't like each other; don't enjoy sharing anything; want to keep the territory for oneself. Cobra is extremely venomous and frightening. In its fighting posture, it stretches the throat wide and flat in a towering, angry and menacing killer-dance. The mongoose is a predator itself, fearless; has sharp teeth. It is quick; has slender, agile body; but, dwarfs that of a cobra. If you watch them fighting just for the initial few minutes, one might think it's a foregone conclusion that the mongoose will soon be dead meat. The entire fight is long, as each waits patiently for its opportune moment to strike, although continuously streaking, evading, and taunting. Each look for an opening, much like the 'cage fighters'. The duel is one of ups and downs; attack, retreat, fake, and wait; decide the best next move. It made me wonder if boxer 'The greatest of all time' Muhammad Ali learned his boxing style of 'jabs-shuffle, evade- bounce off ropes' from these animals. In this fight, at any moment, one might wrest an upper hand, but fails to take a decisive advantage. It's going to be a fierce fight with a series of uncanny tactics and clever strategies, agile movements; jabs, spitting venoms, and taking evasive actions. Both are dogged, and determined. Until the very end, it is quite unpredictable which one shall prevail. Incredibly, statistically, the odds are heavily in mongoose's favor, as it has certain immunity to the cobra venom which other species lack!. Our bettors here would have made a clear bet on mongoose!

Now, from a vantage point at the venue

The bookie is calling out:

“The initial report from the start location indicates **Thayankari** as having the best start.

Just a few moments later, another comment adds:

“Oh! Look.. ***Karuvatta*** gains some momentum, now catching up with Thayankari”.

The race starts in great earnest, intensity, grit, and dogged determination

Sri. Paul Jose Praavady, the color commentator:

For this commentary assignment, I consulted Dr. Jagmohan Bhandari, M.D. an Orthopedic Sports medicine Specialist. According to his research, modeling and findings, the paddling is a full-body exercise. The hydrodynamic forces felt on the paddle are directly transferred to the arms, shoulders, back, and legs. The main muscle groups involved here are:

1. *Abdominal Muscles* which are constantly used in the action and for balancing the body.

2. *Biceps and Triceps*- They are located on the front and back of upper arms; biceps in the front, triceps on the back. They are used in the movement of the arms.

3. *Shoulders*- In paddling the arms have the most flexible and exaggerated movements. The shoulders support and help in the 'cranking' movements involved in the paddling actions. They move within the "rotator cuffs", a ball-socket joint by four ropes of muscles. [similar in action to our ubiquitous kitchen accessory, the "mortar & pestle" grinder used in making "Urad-dal mortar' for dosai & idli.]

4. *Trapezius Muscles*, and the Latissimus (the broadest muscle of the back, or 'lats') just below are back muscles that provide strength and power to the body, shoulder and arm actions.

5. *Leg Muscles*- It's a group of muscles notably thigh and calve muscles where **quadriceps** are the strongest, and calf muscles having the largest muscle mass. When the longer muscles in them contract and relax, they activate movements while the smaller ones help stabilize other joints and bones to do delicate movements by the so-called kinematic (mechanics that deal with the motion in an interconnected system of levers and other moving parts in a system) motions. Besides, as a group, they provide a strong platform for the entire movement by stabilizing and anchoring the body firmly on the seat and perform the work.

The paddlers must keep a good body balance avoiding side to side movements or shifting weight sideways. They must anchor the outboard (Situated toward the outside) foot against a firm surface such as the "ribs" or on the floor to leverage the forces exerted by the whole body. The paddles are kept close to the 'gunwale'. The follow-thru finishing action is slightly away from the body to allow the arms to finish the action rather comfortably and keep the splash away from the paddler sitting behind. The finishing action shall smoothly follow the start of the next stroke by making the entry of the paddle blade for its initial "grab". This happens with the paddle entering the water slightly away and bringing closer to the gunwale. 'Pull' & 'push' follow with a faster and fluent 'follow-thru' actions to finish the stroke. Obviously, the application of the force varies considerably during the various phases of the action. The 'pull' is the hardest. The 'viscous drag' on the hull is a stubborn force. (Viscosity is essentially a measure of resistance in liquids to a moving body through it; identical to 'friction' in solid bodies). The paddlers exert forces in "paired-tandem" and must be in synchrony so that all the additive vector paddling forces contribute to the maximum effect. Although the 'turbulence' created by the action cannot be eliminated, it should be kept to a minimum.

Suddenly, the commentator notices some significant developments in the course of the race. He continues with the running commentary:

"Boats on the right and left most lanes have now come even and slightly ahead of the pack

The partisan crowd immediately sees it; surely, they don't miss anything!

They begin to scold the laggards with vociferous scorn and taunt

The fabulous fives get the message

Immediately, they show what they can do, and signal to their mates what they must do!

At the moment, the boat at the near side, *Maha Keralam* is gaining momentum
While it hears the shouts louder than other boats

Falsely thinks the whole crowd is on their side and cheering for it"

Now onto Srimathi. Ambili:

"The rowers have been warned not to look aside, but keep the focus only straight forward. However, the Rhythm makers face in opposite directions perpendicular to the direction of motion; sometimes in 8 different directions, giving a 360° all around view. The positioning gives them a unique '*vantage and direct*' view of the race boats in adjacent lanes, whether they are ahead or behind. They can sense and feel the relative speed and motion with respect to other boats; monitor the changing tactics with their critical pair of eyes.

They adjust the pace and cadence accordingly

Each lane has its own advantage

The nearest to the bank experience additional drag when it gathers momentum

Because of the hydro-dynamics of the water hitting the hard sidewalls of the bank; feels like a mild breaking force

Pascal's law is in play here!

The middle lanes get an advantage or disadvantage,

Depending on their position with respect to the '*wave tunnel*' caused by the boats in the adjacent lanes.

Depending on whether it is 'on the *crest*' or 'on the *valley*' of the waves created by paddling action of the other boats"

This is similar to the bicycle race, or the bird's flight where the one in front bears the brunt of resistance, the followers enjoy some useful draft. The main difference is that they don't change positions for altruistic ideals! The draft gives them an advantageous 'pull to the fore'!

The bookie:

"*Quarter point* calling! The boat **St. George** Ahead by ..m..m a Meter; I'm looking straight ahead, straight on, place the bet! Over!".

Haneef Hidrose, is a Public Relations officer at the Kerala Tourism Board. He relieves Srimathi. Ambili;

He begins:

"The lead changes back and forth

Between the 2nd, 4th, and the 5th lanes, almost at every measured length intervals

Each boat has its own loyal supporters

They cheer for their boat, taunt the others

When they see theirs leading; Go glum when lagging

The rest of the crowd decides it is not worth agonizing

Cheerfully enjoy the race instead"

The race has some ways to go!

Sri. Hidrose invites Sriman. Vananthan Nambiar, a Malayalam poet laureate and Pundit, to describes the scenes as he sees them.

He is sometimes called "*Nimisha-Kavi*"(poet, extempore!); he can compose poetry 'on the go'. He is said to be a distant descendant of the very renowned and inimitable *Kunjan Nambiar*. Gladly and gleefully, Shri. Nambiar obliges;

He starts in a poetic style: in his lilting prosody and cadence that is quite familiar to the public from his radio broadcasts. In some ways, it reminds us of the Sunday sermon in the church where the Rev. talks without interruption, and listened to with reverential attention. Recently, this style has become popular with some orators, and politicians of color.

"This is truly a great spectacle in the water..

In the background of colorfully dressed and boisterous spectators
Present on the green pastures and under the shades of lines of coconut trees along either banks

Looking around, I notice some wearing costumes of mystical characters of Kadha-Kali actors, Protagonists, and antagonists among them..

Some others of scary creatures...Dinosaurs, and fire-spitting dragons,

Scaring infants

Also, characters wearing caricature masks

Resembling hated and deceitful politicians...

Possibly instigated, and contrived by the opposition parties...

Expressing their sentiments, and resentments...

Amusing people all along

Occasionally, drawing mocks and jeers from the crowd

Facetious cartoon characters hop around on stilts like Kangaroos

Often stumbling up on people

Revealing their true faces while sincerely apologizing.....

Children playfully snatch their masks

And fit it on themselves; strike different poses

Ask others how they look

There are many frivolous fun and games

Also funny and happy scenes all around, aside from the races....

Well, we call paddy, the "*Nellu*", the whole grain with its whole husk

Straight from the fields, or as stored in the granaries

Also, the shredded or milled paddy without a distinction,

Cooked and ready to be served is the life-sustaining "*Choru*", (boiled & cooked rice) that's our staple food

Without it, we can't live a day!

The paddy fields stretch far and wide from here, long, and across...

One joining the other

Only large canals meant for irrigation and transport

Separate them, or can tell them apart

They spread as far as the eyes meet the water and the skies....,

The unreachable vast horizon far beyond!

A "Sanga-Mam" (confluence) of the three entities,

or say, the proverbial physical Trinities!.

Just a couple of weeks ago, the fields were bearing golden rice grains

On branched 'inflorescence of panicles' (the arrangement of flowers on a stem of a plant)

Where flowers and grains appear from bottom to top

Raising hope and joy to the farmers, likewise....

Travelers reported seeing them from afar

The rows of golden paddy growing thick in the vast fields

Later harvested and stacked in giant heaps of grains

Like the Egyptian Giza pyramids standing in the desolate desert of golden sands...

They are the fruits of our hard labor, bounty from the soil...

And a gift from our own earth, mud, and water;

Reaped with toil, sweat, and passion paid by deep family debt
Now garnered in the granaries
Waiting to be sold at a fair price!
Helping them to survive through another year ahead

The Cranes, Herons, Storks, the Egrets
Differentiated only by their "carriage' of the heads,
Congregate in the fields at the moment...
They feed on spilled grains and trapped fish
Ducks feed themselves in the vicinity
While the shepherds maneuvering their tiny boats, lead them on
Ducks dabbling and quacking, shepherds gently shepherding
Huddling and guiding the flocks of fowls
By the end of the day, the flocks are full
Make their laborious homeward journey
Crossing the rivers and waterways
They form different shapes and formations
Generously offering others space in the flotilla
Quacking nonstop
One must wonder what they are babbling to each other
So enthusiastically!
Soon after milking the cow in the early mornings,
The female households gladly collect the eggs
They are the '*golden eggs*'; it's a cash, as well as nutrient product!

The birds of the same feather flock together, do the things they always do; render color, sound, diversity, and beauty to surroundings

I see birds spreading wings and showing off their colorful plumage

To the full extent of its wingspan...

Hover high in the air; glide effortlessly

Elegantly in cruise mode; defying gravity with the draft

As they make the aerial survey...

Some levitate over the same spot, spotting a fish to prey on

Flapping wings vigorously...

Playing the role of 'MC', the Master of Ceremonies...

Or the conductor of a philharmonic orchestra...

While the birds on the field play around with spread wings,

And tweedle their bird-sweet song

Others doing the usual '*Turkey*' bird-dance

A couple soar high; swoop, diving in steep plunge...

Descend to the ground, hop around in circles on one leg....

Then swoon; playfully showing ecstasy

One by one, they join the huddle

While another emerges from the far end,

Does the same routine and goes back; on and on...

As performers in a group dance, as the stomping 'Irish foot-dancers'

Or infants waddling around while playing musical chair....

Producing a *buzz* in the air...

It's a dreary drone in my ears

But a melody so mellifluous to them...

Our children would say...

"It's bird's wedding!"

The Woodpecker normally pecks with "Tup-Tup...; Tup-Tup"

Now blends with 'Thi Thai Tho, Thi Thai Thom', the rhythm of paddling

Dispelling the notion,

That only *Mockingbird* (bird specie found in N. America & Caribbean islands noted for the ability to mimic the sounds of other birds) can imitate sounds

Or only parrots can mimic human speech...

The melodies & rhythms in the air rhyme; they mix and mingle

With our pulse and heartbeat..,

Sometimes swelling in crescendo

Other times decaying in subdued decrescendo

It's the rhythm of our own lives

Of joy and despair

Vicissitudes of life and fortunes,

Boom and bust,

High tide and low tides in the water bodies around us,

Dark days and sunny days; bleak and bright days,

Miseries of poverty; and delights of prosperity,

Agony and ecstasy in the living experiences,

The inevitable ups and the downs

...of our fragile lives!....

A small fleet of speedboats thrusts into the view on the vacant side

Seen having fun in the vicinity, frolicking on the cresting waves

Speed up and suddenly comes to a slippery, splashy stop,

Like the skiers in downhill slalom, at the bottom of the slope

Maneuver to a sudden stop in front of the hedge of evergreen tree lines

When the drivers abruptly throttle down the engines,

Like the geese fly in and land in the wavy waters at a low angle

'Planing' as a hydroplane on the surface until coming to a still stop,

Thereafter floating

Now a gentle afternoon breeze blows, fluttering the leaves ....

The leaves of different sizes and shapes

Vibrate like musical reeds

Of "Oothu / Nada-Swarm" (Indian oboe)

Provides music of varying tones and timbre

As they vibrate in sympathy in varied pitches

Forming a 'nature orchestra' of its own

Energizing our spirits and minds ...

Corroborating the sayings in our "Upanishads" (book of philosophical treatises of Vedanta)

"all objects in the universe are in constant mode of vibrations at its own frequency, and amplitude producing energy"

Mind is where our thoughts, emotions, and desires reside...

Brains are where information processed

And where decisions made...

Our body responds with actions

At the command of, and from the brain...

Producing efforts and movements

Accordingly!..

Sometimes, emotions interfere and prevail

Subdue logic, and dominate the decision-making process

Ceding the vital function to the fickle mind

Causing unintended consequences, and regrets later on

Air is what breathes life into all living beings

And what sustains lives...

Happens without our conscious efforts, or intervention...

All happen in mysterious ways, or simply, spontaneously

Quite thankfully, indeed!

The lanky coconut palms of various lengths and posture...

Some bearing thick green; others shiny golden coconuts

In bunches, thick, round and dense,

Lean in different directions, filling in the space

And vying for a 'vantage peak' at the blazing sun above

The life sustaining sunlight, the cause of all creations

Helping leaves manufacture food by 'Photosynthesis'

Using free sunlight; Carbon dioxide from air,

And also life-sustaining water from the soil

Brought on by the branching, probing roots

Are they in someway smarter than humans?

Towers of haystacks big and small; stacked up in different ways,

Rolled up, rounded up, and bundled up,

Raked up, and collected by the farmhands

Like a mushroom cloud, in dome, Wigwam, Igloo shapes
Also in many geometric, and amorphous forms
Seen through the dense vegetation

Of green trees, bushes, water cactus, and banana plants

Providing a starkly contrasting look!

Cattle feed directly off them

They munch on grass and hay, alternating

Children climb to the top; and slide down

To the fallen hay at the bottom

Using it as a landing pad for their tender bottoms...

Yellow flower bunches of tender coconut seedlings...

Wait to blossom soon, after the monsoons...

Healthy ivory-colored *branches bear evergreen leaves*

The long leaves grow in pinnate pattern (having the leaflets growing opposite to each other in pairs on either side of the stem)

Gracefully tapering proportionally,

Towards the tip in symmetric geometry...

The supple leaves yield in the breeze...

Showing which way the wind blows...

Like the flimsy "*windsock*" at a helicopter pad!

Bushes of exotic plants grow, that no one had planted.

Their clusters naturally arranged in artistic patterns of

Inflorescence (specific patterns of branch arrangements in plants holding flowers) of 'umbel,'

As in candelabrum (branched candlesticks),

The youngest at the center, surrounded by the older;

Another in 'corymb' cluster, the oldest flowers at the periphery

As a whole, appears as a convex figure in two-dimensions;

Tiny flower bunches on grasses in 'panicle' clusters

In branched inflorescence

Like the rice grains grow on branched paddy stems,

The panicles bear the rice 'spikelets'

That grow into grains

And ultimately becomes our staple food!

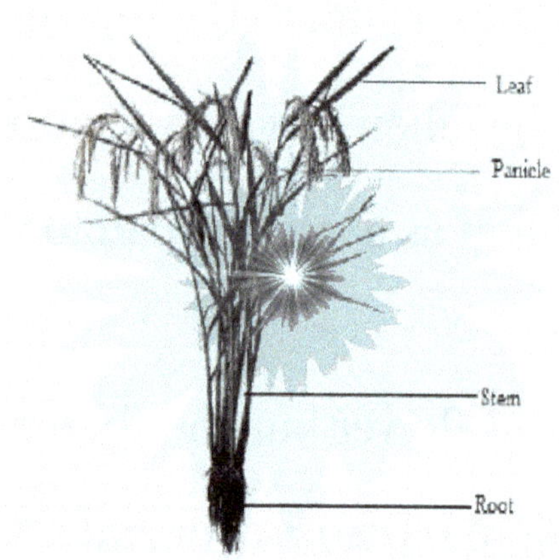

Still others in 'raceme' pattern one above the other

On its own stem, unlike the panicle pattern

But, otherwise, looking the same

The nature giving us clues and instructions

In making pretty "flower arrangements"

Another subtle,

But truly dazzling wonder in nature

Besides rendering environment pretty and vibrant!

The evening sun-rays color the skies...

Clouds form in abstract shapes and figures...

Gradually drifting across the vast skies in the rarefied air

Casting shadows on earth below offering some mercy

From the scorching tropical sun

While also scattering the white light

By the layers of dust and moisture in the air

Into its "VIBGYOR" (Violet, Indigo, Blue, Green, Yellow, Orange Red)

Color spectrum, adding various tints and shades in the act...

Creating different levels of transparency

Painting wavy patterns in the atmosphere...

Raising faint auroras

Of spectacular reflections and 'halo' around the clouds

Building soft and dim replicas of '*Aurora Borealis*' seen near the North Pole...

And the '*Aurora Australis*' in the South ... [The occasional spectacular light-displays shimmering in the nightly skies, visible only at polar regions in the vicinities of the 2 poles of the earth. They happen when the ions in the solar wind emanating from the sun collide with the ionosphere existing at about 100 kms above earth with the earthly abundant elements of H_2, O_2, N_2, and noble/inert gas (He) Helium]

Both, the natural wonders of the world...

Some on full display here

If one spares a minute and cares to look up closely!

"Ma-nu-ja"! (fellow beings!)

Enjoy the sights and sounds you have been missing

All this time...!

None occurs by orchestration

But spontaneously

In some marvelous scheme of things!

Relish the moment; cherish it in your heart and share it

With the loved ones & others...

And live life in peace and in perfect accord

With mother Nature!"....

The cameras were showing the scenes one by one, as the poet was poetically, accurately, and so eloquently describing. It provided a full account of the surroundings and settings of the venue, where the race is now in full swing and steadily heating up.

The commentary resumes by Sri. Paul Jose Praavady in ecstatic fervor:

(Subconsciously, his commentary of the event assumes the tone, style, and cadence of the recital by the poet we just heard)

“Camera images show paddles go into the water and come out of the water

In the trajectory of a circular arc, like a 'camshaft of a piston engine' that rotate the heavy *flywheels,* in constant rhythmic action

Also imitating the accurate oscillations of a pendulum of the grandfather clock

Lends some relevance and credence to the '*Analytical Geometry*' in the real-world application when referring to the '*locus of a pivot*' when in motion

A subject I poorly understood at the time, in Mathematics & Mechanical Engineering classes!

The paddles move in perfect unison, back splash almost negligible to none

The rear heads of the Chundans merrily bounce up and down

Responding freely to the turbulence created by paddling movements

In the majestic style of the tusker elephant's gait or that of a galloping horse

Looking from the front and towards the racing boats, they appear charging straight at, like speeding arrows

Remember, the bow of the Chundan is pointed like an arrow!

Not a suitable perspective for determining which one is ahead at the moment, but pretty nice to view the paddling style!

So far as the paddling goes, well as you are familiar with, they are seamlessly distinguished in 5 contiguous sections. They are actually 5 specialist categories of rowers; have distinct functions, and tasks to perform. All work continuously, but each one assumes importance at certain stages of the race. Accordingly, each has a specific role and paddling style. The largest number of paddlers occupy several rows immediately in front of the arching stern and the helms-men's. They adopt an 'abbreviated' (chipping) style. That's the rear most part of the chundan that makes contact with the water; where the heaviest load and drag is concentrated, and felt. The next section is for the Rhythm makers,- cheer leaders and pace setters all in one. Two of them bang on the 'soundboard' with poles of water cactus and make the rhythm in 'tempo counts' as always shown at the 'head' of the music scores. The rest of the group continuously prompt the rhythm makers by controlling the pace & frequency of the chant- expressed in the articulation of the Vanchippatt that in its turn, is dictated by the current race conditions. It's a long race; the strength and stamina of the entire paddling crew has to be effectively utilized. The second-largest section of paddlers are seated in between the rhythm makers and right behind the "Fabulous Fives". They maintain constant rhythm and medium paddling action; tasked to sustain the cruising speed. The most visible paddlers are the Fives. Their task is to accelerate and keep the momentum in the final stretch assuring a glorious victory. They are ambidextrous and have rather exaggerated sweeping action. They may change sides in smooth and synchronous style orchestrated by the leader; lean forward while making the characteristic 'sweeping" maneuvers. Undoubtedly, their roles are very critical in tight races and always in the final stretch of the finishing moments in the race. Given all the difference in styles and executions, the paddling should carry on uninterrupted in perfect synchrony, and harmony. Every action and arrangement in the boat race has a specific purpose and certain significance. Each position, number, groupings, and symbol bear a cryptic reference to one or more of the followings: event in ancient history, a Vedic citation, a cosmic sign, piety to a deity, or a thing from everyday life, and a certain phenomenon of nature. Supremely, each code is associated with the Sciences of naval warfare treatise & marine architecture.

As in the 'Formula 1' car races, the spectators must know a few things to be able to fully enjoy the races. Otherwise, it remains merely an imagery of cars going in head-spinning circular loops. In Formula 1 races, besides the driver, there are many support groups with specific roles, such as the tire changers; refueling man, wingman, jack-man; a lollipop man to safely conduct the 'pit box' activities. Importantly, there's a mission command center team who monitors many race situations and performance data, and accordingly, give instructions to the driver. They are all in constant communication with each other using telecommunication technology.

In Vallam Kali, most of the communication takes place in the rhythm of the Vanchippatt! Most of the adoring fans understand the full details, situations, and nuances about the Vallam Kali inside out! The tourists are educated about them. They are also pleased to see their country-women enjoy and participate in the Vallam Kali with the natives.

Nowadays, the 'drones' supplant the so called 'spider cam' (aerial camera on 'puppet strings')

Precede the way of all, hovering in the front

Floating like a *Ruby-throated Hummingbird* in the garden in Spring times

Vertically above and high, suspended from thin air

Carefully avoiding any adverse aerodynamic thrust,

From its multiple, rapidly spinning rotors that

Inevitably causes momentary ruffle of the headscarves of the paddlers

They do feel a '*fanning*' effect from above and

The soothing sensation resulting from it

The camera is staring straight down; provides

A steady '*drone view*'! (a characterization borrowed from the phrase "bird's eye view')

They give a sense of the actual dynamics of motion and speed of the boats

As they let the entire length of the Chundan Vallam from the bow end to the stern pass by under their steady, stationary vantage position

Also, provide a view of each row of the rowers

Other times, the drones stay still and low at a certain point

Square (perpendicular) to the direction of the boats' motion

Showing the speed of the boats

Holding a steady reference point

A lead drone with a very high zoom lens is ahead of the pack

It shows the aerial view of the entire racecourse

All the race boats, and the drones as well,

For the benefit of Television viewers

One by one, the drones make certain gliding maneuvers

As the special forces of the Air force do in exercise

Providing another spectacle for the cheering crowd

The camera image '*dissolves*' into a '*close-up*' aerial view

A full view, highlighting action details

Shows the paddling styles of the entire crew

Magnifies the views of:

The pivoting shoulders of the paddlers

The trajectory of elbow motions (similar to the to-and-fro straight-line movement of the connecting rods of a locomotive)

The shoulder muscles bulging,

The tendons compressing and elongating

The rhythmic tightening and relaxing of the biceps

The protrusions and undulations of limbs and body

In isometric, concentric, and eccentric contractions

The 'kinematics' of body movements that

Leonardo da Vinci illustrated in great details centuries ago;

Also, the determination, strain, and grimaces on the faces.

The water speckles kicked up by the paddles

Seem to linger in the air for a while, defying gravity

Then collapsing suddenly, joining the waves

Like the Wemmer Pan's musical fountains in Johannesburg!

The boats cross the midpoint

As the camera maintains the shot

Produces a clever visual effect

Illustrating a fine image, telling the whole story of paddling,

The story of competition, tenacity, and the 'will to win'!"

The commentator changes to a more moderate tempo and to a lower '*octave*' range from tenor to bass:

As a distinct technique of the paddling action slowly comes to view

The paddle blade enters the water at a subtle '*wide to close*' motion at about 80° angle (the pitch); then a series of movements: the '*catch, pull and push*' with a fluid

Follow-through motion out of the water,

Forcibly pushing water strongly backward

Compare it to Freestyle swimming, or to a well-timed,

Baseline ground strokes in tennis!

The paddlers take their paddles out with open faces, facing away from the rim.

They don't want to drench, and blind their own mates sitting behind in tandem!

They paddle in such a way, and style that the resultant force or the vector sum of all the forces is precisely parallel with the keel,

Producing the maximum efficacy and speed

The paddlers maintain a fluent rhythm and style

All is well!". Everyone is doing great so far!

The bookie from 'Bet-to win':

"Yeh Bhai Saab, su-now, Ji! (please listen), this is 700 M point,

The boat Anari is moving slightly ahead"

"Where is St. George?"

"A bit close behind!"

'So, bets are off! Good job...Ye guys watch the event carefully...sure to have a straight-out look, as I told you, without any parallax error, that's the science, the one and only science lesson you have to remember; keep it always in mind!"

"Maha Keralam seems to be struggling a bit. Being new wood, the grains haven't aligned smoothly enough in the optimum direction of motion; maybe the fine wooden surface is still a bit too rough. As often stated in racing terms, "the wood grains need to learn". This deficiency can overcome only with some aging as well as more frequent contact with the water. In fact, in preparation, the Chundan was tugged up and down the river with only two helmsmen on board to minimize the stated inherent disadvantage. People had seen it frequently plying-by, but couldn't decipher why. Some spectators, who are squatting on the quay with feet in the water, juggle water with their feet, like the football players taking 'free-kick', hoping that the current thus produced may help the boat.

Over to Srimathi Parvathy!

"The out-of-state guests, VIPs, and celebrities on the decorated "Mandapam" (Pavilion) remain somber and stolid, struggle to comprehend everything that is taking place in the race, not taking any side. The polite hosts animatedly describe to them what is really happening, the nuances, subtleties in great reverence and in ingratiating 'Ocharam' (obeisance. a local phrase for excessive reverence, an expression associated with royal sycophants). The listeners nod their heads in consent with the typical 'head oscillations' response (an animated physical gesture of agreeing or being properly understood),

Adding a faint smile in courteous agreement.

Sri. Paul Jose returns:

The crowd sees one boat advancing, making them think that it is picking up speed; they become overly excited and cheer that boat; the next moment they see the same stalling, then they curse the paddlers and call them names. They are confused by the physics of motion. When the paddles are in the water and in its '*catch and pull*' mode, the boat is in its propulsive phase; it accelerates; when the paddles come out of the water, it decelerates due to the lack of applied motive force and the stubborn resistance of 'viscous drag' acting on the hull. Not all in the crowd are science enthusiasts; they just believe their eyes- "What You See Is What You Get" (WYSIWYG!, an old computer jargon) hah.. hah.. haahh...!"

The lanes are marked through the entire length of the course by vertical posts at regular intervals carrying "distance marking to the finish line" just as the Highway/Autobahn road signs.
"Behold, our esteemed Television viewers! Please take a good look at the lines you are seeing now!". They are shown as superimposed or 'keyed-in' "virtual lines" across the lanes that are physically not there, as indicated by the term 'virtual'. The applied computer technology is called "Virtual Reality" (VR)

Similarly, a different colored 'moving line' can be seen in front of the leading race boat, showing the relative positions of all other boats. Though they are called 'virtual' they indicate real-time situations at that moment! The direct viewers cannot see them as it is not there. They miss a lot of vital race data and 'Info-graphics' displayed by the "TV Magic' seen by the TV viewers!

They tell you accurately which boat is leading at the moment, and by how much; also, in other words, currently how far behind each of the other boats are. Hopefully, next time, we can provide you with real-time measurements, as in Formula 1 races, the actual measurement in meters and the distance each boat is away from the leading boat, what is its current speed in m/s, as well as other interesting parameters.

The bookie:

"Hello! From the sprint location, only the final 100 M to go!

The race is still tightening, no clear leaders, even now, ..., not just yet!"

Haneef Hidrose (commentator):

"Every one is a contender at this stage; anyone can win this race

Maha Keralam decided to adopt the sprint style of the ace sprinter *Usain Bolt*

Sans his tendency to '*look around*'

They have gone through the drill in the training sessions

Shoulders down and relaxed, maintaining a strong mid-section

Pivoting at the hip in the "*catch & pull*" phases

Avoiding any lateral torso movement at relax phases

Arms held loosely, elbow taken back in quick 'throw-back' motion

Keeping center of gravity within and in-line of motion"

All executed superbly!

The bookie:

Well, now the boats Kavalam and Karuvatta look even, just passing by my current position, going 'neck & neck';

Two others, Anari & St. George, in close pursuit...still too close to call anything!"

"Use your binoculars!. Come on, Yaar...! Make a call !" yelled the personnel in the '*situation room*' at the betting center.

Srimathi. Parvathy: (in hyperactive tone)

"The Rhythm speeds up, they are in sprint ...oh,.. there,...look!.

The first of the trinity helmsman on the lagging boat does his trademark kick

The others follow suit

Spectators become excited,...they are ecstatic!

At the moment, ...arriving at 50 m mark....!

When the boats pick up speed, and accelerate, the bow streaks and splash water forcefully aside to both sides challenging others "follow me, partner!". The streaking indicates the speed of the '*dash*'. The ideal situation is that the boat crosses the finish line at maximum speed; it should happen at the propulsive stage of paddling. But it is difficult to ascertain such a situation. The logical option is to increase the stroke rate and sweep action; shorten the time when the paddles are out of the water, or idle. As a matter of choice, some teams prefer chipping as learned from the chickens in their own hatcheries, and others the pecking imitated from the woodpecker perching on the trees in their palm grove". [the bird can peck at 10-12 times/s; awesome!]

All have rigorously practiced, timed most efficient techniques; had come up with several options too. The maxim is: "not to relax" until at least half of the boat has crossed the finish line. Several video footages of sprinters have been shown that after having built up a seemingly unassailable lead until the final couple of meters, some took it easy, and relaxed; only to lose it all to an on-rushing follower behind whose momentum carried him/her over the finishing line ahead, in stunning upset.

The conventional wisdom says "a race doesn't end at the finish line; further beyond it!". Some boats are told to follow a specific 'whistle-cue' from the captain who is standing midway between bow and stern.

The captain decides on the tactics and strategy; he is also the coach.

He carries a whistle which he always holds in his mouth, ready to blow instantly; as and when needed. He must be very serious and dutiful in his role, as his decisions are too critical and deciding!

The rhythm makers change the tune accordingly

The paddlers must listen to the chant carefully and execute accordingly,

Lest the paddles should clash, no time is left to recover and regroup

The execution must be smooth and fluid, as an 'automatic transmission' in automobiles

They cannot afford to lose momentum and speed of motion they have so strenuously created so far

Maintain everything as it is, except for the frequency, and number of stokes

Chapter 35

Trip to the VallamKali nostalgic reminiscence

My paternal hamlet is Kainady in lower Kuttanad at the very edge of "*Pata-Sekharam*" (large blocks of Puncha Paddy fields) laboriously reclaimed from the vast Vembanad lake basin. It lies equidistant from the nearby towns of Alapuzha, Kottayam & Changanacherry. It doesn't own any race boat; has only gigantic '*KettuVallams*' (Cargo boats to transport paddy harvest, water pumps, farm equipment, tools, fertilizers, and labors) or other smaller boats for family & personal travels. Today's '*House- boats*' are the outfitted models of the '*KettuVallam*' of yester years. Most of my dad's cousins and other second-generation relatives live in the neighboring village of Kavalam. Usually, we adopt and root for the Kavalam chundan in the races. My maternal village is *Karuvatta* in upper Kuttanad further south, only 5 Km away from the shore. The region and its vicinities are the most prolific in fielding chundans in races, thanks to its central location in the erstwhile Chembakassery kingdom. As mentioned before, the race boats are relics and models of the Kingdom's marine fleet. My cousins had taken me several times to watch the local races held in the rivers and canals in the vicinity. My maternal grandfather, affectionately called "*Vallya-Appachan*" was an avid VallamKali enthusiast. He used to take his grandchildren, boys & girls, to VallamKali. It shall be proudly stated that he treated everyone equally and with respect, as well as equal affection and attention. To begin with, it was a long and boring journey to the races. The route stretched along rivers and canals, waterways around large paddy fields. Sometimes, we took shortcuts, where the locations of breaches of the levees can be spotted or known from other frequent travelers. Every time, he took his violin he acquired from a German missionary, who had to leave the country on short notice on '*Rückkehr-Ordnung*', (a peremptory order to return forthwith), from their *Führer*, Adolf Hitler. The missionary was only able to hurriedly show him the 7 keynotes in syllabic notation: "*Do, Re, Mi, Fa, Sol, La, Ti*" (though Germans & Scandinavian countries used corresponding alphabetical notations). The rest of it was improvisation, practice, and perseverance on the part of my grandfather. We sang along in annoying cacophony, as our primary intention was to drown out each other's voices. It is fair to say that he practiced the instrument regularly and progressed well with it. He organized church choir and orchestra, sang devotional songs in church feasts as an "*ersatz*" (substitute) for the cantor. His eminence in the congregation and leadership inspired us quite a great deal. We pursue his legacy of music today at varying levels of competence, proficiency, and passion. [For a while, I tried relentlessly to locate and repurchase that sentimental violin, he donated to his talented granddaughter who had joined a convent. To our great chagrin, it soon became a common property of the religious order and my cousin seldom got a chance to use it. Now, it remains lost forever. It would have made a truly memorable heirloom possession that would have remained in the family much longer. You will never recover anything you give to the godly people in goodwill and faith, a very painful and resentful lesson learned quite early in life and with other periodic reminders. I wish there were some laws pertaining to community donation that a gifted item shall be returned after a certain length of "grace period' simply because a family property belongs to all of its offsprings though given by an ancestor.-something similar to the "lease of Hong Kong"!]

Some of my intellectually inclined cousins took along chess boards, made from wooden pieces. Little ones were given pieces artfully fashioned by our beloved grandmother "*ValliaAmmachi*" out of banana branches. Inevitably, some sort of dispute would break out soon, on an unauthorized, or false move. We were never able to complete a game. It was frustrating! The artistically gifted performed "*Kadha- Pra-sangam*", a musical 'storytelling- performance' by the protagonist while the rest of us filling in the role of a chorus. They were all fans of Late. Sri. Joseph Kaimaparampan, a distant relative and the most renowned performer of the art-form at that time. Mothers had carefully packed lunch & dinner wrapped in flame-treated, pliable plantain leaves with rice, Idli, Dosai, Sambar, vegetable Medley, fried beef cubes mixed with coconut slices, and fried Mackerel. Freely, did we share items among ourselves. Some attempted to 'pick from others' portions and often start altercations; disputes; crying would soon follow. Often, the little girls were the easy victims. Older ones would try to broker and restore peace. We enjoyed the chlorophyll flavor of the plantain leaves; liked the fact that, it being biodegradable, can be conveniently disposed off without causing environmental harm. It becomes an organic manure & fertilizer for the trees!

Mothers entrusted younger ones to the care of designated older and responsible siblings. The innately naughty and the habitually mischievous received an earful of advice, and cautions beforehand. To arrive at the race on time, we helped the punter by pulling the boat with a coir rope while walking along the embankments, as he steered the boat safely. It is the same technique used to pull barges by steam locomotives using steel cables to aid navigation against heavy tide/current in the Danube river at the *Iron Gate*. (*Now,* a hydro-electric dam stands there, jointly built & owned by Romania, Bulgaria, Moldova & Yugoslavia, currently Serbia at their joint border). This was done to overcome the forces of strong current and turbulent waves acting against the movement. The chums in the Vallam boat cheered and encouraged, asking for more speed. We would get tired soon and retire back into the Vallam boat, drained and sweaty. Some Vallams tug along behind motorboats; others tried to ride on the crest of its waves. Some 'KochuVallams' (small boat) could be seen entering the wave stream in the wrong way (parallel to the waves), instead of going across, and capsized. Onlookers feared for their safety, wondered what would happen to them next, whether one should attend a funeral, or go to the VallamKali. But, they were experienced and fun-loving daredevils. They were actually making a show of it. Skillfully, they straddled in the water like the goalkeepers in water polo games; promptly re-floated the Vallam right on the spot, and carried on, on their way, although fully drenched from top to bottom.

Along the way, we learned some lessons, gained new insights, and experience; some notions were falsified; some beliefs were held in abeyance; formed bold new ideas. One is exposed to the outside world from the protection of home and the company of readily tolerant, and 'spoiling' relatives. The world can be friendly and hostile; considerate, and inconsiderate; magnanimous, or mean; kind, or cruel; helpful, or indifferent. The world consists of people with varied outlooks, perceptions, and tolerance levels. All want happiness and prosperity; choose different ways and separate pathways to achieve their often elusive goals, and dreams.

Time is fleeting! Before we knew it, we arrived at the site of the regatta. The sight & scene felt different, the ambiance was scintillating, the mood buoyed, a spirit of adventure prevailed; images & memories of the past rushed to mind. As children, we felt a soaring sensation of spirits. Collectively, we saluted one of our uncles, *Thommichayan* for manning the race boat as its helmsman in several races, albeit mostly in losing contests. But he excelled in Volleyball; played competitively in tournaments. Unbeknownst to our grandfather, he was the family's elite sportsman, and hero, although his performance adventures took place before any of us were born!

Enthusiastic spectators make their journey to the regatta by whatever means they can afford to find

I am regaining my presence of mind, returning to the present. Today, I am alone, without the company of my grandfather, siblings, or cousins. Sadly, some of them have departed from this world; a few too early; now surrounded only by strangers. Standing here, I was contemplating the events of the past just the way I described. It is crushingly nostalgic and sentimental. My loyalty lies unequivocally on the maternal side, mainly as it means much more to them. I feel quite thrilled at the prospect of her winning this race.

Now, back to the races!

The bookie:

"Listen, Mukesh Bede Bhai" (big brother, Mukesh!)

"Bolo! Bolo! Ja-ldhi se, bhai! (Go On! ...make it quick, my brother!)

The race is heating up right from the start; wresting and losing the lead frequently in the long race

“Kavalam is the smallest boat, I guess it has a distinct advantage; but others are bigger, and carry more manpower on them, just a thought,...well, it's your decision, Bhai, Saab!”.

Sri. Thommichan MalikaMukalil with his commentary:

“Now boats are at the *30 M* mark..!

Four out of 6 boats are at ‘neck & neck’; in close contention, all with explosive momentum. It's so well, so expected. After all, they are the final six,...6 fastest... qualified out of 36 other Chundans. They must be very competitive indeed!

Good sprint action can spring surprise results, you know...?

Paddlers know this for a fact, they have experienced it before

They also know the crowd is thicker near the finish

The shouts sound louder and exhilarating

There, there are fans of all the teams; most have filled up the places the fans of the non-qualified Chundans have vacated in disappointment

The closeness to the finish line raises desire, and adrenaline;

Adrenaline gives strength and courage and energy to do the unthinkable!

The desire sparks passion; passion makes efforts

And, efforts bring good results!

Suddenly arrives the *20 M* mark..!

As I speak,

St. George & Kavalam find themselves falling behind slightly

The helmsmen throw caution to the wind

All of them use their big paddles, their ultimate weapon...also the strongest

The First of the 'Trinity' does the Cartwheel kick, oh, my...!

In flamboyant style, and with striking effect

The Second ones also join in

Both in perfect timing and synchronization.

Karuvatta Chundan quickly spots it; do the same

Now, the Third ones of all other chundans venture to do the same

Leaving the two of the 5 helmsmen left with the task of steering.

The first paddle is high up in the air

The second midway in the air; the third in the water,

Each 120° apart, as in the 3 leaves of a pedestal fan, or in farmer's parlance the wheels of '*Chakram*', the water wheel.

Performing the "Kuthi-Eru" in smoothly executed rotation, and cyclic action

In this routine, one big paddle is in the water and in action every moment

A lot of force is generated continuously

One could see the effect in the boat's sudden jolt and kick in acceleration!

Maha Keralam has been always a renegade (a rebel)

Following the personality of its audacious captain

He is brash, bold, and adventurous, a risk-taker

He calls out to his fellow helmsmen

Disregarding the rhythm makers

To do the paddling all at the same time; 5 rudders; each at 72° apart

Surely enough, they throw their weight behind the leader

Together, carryout the swashbuckling *Kuthi-Eru* with exaggerated follow-thru

Occasionally using all the paddles in the water in desperation, pulling at the same time, creating a combined force, and a tremendous '*shake*'

The head of his Chundan jumps up and down as in a '*horse-back ride*'

In the turbulence of speeding waves so created

The *10-Meter* mark.., Right here now!

All boats are at the same mark at the same time!. Incredible..just awesome... Are, Bapre!!!"....'Common..."Kavalam!"

Srimathi. Parvathy grabs the Mic in great excitement.

She continues the commentary in her high-pitched soprano voice:

"The crowd goes hysterical, shout from the top of their lungs

Throw their limbs widely around, hitting other people nearby

But they don't mind, accepting that they might have done the same to others too

Thayankari is not to be out done!

They know the technique as well; they were here in the same place 3 years ago

They have helmsmen with strong shoulder and rotator cuffs

Also do the cartwheel kicks,

Like the gymnast on the mat in the floor exercise

Attempting the final '*dismount*'

Gathering momentum

The sight is so pretty, elegant and awesome

All are in the critical stage of the race

Frantic actions add to the hype and speed, all at the same time

It is exhilarating to the spectators!; Challenging to the paddlers!

Some interject exclamations

In their minds they hold tempers,

But outwardly tempered by civic consciousness

The easily irritable and agitated yell '*expletives*'; the broadcasters are too late to "bleep" the words, nothing can be done in such 'live' broadcast situations!

Right now, nothing else matters, but the race alone!

Fans erupt in 'Kurava' (rapturous 'ululations')

The decibel level reaches its climax and crescendo; the loudspeaker distorts words; affects clarity and intelligibility, also some howling audio 'feed-back'

The echoes resonate far and wide

Everyone joins in the ecstasy and in the display of unbridled exuberance

In their own natural way and manner

The moisture in the air propagates sound faster and further, '*colors*' the sound

Sometimes the sound waves, imagine, sinusoidal waves

Arrive at a certain point in phase (its crests coinciding) amplifying the sound

Other times, come in out of phase (crest and valley coinciding) neutralizing each other

Creating frequent dead or silent zones

Listeners feel the sound is hopping around the entire racecourse

Approaching towards them and then passing by, without stopping

As the Kangaroos hopping by

A sonic '*mirage*' of, sorts!

Boatmen see the approach to the finish line

The race will end very soon

Now or never; a time to 'do or die'!

They bent down to gain some aerodynamic advantage

Like the cyclists in "Le Tour de France" in their final sprint; redouble their efforts

Hope the momentum will carry theirs through the critical final length

At maximum speed, crossing the finish line

In a dream finish that they will relish the whole life long!

Some disregard the rhythm, tend to overdo a bit

Few paddles clash; some slash; some splash, drenching the ones behind

Miss a stroke or two

But, quickly rejoin the flow

Spectators weigh their stakes and chances

For many, it's not only a race

A matter of great pride and prestige as well

Not to mention the high stakes of prize money

The many things money can do for their native hamlet; better walk ways, schools, clean water, safe bridges, public comfort facilities,.. etc...etc..

On and on...reasons are aplenty!"

Some agonize, others tense, none has reason to be overjoyed, or despairing

As it is still '*too close to call*'; it's hung in precarious balance

All options open, and possible!

Impulsively, they build up some internal anxiety, and wishful expectations

Conjectures and scenarios of many kinds brew in their restless minds

Their faces reflect, and show all the gamut of acute emotions

Regardless, the spectators give an exultant cheer, applause, and shrieks of *Kurava* ululations in the top of their lungs to all the competitors. They earned it; fully deserve it; putting on a magnificent race. “Bravo!”

It's turning out to be a thrilling finish in great drama, uncertainty, controversy, intrigue, and great suspense

The race is just less than a meter to go!

If at all anyone is ahead, it’s only by a *whisker*,

Question is which one that is...

...Oh,..My... Oh, God !... h'm...h'm !

The race is not yet quite over
There is still hope left for any Chundan to win it all!
It is "anybody's *race*!"
Nobody ventures to call it one way, or another
It'll be known in the next second, or fraction of it!
Rightfully, leaves the task to the referees and judges.

Some of the young spectators jump up and down,
Like '*the greatest of all time*' boxer Muhammad Ali,
Doing his trademark '*rope skipping*', shuffling feet in rhythmic speed
Some lean over precipitously, others tiptoe for a fleeting glance of the 'FINISH'
Over the shoulders of big men
Everyone seems to do the same
Feel a gentle lean gathering force from the rear
A bump here; and another there
The thrust builds up: one trips, loses his footing, slips and stumbles,
On top of another
Find no rails to hold on to
A domino effect ensues
Suddenly, a section of the front row of spectators
Find themselves in the water
Drenched and embarrassed
Fumbling for wallets, phones, and valuables
Others ignore the fallen comrades

They wanted to see the exciting finish

That's what they came here for

The culmination and the climax of it all!

They came to see the critical moment

The incredible moment, as the glorious goal scoring moment

In World cup finals

Want to be the first to know the result

And savor it all for a long time, and brag about it all along, forever!

The spirited action reaches its climax

Heart beats grow faster and louder, raising pulse and breathing at the same time

It's awesome, exhilarating, maddening at the same time!

"Oh!my..God!,..what 'a race! Bra..vo! MaMma, Mi..a! Heileger Strohsack... Wunderbar! Faan-taastic! Maa...Shaa...Allah! Aab-so-lu-tly Sooo-perb! In-cre-dible..! A humdinger,... indeed!"

Exclamations & phrases of amazement shouted out, as in musical ornamentations of "*trill*" (vibratory or quavering effect) and prolonged "*fermata*" (music; a long held out, sustaining note) emerge from the foreign broadcast booths of BBC, Australian ABC, Sky networks, ZDF Germany, Italia network, Arabsat, RPN Philippines, and Deutsche Welle.

Emulating the blare of the Brazilian football commentator's

oo **oo**
"Go..o oo... o..l!"

The bookie:

"Its over,... finished....all over, *Bhai Saab*!

"They crossed the finish line, by a whisker from each other!"

"Who won?"

Silence... (no answer!)

Pin drop silence at the venue, even the breeze stays still

"What did you see?"

"Don't know!"; 've no comment!; that's the first time in my life!

"All crossing at the same time?

Listen to the live commentary!"

There's no audio, nor running commentary.

Everything stopped;...frozen,... abruptly!

The stubborn and sombre silence persists; it turns mysterious, and lingers on.....!

Chapter 36

THE SOMBRE SCENE

Parvathy has been talking at steadily increasing tempo, and crescendo;

And people had been intently listening to her commentary.......

“What happened, then?”

Nobody knows!...

“She ran out of breath?

Poor she!”

“Par...vathy!...?...Par...vathy!...?” her colleague is heard over the speaker system asking in grave concern for her in trembling voice

She needs to catch her breath and recover before she can utter another precious word.

It was extremely exhausting!

Only heavy panting can be heard, over her open microphone,
Like puffs from a misfiring engine, or heavy breathing of a Marathon runner who just crossed the line after continuous running for 2 h. 10.27, and collapsed on the running track; finishing the finish-line.
“Is she sobbing...?”
“Did she faint...?”

Concerned listeners and spectators are heard asking among themselves.

This is more suspenseful right now; the only subject people are talking about. Not the results!

Showing concern for human welfare and life above all!

The bookie: (in stuttering speech)

"It's total silence here;... hear, talk or see nothing,... dead silence...

Nobody knows anything;... nothing at all"

The only person amused is Lord Karthikeyan, the god of war; he has reason to laugh his head off having conjured up such a tight fight, and a close race to the finish! Simply awesome! Truly, 'The mother of all races!

Everyone seems to behave like "*the three mystical monkeys*"

Looks around blank and blurred in utter disbelief!

"Checks the TV screen!"

Suddenly, the euphoria is replaced by concern, suspense, and anxiety

A collective sigh rises in the thick air of mystifying silence

A '*deadening silence*', to say in an 'epigrammatic' (as in a pithy statement, often containing a paradox) expression.

No one ventures to speculate

Here at the present time, silence is golden!

Some commotion and ruffling of papers are heard at the finish-judges' table too.

That's all!

No one knows what's just happened, or what's really going on

Everyone sharpens their ears, like Meerkats in Southern Africa do, standing erect, ears raised and pointed; listening to any sound of an oncoming predator.
All wait in grave anxiety for the result to be announced.

Meanwhile, after crossing the finish line, the boats, make a wide roundabout turn one by one; paddle back and lineup in front of the crowd as regiments in the Republic day, performing the Parade of Honor. Now, the flag-staff and nameplates on the boats come into clearer view, majestically revealing its personality, poise, size, and shape in full 3 D respective. The paddlers greet and acknowledge the cheering crowd for their presence, and enthusiastic support by holding the paddles up in two hands across their bodies. They are exhausted, but cheerful; feel they have done their part admirably; feel sanguine about a victorious outcome. It's all done and over with. Time will tell, rather soon!...

The paddlers are overjoyed to spot some familiar faces; gesticulate to the crowd and convey various cryptic messages in their own 'sign language'. There is no hugging or embraces like in other sports.

The water creates a dangerous barrier, and prevents any physical contact!

Here, the runtimes of the races are not shown instantaneously as in Olympic sports. The traditionalists are technology and Science averse like the conservative political parties around the world. They cling to the past customs and beliefs. All the rituals, ceremonies, and practices have to be faithfully carried out one by one in a prolonged, timely manner.

Time shall also have to wait in line for them!

How was it then, shall be now, and ever afterwards!

Their hold on public matters can come lose only with the passing of time, or by other extenuating and consequential events. Often, they seem to be impervious to logic, reasoning, or persuasions. They don't listen! Media broadcasters and sponsors have been arguing for automation and innovations, as they are very sensitive to time schedules, and business/profit concerns. It's like passing legislation in both houses of the legislative bodies. There is pre- & post-negotiations, reconciliation talks, debates, before voting and fulfilling many procedures in between.

Changes can occur only at 'turtle speed' not in rocket speed!

"Things have to change!" people wish in vain.

Chapter 37

A Futuristic Master Plan

People become tired of waiting. They notice a number of people are seen huddling around a '*jumbotron*' in front of a booth. Almost everyone is curious as to what is happening there. A *herd mentality* pervades; spontaneously, everyone converges to the same spot. The place is quickly thronged and crowded.

A large sign identifies the booth as "*Bodoll Multimedia consultants*" a JV (Joint Venture) defense contractor. The company senses this as an opportune moment to divulge some privileged information; gain some exposure, and interest, as well as potentially favorable public support for its pending business proposals. They decide to play back a slide/video presentation given previously to the executive committee of the race organization, state administrative officers, the '*RDO's*, District Collectors, business entities, members of the Society of Engineers, and the race-boat captains. At the outset, the booth manager Mr. Balan BalaRaman assures that "the presentation will be paused or preempted as and when the race results become available; if you like, you can watch it on the same screen without turning away!" The crowd is reassured and settles down quietly with the curtsy refreshment snacks served by sari-cladded, smiling 'booth hostesses'. The presentation is entitled "A FUTURISTIC WAY FOR THE RACES & THE REGION".

Under the prevailing circumstances, it generates keen interest. In fact, anybody who opens the mouth will receive a reasonable hearing.

The project engineer, Mr. Jolly John Joppan, a graduate in (M.Sc.) Tech Engineering Technology from the Birla Institute of Technology & Science, Pilani, also a native of Kuttanad opens the presentation with a '*3 D animated*' slide show illustrating each concept. At the outset, he apologizes for not being able to show several other videos and footage due to time constraints. Politely, he announces that the entire presentation will be posted on the company's website in the next day, or two.

"This proposal contains privileged information; somewhat complex and highly technical; includes futuristic, electronic, display, computer, *geodesic*, and architectural components. Immediately, it spikes the interest of the technology-savvy and the ambitious entrepreneurs. Firmly, they settle in their seats.

"We distinguish between two groups of viewers. They are on-site viewers and the other Television/Internet online viewers. Although both see the same races, some would get to view additional data, info-graphics, and animation. That's because of 'virtual reality' that can be presented via the media only. The 6 lanes are marked by 'LED' light poles at every 100 m. Each lane is assigned a particular color and a number, which suggests that each pole is sandwiched between two distinct colored LED light poles with a thick opaque partition in between. Each pole is marked in sequential order. Additionally, it has a distance marker, like the scorecards hanging from the net posts in tennis courts. It displays '*Distance to the Finish-line*'" information.

Some people nod their heads in firm approval, indicating it's a smart idea indeed. As it is right now, the paddlers don't have any idea where they are, how far to go, and so on. This feature enables the spectators and paddlers to know exactly where they are. *Fantastic! Go, on...!*

"In the final 100 m stretch, there will be poles at every 20 m, marks. The light will change its color to green when a boat passes by its designated lane. The poles also serve as background and reference marker so that the fans can better determine if their boat is leading or lagging, besides providing an interesting sense of motion, and speed against the marker. At the starting docks, a 'Traffic Light' system is adopted. An individual Green laser, at wavelength 510–570 nm range, is installed for each lane. This is used as an Image detector that reflects light when incident on the tip of the bow of the boat. At this position, the boat is on the 'official start line'. Each participating Vallam is fitted, ornamentally of course, with a plate of photo-sensitive material at the tip that scatters light. Any intruding boats breaching this line will be retracted and repositioned. A red ruby laser emitting a red beam of light at 694.3 nm is installed 5 cm ahead. A Yellow laser diode emitting around 589 nm is installed 5 cm behind the Green laser. With the help of these color lasers, the race boats can be precisely and quickly aligned at the start. Electromagnetic wavelengths below Red (infrared) or above Violet (ultraviolet) are not visible to the human eye. The amazingly narrow visible spectrum range is limited from Red at 700 nm to Violet at 400 nm wavelengths.

Any radiation below Red is infrared and above Violet is ultraviolet (Frequency is inversely proportional to wavelength!). The lasers are inherently intense directional beams and form a fine straight line of propagation or path. Everyone is cautioned against looking directly into it, as it may damage the retina of human eyes, although in this case, very low power diode lasers with less than 50 uW (micro Watt) are used. When a boat is lined up, a big bright Green light corresponding to that lane flashes. A piece of pleasant, cheerful music plays, a major key of 8 notes is played in ascending and descending order. Each lane has its own musical key and its own instrument. The 1st lane plays C major on Piano; the 2nd G major on Violin, the 3rd D major on Flute; the 4th A major on Sitar; the 5th E major on Xylophone; the 6th B major on Clarinet. When all boats are lined up it plays like an orchestra in an 'Andante' (a moderately slow tempo) from one end to the other in a sequence of musical harmony!

The start-judge needs to have a musical ear! Ha ...ha.haa...!

The idea is that, potentially, the musical notes relax the paddlers and help relieve their anxiety. Well, now on the other hand, when any boat overshoots the start line, a short, but annoying discordant chord of adjacent notes is played on a blaring horn. If a boat is behind the start line, an intermittent and encouraging sequence of notes in ascending and descending arpeggio is played until it is brought to the correct start line. A laser & musical alignment of accuracy!

Having all the boats aligned up precisely, a green traffic lights up corresponding to each lane. The boats are held in position 'On the Mark'; all paddles must be out of the water, resting flat across on the laps. After a brief pause of 3 sec, a yellow light comes on for 'Set'; A 'count down' display starts with 3 seconds on the board; it is triggered, the venerable '0' is for the "Go!". At the front, a H2 (Hydrogen) torch is triggered, which spits out 6 spurts of flames in quick succession. Normally, the H_2 flame is colorless, but chemical impurities are added to make it colorful and visible from a distance, all the way to the finish line. The start signal triggers the torch, acting like a rapidly rising 'flare' going up and down in flashes for 5 seconds. Associated with it is a prolonged sound of 'horn' for 2 seconds. Thus, the start signal is both audible and visible; it causes no signal delay whatsoever to any of the lanes.

That's the glorious start of the race!

“The signal triggers 6 digital stopwatches, one for each lane. By the way, our website will post an 'App' that can be downloaded onto your smartphones 'free of charge' of course, which can trigger your own timer/stopwatch. It also displays the speed of each boat and current placements in the race. As this is a 'GPS' / satellite / Internet & communications application protocol, it triggers all devices with the installed App, as well as the 'smart' TVs installed around the racecourse. The panels can also show various information crawls for breaking news, stock market, sports, weather, billboard advertisements, etc. much like the familiar ‘Times Square display’ in New York City. In this case, the crawls are meant for emergency, security, and crowd management matters of the race.

At every 100 M mark, intelligent terminals are installed. When the first boat crosses the mark, it generates a text display identifying the distance to the finish line, similar to the navigation display on passenger airlines showing the distance to its landing destination. It looks like: (A slide is shown color-coordinated display device). For example:

Standings at the point of 900 M to Finish

Rank	Lane	Name of race boat
1st	Lane #5	ABC
2nd	Lane #2	DEF
3rd	Lane #1	JKL
4th	Lane #6	MNO
5th	Lane #3	TUV
6th	Lane #4	XYZ

The current text based data is automatically inserted in a narrow vertical column to the very left of the TV screen. In addition, it shows the leading order of the race boats at the time (or programmable as lagging order of each boat by its time difference with a ‘minus’ sign). The actual sequence is as follows: first, it sounds an audible alert, a chime; then the picture dissolves on to the full screen momentarily; then quickly squeezes itself proportionally and placed to its assigned left flank location. The data remains on screen, until it’s refreshed by the new data at the next ‘distance pole marker'. The automation virtually eliminates any human and visual error. It's an up-to-the-moment status display.

Similarly, at the finish line, Laser beams are aligned and installed for each lane following the color scheme at the start. They are modified/converted version of the invisible spectrum channels. Each forms a 'virtual Intrusion plane' as if it is a sheet of thin glass that can detect any intrusion through it by any part of the boat. The principle is the same as that in the Burglary alarm systems; the difference is that in them, they use LEDs or other electro-optics in place of Laser beams. When a finish plane is intruded, a flare-torch is triggered; it goes off on the last pole corresponding to the lane. All finish point systems are connected to super hi-speed microprocessor devices & computer systems. It can identify which lane is intruded and in what order/sequence. A, '3 D cube' shaped, slowly rotating lane display assures an all-around view of the order in which the breach occurred. In other words, if the boat in the 3rd lane has arrived first, it breaks the intrusion plane first and the corresponding lane display shows " 1st " & the boat's name; if the boat in the 6th lane comes next, it breaks the intrusion plane as second, its display shows " 2nd " & boat's name, and so on. The same trigger signal issues/transmits a stop command to the timer, simultaneously showing the 'elapsed' time or total time of the race for each of the boats. The finishing order will also be 'auto magically' displayed on the broadcast TV screen. The beauty of it is that the results are shown in real-time as, and when they occur!

In order to provide a realistic and 'immersive' viewer experience, a suite of Virtual reality features is incorporated with the existing Television production techniques. The lanes are continuously marked as in the athletics tracks, but with virtual lines. Each race boat is tagged with two 'tracking video stickers' one at the tip of the 'bow' (front of the boat) and the other at the top of the 'stern' (rear of the boat). As they are virtual only, and superimposed on the camera signals, they are visible only to the Television/streaming viewers. The stickers, seamlessly adhere to its spots and faithfully tag the movement of the boat. Generally, the sticker at the front is shaped as a visible crossbar spanning across its lane. For the Television viewers, it would seem like the boat is pushing a colorized beam when it moves forward. Each lane is assigned a specific color from the main color spectrum -Red, Yellow, Green, Cyan, Blue, Magenta. Meanwhile, the sticker at the rear is shaped like a solid flag and would seem like holding a flag that doesn't flutter. All the race data such as speed, boat parameters, dimensions, and other details can be displayed, and refreshed on it in 'real time'. Speed and other dynamic data are measured using 'GPS transponder'. This will also help to determine the leading boat to the accuracy of 2.0 mm. The leading boat is shown holding the horizontal beam spanning whole cross the entire 6 lanes. Now, the TV viewers can visualize the relative positions of all other boats. It's possible to measure, calculate and display the relative distance of each boat with respect to the leader in the race. It can alternatively show distance, lag time, speed, etc. as desired, and so programmed. Drones totaling 7 in number are fitted with multiple remote-controlled cameras, that will be used for the Television coverage. They track each race boat; frequently vary their elevation and perspective; cover the actions and excitement of the races; on-board electronics calculate the speed of each boat; can also show the view of the cheering spectators along the course. The current methods of using the noisy helicopters, their adverse wind effect on the boats will all be the things of the recent past; so also the motorized speed boats tracking the races and making waves that adversely, and unequally affect the boats. These external, obtrusive forces and elements will cease to exist. Now, onwards, it'll be all transparent; makes the races as purely' *performance-based*' as could be imagined!

Still, there is no announcement. Over the PA (Public Address) system, the organizers apologize for the delay and ask for the viewers' kind patience.

Kochu Moan Antony, comes on as the next presenter to talk about a "Futuristic plan for the Kuttanad region". He has been lauded for his insightful knowledge of the region, his researches in its vast and delicate ecology. Right from the beginning, he shows his characteristic passion, flair, and eloquence.

"A company by the name '*Roltaz*' will be commissioned to supply geodetic data of the region and to build simulation studies. They provide the critical geographical coordinates and design data for the future projects for the building of quays, promenades, public parks, restaurants waterfront residential, hotel complexes, etc. I would like to strongly recommend you to take a look at the pictures of the pristine Dutch village of *Giethoorn*, a waterfront village located in the Dutch province of Overijssel. It doesn't have any roads; is built with a system of crisscrossing canals and waterways; seems like a replica and model of

Kuttanaad we used to know until quite recently. When I saw it first, I was obsessed with a curiosity, as to which existed first, and which one was the original; who copied whom, and so on.

My boss retorted in a wry sense of humor: not be too obsessed with it; simply do your job!

A cluster of floating islands in the Vembanad lake is also in the conceptual design, each serving a particular purpose for marine research, Fisheries, Eco-tourism, Scuba diving, water recreation, and electrical power generation. The sprawling Boardwalks will have canopies with solar panels on them (slides show boardwalks in Coney Island in New York City, Atlantic City & Ocean city in New Jersey with '*photo-shopped* canopies superimposed). The standard solar, photovoltaic (PV) cells have only low '*sunlight to electrical*' efficiency of the order of 15-20%. The more advanced but costlier 'layered/stacked cells' which absorb different wavelengths (e.g.,.: blue, green, yellow lights) could potentially double the efficiency. An identical technique is already widely used in making 'multi-layered' circuit boards, *3D NAND* memory systems, and semiconductor battery packs. We must be aware that toxic chemicals such as *gallium arsenide & gallium phosphide,* among several others are used in them. Battery storage units with the 'non-combustible' **V**anadium **F**low **B**attery (VFB) (slide shows Kopsla's Lithium-ion polymer, and a pack of safer *Valance*'s Iron Magnesium phosphate, 150 kWh batteries) will supply electricity day & night. All the new buildings will be supplied by this independent, gridless system. VFB has huge storage capacity, long battery life, long life cycle, and relatively smaller size. There are other emerging storage devices called '*ultra-capacitors* rivaling batteries in charge/discharge time, light weight, and close to 98% efficiency.

Its characteristics make it very suitable for energy backups and application where quick power release is needed as in acceleration.

A *Renewable Energy Complex* (REC) will be designed and built. It will consist of various, robust, 'tried & tested', proven technology combinations of solar, wind, waves, tidal current, and salinity sensitive devices. An array of wind turbines will be installed to take advantage of the constantly available wind forces coming from the Arabian Sea. Wind produces waves too, which is stronger and denser than the wind! The waves and tides are harnessed to drive turbines. Additional electricity can be produced. Interestingly, a Danish/Norwegian company has shown some prototype products in this category. In addition, there will be offshore floating solar panel dinghies anchored in the sea as well as in the lake. *The Maldives* is said to have plans to implement it. Our neighboring little island nation has demonstrated certain interesting steps to protect and sustain environment. Obviously, they are most vulnerable to the 'climate change' disasters. We must all salute and credit them for taking precautionary and proactive measures!

The platforms will serve multi-purposes of producing electricity, housing Tsunami warning sensors as well as beacons for fishermen. The panels will have an anti-glare coating, making them seamlessly blend into the landscape. The electricity produced will be brought ashore by marine grade cables laid and anchored to the bottom of the sea floor, much like the undersea *fiber-optic* cables already in use for Tele-communication networks. Solar panels produce electricity only when the sun is shining. Any excess of energy produced in a day, over the normal consumption, can be stored in batteries for nighttime use or during cloudy/rainy days. Just imagine the scenes of many paddy fields with floating solar panel platforms when not growing any crops, instead of collecting "*Payal*", Hyacinth/moss, and invading the entire area! The electricity producing platforms could be used as electric plug-in points on the spot; can be used to charge electrically operated agricultural machinery, water pumps, electrified trolling boats, dredgers for collecting 'Payal' & algae using large nylon nets, etc. In the process, we may catch some fish too! Housewives will love to cook them as much as the family love to eat them. There are numerous natural resources readily available here to put to multiple uses, clean environment, better living conditions, and for the public good. We only need to apply our imagination, and allow technology to work for us. We can achieve the goals of carbon reduction & carbon neutrality helping us to ward off the ominous 'existential threat' facing the world at least by some more years! Furthermore, we can temporarily shut off our hydroelectric dams one by one to undertake the overdue maintenance & rebuild plans.

The much cursed "*current cut*", or power outage, will be a thing of the miserable past. No one is going to miss it!

Burglars, prowlers, and criminals are the only people who will resent these changes; certainly, the general public won't. What would you say, Ladies & gentlemen?"

" Like it, so far; we'll drink to it sir!

But, who'll 'bell the cat'?...we mean who'll convince the hopeless ministers?". The front-row of listeners murmur among them thoughtfully.

Renewable energy schemes to alleviate the much cursed "**current cut**"

Once, there were fishermen tribes who survived through generations solely on fishing. Lately, due to industrial effluence, pesticide usage, and indiscriminate polluting activities, their occupation is seriously threatened, if not entirely terminated. Practically, they are driven out of their livelihood. Now, we find an undesirable new breed of aquatic vegetation and 'water Hyacinth' drifting around, polluting water, killing fish, affecting agriculture & cultivation, as well as impeding regular water traffic. They carpet the surface, cutting off vital oxygen supply from the atmosphere, and blocking the sunlight. The much-needed oxygenation process for marine life is severely inhibited. The shutters of ThottaPally & ThannierMukkam prevent the entry of salty water into the lake, which did in reality restrict the growth, and spread of '*payal*'. The freshwater fish needs some bit of salinity in the water for their prolific breeding. Under the current conditions, the fish move away; many perish; some species become extinct; fish population dwindles, there is not much left to catch! Fishermen and their families go hungry. They have to go elsewhere looking for work and subsistence; tragically, become migrants in unknown cities facing hardships. We can prevent this human catastrophe, if we take the proposed actions. We can build sustainable ecosystems that promote health, wealth, jobs, and prosperity! It can revitalize traditional rice cultivation and make farming profitable and appealing again. But, if only we act, right away, and maintains the intensity of purpose!

Pollution kills coral reefs, fish, and many marine lives; causes environmental degradation & a number of health hazards

There are certain severely debilitating diseases caused by industrial pollution. One of them is the horribly dreadful '*Minamata*' disease that prevailed in the North Japanese coastal town some 50 years, or so ago. It must serve as a grim reminder to the whole world. It was caused by severe mercury poisoning from a chemical factory on land. The local populace consuming contaminated fish (Cod, Tuna, Crab Mackerel etc.) varieties caught from the nearby seas were seriously afflicted by muscular, and neurological diseases incapacitating many of the inhabitants in the area. Heavy metals such as *Arsenic, Beryllium, Cadmium, Chromium, Lead, Mercury* are contained in gadgets and items in everyday use such as electronic circuit boards; Batteries, construction & automobile parts; mining, smelter, welding operations, etc. besides naturally occupying in soil, rocks, and water. It gradually finds its way, and seeps into the water bodies through different means, paths, and ways.

We hear about a river turning into crimson color in the near-Arctic region of Russia where a Nickel processing plant is located. There are rivers in the world which have become yellow, suggesting an excess of sulfur content. It can even happen in a region where there are few or no industries operating. The CO_2 emission witnessed few decades ago from Lake Nyos in Cameroon is a grim reminder of this. Living beings in the vicinity, including 20,000 people, cattle, wild animals, birds, and insects perished overnight in the catastrophe. The dawn revealed the horrible tragedy never witnessed in human history in a single overnight disaster. These incidences remind us of the delicate fragility of nature!

Oceanographers say the ocean is "*plastic fantastic*", with unacceptable levels of disintegrated plastic particles (microplastics). They enter the food chain via fish. It is an extremely serious health hazard! They also cite the "*acidification*" of Ocean waters which destroys coral reefs. The reefs in the vast region comprising the Philippines, Indonesia, Malaysia, and the many islands nicknamed "*undersea Amazon*" is severely threatened where many inhabitants survive solely by fishing by diving deep into them like our clam fishermen in Kuttanad. These findings and discoveries only emphasize the tremendous need for monitoring, vigilance, prevention, and proactive strategies. Hydraulic '*fracking*' in the US utilizes cocktails of a few hundreds chemicals depending on the geology, metallurgy, and other considerations. After use, they lay exposed in open vats or wells with almost no cover or shield. The noxious fumes permeate into the atmosphere. A good quantity seeps into the aquifers. There are several cases reported in W. Virginia, USA where pipe water becoming combustible, posing potential threats to humans, dwellings, and possessions. Suddenly, people fall sick with dreadful diseases. The fracking regions experience unusually heightened incidences of earthquakes. It is confounding that the corporations and the governments undertake only minimal, token mitigation efforts just enough to justify, mollify, brag, and deflect attention from serious effects & consequences for the environment. Generally, they turn a blind eye; justify as needed to create jobs, totally forgetting the fact that their children will

also need jobs. This group of people treat guns and gas as the most vital things in life, ignoring air and water. There are some irresponsible political leaders who vehemently deny that "*global warming*" is true and happening. They feign ignorance to Science, logic, and clearly visible signs. They are deceitfully serving their masters, the fossil fuel based energy corporations. Pharmaceutical industries prosper as they are presented with opportunities to develop and market drugs for entirely new diseases. Momentarily, they forget the fact, that they also breathe the same air, drink the same water, and consume produce from the same environment. During the Cold War, people feared that the earth would suddenly become a '*nuclear wasteland*'. Now, the overriding fear is that the planet earth is quickly and painfully transforming itself into an '*industrial wasteland*' wiping out the entire humanity, and posing a real '*existential threat*'! The animals would sense the danger and move away; the birds would fly off to relatively safer locations, while mankind wait to sell their houses, close bank accounts, and wait till the end of the '*playoff*' games. The '*modern world*' loses sight of what's important, and what is trivial. An old saying might describe our collective attitude: "W*hile Rome was burning, Nero was fiddling*". Here, Nero is each one of us! Besides persecuting and executing Christians, Nero set fire to a large piece of land in Rome to build his palace!
Attitudes of some of the world leaders are not any different!

We should not allow rich countries to form our consciousness and dictate our lives. Any, who trashes the environment or nature are heading for certain doom, and disaster. The power of nature is infinite. It survives everything and lasts infinitely longer than anything else. It existed much long before humanity arrived. It will last much longer after the disappearance of humanity.
We must always keep that absolute fact in mind; live our lives accordingly!

Excessive toxicity affects all biological systems and enters into food chains, organisms and thus to humans. The toxicity of water needs to be continuously monitored. This is an urgent matter of health and survival; undoubtedly,it merits everybody's immediate attention. Hopefully, this project and research will focus on these important matters. It may help to find expedient solutions. If we succeed, we will have clean air, water, and better health & living conditions. People will become healthier; live longer, and happily. Fishing will be revived; fishing industries will return. It may also add to the beauty; help maintain and preserve the biosphere of the region which is good for everyone and everything. A recent serendipitous discovery showed that CO_2 can be directly converted to '*ethanol*' using a particular '*nano spike*' catalyst. It will be a boon to the efforts of controlling atmospheric as well as '*Greenhouse gases*'. The CO_2, Methane (CH_4), nitrous oxide (N_2O), fluorinated gases, etc. are all harmful greenhouse gases produced by rampant and inconsiderate human, as well as industrial activities.

A 3D scaled model will be kept at the municipal office and shown on its Website. The public, developers, and business community can freely avail themselves of these resources for planning and building in a responsible, and sustainable manner".

"Well, As we are still waiting for the results, I guess we could take a few questions in the meantime!"

A gentleman stands up; introduces himself as Mamman Marthomachan, a writer for the "Business Economist". He inquires: "sounds very interesting, how would you propose to finance this?"

The presenter: "oh!, well!, that's on everyone's mind; is also my CEO's favorite question; he includes the same question, inadvertently of course, even in his greetings, also the subject matter of his Emails. Our CFO (Chief Financial Officer) Mr. Ramesh Phulwari" is the best person to address this question". He hands over the head-worn microphone (similar type popularized by the Silicon Valley executives during their company's annual investors meetings, and subsequently became the de facto trademark of stage performers).

"Well, thankfully there is a number of ways-public, private, a combination of the two, lease, outsource, "*Build–Operate–Transfer (BOT)*" to name a few. This is an opportune time to borrow and invest, as the global interest rates are low, steady or declining. Since the Dollar/Rupee rates trending favorably in a way that for the same amount of Dollar more rupee is available. Alternatively, less Dollar amount needs to be borrowed. Money could be available from international financial institutions (IFIs) such as World Bank, IMF, NDB (New Development Bank, recently created by the BRIC countries head-quartered in Shanghai). Issuing tax-free bonds is another option. It will be quite attractive to citizens for taking '*part-ownership*' and realize a steady interest income from it. It will be a good investment boon for Provident Funds/Insurance companies as well. Before you borrow, you must consider how it can be repaid. In this case, a stream of revenue sources are created, namely, utility charges from the electricity produced; taxes from additional business & commerce, employment, Real estate, Banking, Investment transactions; excursion & Eco-tourism, wealth creation; excise duties from the export of Fish & marine products. A host of ancillary products and services will be formed; operate and proliferate in a symbiotic manner. Facilities for marinas, boating, sailing, water sports, scuba diving, etc. can be built. Citizens will be gainfully employed. All these mentioned activities can be built and served in a very environmental-friendly, and sustaining manner. Prosperity will return to the region and state!"

Wonderful ! A murmur escalates in a section of the audience.

People sense a rare opportunity for an engaging, and exploratory conversation here; begin to voice their candid opinions:

"It will be a bonanza for our ministers and politicians. Right now, they make money out of nothing- liquor licences, permits, and road contracts only". "I am sure they will go all out for it; will use their power and influence". "They are writhing in mirth at the vast potential for their '*cut & share*'; gain & favors"; "must be gloating at chances for widening their influence; tightening their stranglehold on people". "Oh, Lord, the magic of the 'rules' always favors and works for them!"

"They have been always cynical and deceitful towards people...except at election times!" "There will be a huge pool of money for everyone to take home, from the ministers to the peons, the doormats, and anything in between, Ha,, ha ,, ha..".

"The horse-trading, pardon me! Rather 'cattle trading', nepotism, betrayals, switching party affiliations, deal brokering, influence peddling, cheating, vendettas, etc..etc... It'll be another "Great Eastern Circus!"

A banker interjects: "You don't have to borrow from banks. It's all right here. Imagine! The black money, bribe money, unaccounted money, extortion money, contract '*execution*' money, drug money, investment money, NRI contribution, etc.. etc. It will all be oversubscribed the same day, like a blockbuster IPO (Initial Public Offering) of recent times".

A senior citizen who was an active freedom fighter quipped: "This country, our God's own will instantly become '*Devil's own*! Oh..hoh...Before it happens, I want to remind myself of Mahathma Gandhiji. May I invoke God's name just one more time.. "Oh!, God!". He uttered his opinion in obvious skepticism.

A former official of the Kuttanad Karshaka Sangham (Farmer's Union), a stalwart organization for the farmers' causes, Sri. Aappachan Ambili-Ammavan, a prominent and traditional farmer, who is known for his fiery temperament &

outbursts, stands up and says: "My son, who is a Chemical Engineer in Switzerland told me that we can produce electricity from saline water by something called "*osmosis*". The ideal place is a confluence of saltwater and fresh water. It requires a '*nanopore*' membrane or isolator in between. He pointed out that this condition exists at the estuaries of the Vembanad Kayal and Arabian Sea. As I understand, the salty ions have electrical charge, which can be captured by this chemical process, and be harnessed. It is available all the time; day, or night; wind, or no wind; in that respect, better than all other forms of renewable energy technologies. According to him, ... God bless him! There is energy everywhere in nature in different forms. No wonder our forefathers worshiped each of them. The key to make them useful, is to harness them! From what I heard so far from your presentation, everything is at our doorsteps. How could we **not** see it, how could we **not** feel it...and ignore it all this time?... Yeh?.. Now, look, I have some acreage in "*R-Block*" (a reclaimed Pata-Sekharam, intended for all- season cultivation) invested borrowed sum of monies and there are folks like me;....it is 1400 Acre block, all submerged, crops destroyed, the coconut palms look like trees in cold countries in winter, pale, and yellow; without leaves. Since when have coconut trees, the tropical trees became *deciduous*?; also the nuts too; they fall like rain from above. There is utter devastation; hard to bear the sight. There is no electricity available to run the very few pumps left that may still be working. Remember! We had electricity for over 70 years now. Obviously, we regressed, worse than stagnation. You said it is all available where we need it, but these guys don't know that; if somebody tells them they won't understand it. We had had almost 20 governments with parties alternating. What a calamity they have presided over so far! Instead of running around crying, shrugging shoulders or asking "*what can we do?*; *it's the nature's curse!*" We must begin to ask what humans can do. Just look at the technologies you have articulated so far! If we do something, perhaps not us, our children may start to benefit. We need food to eat as long as we live,... Don't we?. We are too eager to build the so-called "*Info Parks*' in every town at the expense of Crores of Rupees, while all these years, neglecting to build storage facilities, sheds, pumping stations, and transport facilities for farmers at a minuscule of these outlays. They levy taxes for everything; even for breathing nature's air! '*Gift of the gab*' is the only gift our leaders posses from all the plentiful bounty our good Lord has bestowed on humanity; they only talk a lot of nonsense;...utter gibberish! no substance, or solutions!.Ignorant! All of them must be summarily dismissed! Let them go to hell!... They serve no purpose to humanity!

I loath seeing them, no point in talking to them,.. Humbug...they don't care!"

Sri. Avirachan stated his case candidly expressing utter disgust; his eyes visibly wet; voice trembling towards the end; that only stopped his outburst! People who are listening feel a great deal of sympathy and utmost empathy towards him. In fact, many are suffering from the same woes and predicaments. They are all farmers! Agriculture is the only occupation they know; today's boat race is their sole fun, and pastime.

Everyone stands up in loud encore applause. They fully appreciate his emotional, valiant utterances, and speech, as well as the poignant points he raised. Pity him and his plight!

The booth organizers suddenly become alarmed, they are astute business people. Their business interests weigh the foremost in the whole exercise, but it is taking a bizarre turn. It has disintegrated into a passionate public discourse venting frustration against the powerful government. Now, the audience is stealing the show! They know that their proposal & project shall be ultimately decided by the ministerial cabinet; not the public. Lots are at stake here. The prevailing situation doesn't bode well for their cause and intentions; quickly consult with the Headquarters in New Delhi. They are instructed to wind up the presentation in a civil and '*face-saving*' manner, adding that the company needs public support as well. It's all a part of typical, cunning, and contriving management tactic, dubbed '*doublespeak*', a subliminal comment. If something goes wrong, the hapless junior executives are victimized; on the other hand, if the opposite happens, the top management cadre claims victory, and cite their superb vision & insight, exceptional decision-making skills. Immediately, they will ask for a bonus. As usual, the subservient board will promptly accede to their demands. That's how corporations operate, everywhere; no exception here. The announcement of race results cannot happen too soon for these young, bright band of junior executives!

Mr. Phulwari continues: "Your point did resonate well among those who are assembled here, as we see, loud, and clear, quite dramatic, but horribly tragic! We do sympathize with you and all the others affected so badly. The case may best be addressed in a different platform, or a media event. I might hasten to add that, in a different way, you have also highlighted the benefits of our proposals. We are gratified! We are truly hopeful that they will bring about the kind of positive changes you all want to happen, as we have painstakingly, and clearly envisioned and presented here. Furthermore, we must strive, adapt and constantly innovate!"

Let's move on, now!

The audience feels the elevated excitement; begin to show keen interest and want to understand all its details. The '*naive*' see it simply as an opportunity to shine in front of their peers and be the first to divulge the plans to others. The business-savvy view it as an '*inside information*' of great value; can make money if acted upon, instead of ignoring it. The first instinct is to buy some properties in the area as soon as possible. They know it is a sure bet, and for a fact. Some ministers who gave approval to the new Kochi International airport did the same, built luxury hotels and resorts; made ridiculous sums of money using the privileged, and confidential information. It is still fresh in their minds. They believe that they have grown smarter too; have learned from the master swindlers; fully understand the aforementioned paradigm of the new *crooked 'world order*'. Within, they gloat in glee and are excited about the new opportunities in store for them. No one shall know the secret plans until they are fully executed! It's a sure bet. That's the plot, a pot of gold! Ha...ha...haa... Come on, bring it on! Let's play it fast; make some money like the ministers & party leaders do!; yeh, we've learned it too!

"This region has an extensive network of estuaries, rivers, lagoons, and canals with scenic settings, but has a fragile ecosystem. Well, ladies & gentlemen, we are not aware of any active water management, dredging, or flood mitigation/ remediation measures undertaken around here. During monsoon rains, erosion of soil from the highlands flows down constantly. The gathering of 'alluvial soil' (a deposit of sand, mud, silt, or gravel formed by flowing) makes the river beds shallow; pollution accumulates, species die away. Life stagnates! Dredging is needed; absolutely! It takes 'Water stewardship' to manage the situation. All river managements do. It is a common sight in the Danube river. Stones, sediments, and other materials descending from the Alps and surrounding areas, flow far downstream into the river clogging waters, causing harm, and affecting navigation. The dredging regenerates life-sustaining forces, cleans the lake bed & water; deepens the floor, helps to hold millions of liters of more water, thus mitigating flood, by raising the tipping point of catastrophic level a bit higher. If we begin to do these things here, it will become a 'freshwater sanctuary'. The ecosystem heals. Stocks of Lobsters, Mullet, and marine species will return, magically!

If they don't come, we can deposit some; they will spawn their eggs here, and proliferate. Fishing will once again become profitable!

Underwater fauna and reefs will grow back, and thrive!

Remember! there are lives beneath the water. They are most essential too. After all, life originated on earth because of water, shown by the fact the scientists are always searching for water in other planets to determine its viability for life.

The flourishing flora & fauna of the ecological community of the region

Within our sight, lie blocks of paddy fields, groves of palm trees. Some of our ancestors with great imagination, plans, and determination embarked on these massive adventures, and undertakings. They helped to develop and cultivate new land all across the vast lake basin. Surely, it didn't happen easily and overnight. Ladies & gentlemen, we have to draw inspiration from it; should

not let that spirit of adventure and entrepreneurship die away! We need to employ these qualities to survive, live and prosper. Each generation is responsible to preserve and maintain the legacies of the previous ones!
"Any more questions at this point?"

A person has been eagerly waiting in line. He introduces himself as Haridas Kuberan, an environmental activist. He had participated in the 'Save Narmada Movement' alongside Miss. Arundhathi Roy. He asks: "Will there be any displacement of people?. What are the undesirable ramifications and ill-effects?"

"Quite frankly, None! Also, quite emphatically, none; none at all, Mr. Kuberan! This is a feature-rich, futuristic & Eco-friendly Master plan, a project with a specific goal to protect, preserve, enhance, beautify, and sustain our ecosystems. Right now, it is in dire neglect; in disrepair; polluted, and in appalling conditions. This proposal addresses major issues and offer expedient solutions. The quays and boardwalks are built into the water just the way some homeowners now extend their bathhouses /cabanas from the river banks. No displacement of people is envisaged. It will promote health, hygiene, sanitary conditions, and quality of life. It may help to eradicate many waterborne diseases. Firstly, the lake-banks will be fortified with non-toxic, carbon based, organic materials. This project is designed to utilize locally available, earth-friendly materials just the same way as our ancestors did, without competing for raw materials from other daily needs. Thus, the effect of inflation will be minimized. It will be a fabulous area or village complex for living, and working; offering many amenities, pastimes, and convenience of accessibility. It will have a great impact on many aspects of life and work. Further, it will revitalize the entire region".

Miss. Diana Ann Danapady, a charming lady, is a graduate engineer from the College of Engineering, Thiru-vanantha-puram. She has a question:
"Would you kindly describe the entire process and sequences involved in a typical project undertaking of this size and scope, for the sake of all the engineering professionals?"
Mr. Kochumon comes forward and delightfully takes the question.

"Well, the project is first conceptualized taking into consideration the objective, scope, budget, complexity, components, and resource requirements.

We adopt a concept called "Da Vinci doctrine", not to be mistaken here with the enigmatic '*Da Vinci Code*' movie film! Leonardo da Vinci is still one of the greatest persons in history. He invented many concepts in Engineering, drawings, and modeling, among a host of other incredible and valuable contributions to the world civilization and modern-day innovations. He analyzed many things, including the anatomy of the human body; organs, limbs, ligaments, their geometry, and the animated sequence of motion. Likewise, he conceptualized devices and contraptions, as to how they might work and function, including those for flying, machines, and aqua-ducts. He made geometric & mechanical drawings; created models and worked from those starting points. Thanks to his enormous contributions, CAD (Computer-Aided Design), CAE (CAE or Computer-Aided Engineering: a term used to describe the procedure of the entire product engineering process, from design, virtual testing with sophisticated analytical algorithms to the planning of manufacturing), and a specialized branch of Mechanical Engineering called "Kinetics & Kinematics", came into being. Originally, Leonardo da Vinci studied relative motions initiated by the tendons, ligaments, and muscles in our body, later applied to levers, gears, motion couplers, biomechanics, robots, prosthetics, and ultimately to any mechanism involving a motion. Led by his inventions and inspiration, design, fashion, art, and handy-work flourished in Italy. Another person we regard in equal esteem is Pythagoras. You, see, it's drastically different from the concept of quick wealth-making by 'financial engineering' schemes on Wall St. We need to know that these great people examined, studied; exercised diligence; made models to fully understand the Science and technics. It wasn't a divine revelation at all, but a lot of perspiration and determination all the way in their pursuits of knowledge, and discoveries.

Time to time, the world produces a few genii, and visionaries with such supreme qualities!

Diligently, we apply the methods & concepts shown by these great people in our work to greater success, so much so, we honor them by hanging their portraits on the walls of our design studios.

Well, most of the work is done in the office on paper and computers. To begin with, we do design calculations; compile specifications; create drawings using 3D CAD (Computer-Aided Drawing) as well as CAE (Computer-Aided Engineering) programs; build prototype models on computers; then physical models with simple craft & modeling materials; compare various designs; make equipment and material lists; write documentations & RFP, ('Request For Proposal'), coordinate various departmental activities; find outsourcing partners. A timeline is developed considering various delays, environmental impact clearances, and complex approval processes. We do time and capital outlay allowances for 'Force Majeure (events & effects that may occur that are beyond control or imagination) scenarios. We ensure that the 'ESG' [Environmental, Social, and Corporate Governance] criteria are met.

A feature rich, futuristic & Eco-friendly Master plan to support large scale developments in Vembanad region

Besides, we have to do some testing of materials and components in labs, although the manufacturers usually provide these specifications. The goal is to do as many things as possible in a controlled environment, i.e, in the comfort of the office itself. We plan, develop, and document a systematic sequence of operations, step-by-step procedures, much like writing an 'Operations Manual' or writing a computer program application. It is quite important to pay attention right from the design stage to all the details such as diagnostic procedures, periodic maintenance tasks, as well as facilities for the upgrade and expansion in the future. We don't build anything which cannot be repaired, replaced, refurbished, or recycled. Wasteful is sinful! As taught us all right from childhood. The final work is carried out in the field. Every conceivable and minute detail is addressed beforehand, described, and illustrated in elaborative drawings and documentations. We create animated sequences and figures to simulate conditions and aid in the installation, assembly, and construction. Further, we adhere to and comply with all ISO / industrial & (UL) specifications, and requirements. By the way, UL is a standards' developer with versatile scientific and testing expertise. DIN (Deutsche Institute für Normung) is the German equivalent. These standards help us to build a safer and sustainable world and freely interchangeable parts. Generally they refer to safety, compatibility, by setting specifications of: materials, components, and accessories. We choose the best available materials in our projects, such as Vanadium steel that is lighter, stronger, durable, and rust-proof, for structural works. Besides, we conduct our own exhaustive "Strengths of Materials" tests in stress, strain, tensile, maximum & ultimate strengths, destructive loads, considering natural, dynamic, and constant forces each component must withstand. We prefer to use Additive Manufacturing (AM- a Desktop 3 D printing method of adding materials as needed, in layers rather than by cutting off from a large piece, and trashing the remnants); use standard replacement parts as much as possible. Theoretically, everything must be clearly spelled out; exactly what is to be done, how it is to be done, and in what order. Many meetings, discussions, and brainstorming sessions are held at every stage. Our axiom is also the same as that of a good tailor, "measure twice; cut once!" Then, the Engineers and lawyers collaborate to write bids & contracts. The lawyers include major clauses for 'F*orce Majeure'* (extraordinary occurrence beyond control) scenarios, 'indemnifications' for delays, change-orders, damages, liabilities, litigations, etc. Cost accountants and auditors recalculate costs, item by item, and 'future market contract rates'. Bids go out; submissions are carefully evaluated; costs & expenses are tabulated for comparison. A capable winning bidder or several specialized entities with impeccable track records are selected, vetted, and entrusted with the work. Our engineers closely supervise the work, coordinate different aspects, and specialties; finally, commission the project.

In the end, one of the ministers is brought in with great fanfare, and entourage to '*cut the ribbon*' in front of the cheering public. He/she takes the whole credit! Before leaving the scene, he/she makes sure the plaque is affixed showing their name immortalized on the structure, correctly spelled, and several photos taken in front of it!

This is what happens in a project, in a '*nutshell*'.

Was the explanation helpful?"

He gets a response with an affable smile, and continuous nods of satisfaction.

"By the way, over my headset, I hear; our HR would like to talk to you, if you will, Miss. Donapady!" Kochumoan added in a follow-up request.

HR is always looking for people with initiative, good imagination, and intellectual curiosity.

Dolly Ann is pleasantly surprised, judged by her body language and mannerism.

Mrs. Nalini NallaThumbi, a research scholar at Kerala Agricultural University, Vellanikkara, Trichur, asks "Agriculture plays a vital role in our local economy, especially the coconut tree. It is baffling and frustrating that all these years, why haven't we invented any contraption or device to pluck-out coconuts from the trees?. Our researchers have not found a cure for diseases like "Mahali", characterized by the shedding of female flowers and young coconuts. Spraying of '*Bordeaux mixture*' on the crown does not seem to be effective. Various bacterial, fungal, and Phyto-plasmic diseases are destroying our '*money*' trees! Have you got any bright ideas in this matter?

"Well, '*Necessity is the mother of invention*'. Necessity is there; inventions have not yet caught up with it! It's disappointingly lagging.

We have to acknowledge these simple facts first. Innovating is a spirit and a culture that seeks new things, and methods. It has to be cultivated, promoted, and nurtured. Innovations and inventions are the products of research, incentives, and dedicated work. It takes capital, know-how, time, recruiting uniquely talented, and dedicated personnel without showing any nepotism or favoritism, providing funds, building prototypes, testing, trials, and outsourcing specialized/expert technologies. We must convince ourselves that these are the necessities. It takes visionary leadership with a good imagination, purpose, focus, determination, self-motivation, and conviction to pursue something of this magnitude. It goes without saying that these essentials are dreadfully lacking in our society. We need to collaborate and partner with other coconut-growing Polynesian, Caribbean islands, East-African, and Central American countries. Almost every one is affected. We need to share their experience and remedies; try out different seeds, breeds, and hybrids. By the way, I heard recently that a hybrid of 'Malayan Dwarf' & 'Panama Tall' shows some promise. It is somewhat clear that this plant disease is spread by pathogens such as bee, wasp, hornet, or Phyto-plasma (tiny, parasitic bacteria growing within the plants carrying infectious pathogens) and spread by 'sap-sucking' insects. Incidentally, it occurs to me that once, I was having a conversation with an American ornithologist on a flight. He was on his way to Kochi, and onto the Sahyadiries to do some study on Indian woodpecker. He told me that the 'pileated woodpecker' in some N. American forests is relatively a massive bird, chisels out holes pounding its long beak into the tree trunk, looking for larvae of insects, beetles, and ants for its food. It may be interesting if our researchers can bring a few and keep them in a tight environment of large nets, or enclosure within a small coconut grove, and find out if they can help. Quite funny as it may seem, at this point, we must not leave any stone unturned, Mrs. Thumbi!

Well, in any case, research and experimentation are the prime avenues to be explored, and it has to revolve around *molecular biology,* and *plant pathology*. All the findings must be published and widely discussed in depth. It may result in developing effective pest control formulas or methods. Besides, soil tests must be undertaken to determine any deficiency in some way that it may not be getting enough nutrients to keep them healthy and fight off diseases. Is it not the case with humans too?. Well, everyone is as frustrated as you are! By the way, Srimathi. Nalini, thanks for bringing up this critical issue. Certainly, it has been an unforgivable omission, and a big shame for all of us. At this late juncture, we cannot afford to stand still and choose not to do anything at all. As a country, we have not yet earnestly recognized and embraced the importance of research work. We lag way behind other nations. To remedy this backwardness, we must begin to provide and equip at least selected regional schools with good labs and workshops to cultivate an early interest, and build a culture to do exploratory/investigative work. Our education has to become more *heuristic* (an educational method in which learning takes place through discoveries that result from investigations made by the student) based on critical thinking and experiments, drawing inferences, and so on. In order for that to happen, we need to first encourage fellow citizens to elect visionary leaders with sagacious minds, impeccable integrity who possesses enough intellect, and education to understand the necessities; having an ardent desire to do good things for the people, and country. Hopefully, people understand, that they cannot progress individually. The whole society must progress together as a unified nation to enjoy the goods.

We must persevere to succeed! That's the keyword here, ladies & Gentle men!

A team of a capable, experienced workforce with qualifications will be essential to do this kind of task. The *onus* (a burden or obligation) is on ourselves! We need to *hortatorily* (in strong urging) ask everyone to do their individual part in solving common problems; besides, to exercise their franchise in a serious and responsible way. Well, you see, I must stick to my script! Reluctantly, of course, I must leave it at that, after simply adding that the goals you raised are definitely achievable, within the realms of capabilities vested in every one of us!"

In conclusion, let me recite a few quotes from The Late. President, Dr. Abdul Kalam which I find very powerful, *sagacious* (*showing acute mental discernment and keen practical sense*), and meaningful in this very context. Its power is validated by the innovations and achievements the DRDO (Defense Research and Development Organization) has achieved under his leadership in spite of occasional failures and disappointments. Unfortunately, that's the arduous path of progress and success.

- "Excellence is a continuous process and not happening by an accident
- We must think and act like a nation of a billion people and not like that of a million people. Dream, dream, dream!
- You have to dream before your dreams can come true
- My message, especially to young people, is to have the courage to think differently, the courage to invent, to travel the unexplored paths; the courage to discover the impossible and conquer the problems, and succeed. These are great qualities that they must work towards. This is my message to the young people!
- Never stop fighting until you arrive at your destined place-that is the unique you! Have an aim in life, continuously acquire knowledge, work hard, and have the perseverance to pursue a great life!
- We should not give up; we should not allow the problems to defeat us!"

Quite inspirational, motivational, and forceful utterances!

"Again, the onus is on you, yourself! You are the protagonist; let initiative starts and emanate from you! Lead from the front; champion the cause! Wait not for others! You can do it, and must do it; eventually, others will follow!

That's thc way it is!, Srimathi. Nalini!"

Suddenly, it comes to my mind a song, we used to sing at the end of Sunday mass, a valedictory (a closing or farewell statement or address) and devotional hymn "*Let there be peace on earth and let it begins with* ***me!***" It ends with a word from the priest, "Go *in peace and serve one another!*"

Start with one, and that's **you**; that's how we make our world a better place!

People are seen waking up to certain new-found realities. They find the presentation not only as a business project, but also passionate, inspirational, futuristic, thought-provoking, even uplifting; deeply philosophical, as well as *altruistic* (Unselfish concern for the welfare of others; selflessness).

Remarkable, challenging, and at the same time inspiring!

Philosophy, a pillar of thought, has a significant role in all ancient cultures, Greek, Roman, Chinese, and Indian!

Sri. Thampuran Thamara-Kka-nnan, an NGO, asks with grave concern and skepticism:

"To have an idea and scope of your proposal, what major global project do you compare it with?. We have been hearing about ventures heralded as mega projects in recent memory such as *Al Khobar-Baharain Causeway, English Channel Tunnel, Commonwealth/Olympic Games venues & infrastructures, the Swiss Tunnel deep and across the Alps*, etc. What the world has noticed is, that a lot of cranes, haulers, earth moving and drilling equipment were working day and night, but precious too few people. The mega ventures were completed in just a few paltry years. Only major global corporations, financial lending institutions, and few contractors benefited from them. The miserable majority of the public was invited solely to pay taxes, and foot the bill.
Is that logically what is to be expected here as well?"

"No...! No... excuse me, Sir!", The presenter started his explanation with fervent defense, and passionate denials. He hastened to say: "This is conceptualized as an enduring project. Our philosophy is totally different. The objective is to bring people and nature together. We cannot change nature, sometimes it is so brutal, but we can affect and mitigate its effects; can try to defend ourselves, and our properties. We strive to make changes in a 'non-destructive' way; in fact, in a conducive way for both. The core principles and vital components of our proposals are humanity, and nature resulting in a harmonious, mutually dependent, and enduring relationship, or described in a truly '*symbiotic*' (a relationship of mutual benefit or dependence) relationship!

We must incorporate stipulations for the maximum number of people to be employed, and as long as possible. The design is conceptualized in that way. This is a social endeavor, as much as a business endeavor. This will be an enduring and endearing work for mankind, and the environment, Science & Technology. We rank people in front of everything else; insist that they be offered a chance of participation, employment, and that they can find enjoyment of lasting benefits. Machines are only sparingly used to carry out dangerous, laborious, and monotonous work; in circumstances where structural integrity, precision, strength, safety, and durability are of paramount considerations; eliminating catastrophic human error. Certainly, there will be cases of machines drilling for columns, or underwater spot welding, fastening; certain work performed at dizzying altitude and in pressurized capsules at great depths. They are essential. In spite of that, there will be plenty of work for installers, construction, skilled & semi-skilled; maintenance workers, electricians, managers, civil & marine engineers, as well as researchers. In the subsequent phases also, it will generate additional jobs in services, hospitality, tourism, and theme parks; perhaps also a floating hotel in the lake, similar to *Pichola Palace* in Udaipur. A bird sanctuary for migratory birds is also on the cards. Besides, fishing will be a large beneficiary. People haven't had a chance to taste some delicious fish curry for a very long time because of its exorbitant prices. Imagine, the fish prices come to affordable levels! The bodies of water can sink tons of CO_2 to the bottom; of course, we have to first capture and then liquefy the gas. The movements of schools of fish in water is enormously beneficial to the cause! We could see a thing can cause many other tangible effects. It will be a vibrant region, offering plenty of new and exciting opportunities. In the future, we could also think of building a mono/commuter rail on "stilt' columns using 'Magnetic Levitation Technology'; or Zero-emission, environment-friendly Hydrogen-powered commuter trains going around the banks of the Vemba-Naadu Kayal. Conceivably, it connects Alappuzha, Kochi, Kottayam, Changanacherry, and the towns on both banks in a circular loop; Extending the loop later to many 'Karas' in the Kuttanad region, then on to Wellington Island, Bolgati, and other populated islands. We were disappointed that the '*Kochi metro*' did not choose one of these systems. It will reduce traffic on the highways, NH 47 and KK Road. Businesses will have more room available to expand. People in the suburbs can benefit from it. This will go a long way towards '*De-urbanization*' & de-carbonization; reduce congestion, pollution in cities; sporadic, and contagious diseases. It will mitigate the burden on space & water resources as well as the infrastructure networks. People will have more time to live their lives stress-free, which are otherwise spent stuck in noisy, polluted traffic jams.

By the way, Sriman. Thampuran, it's not comparable to any of the global projects that you have named. In dollar terms, it will just be minuscule in proportion, and budget; an affordable one that can grow in phases. Surely, in spite of it, this will be a successful endeavor, rich in features and usefulness, in par with the other major global projects". Besides, none of the projects you mentioned has addressed 'environment and sustainability' as adequately, and *comprehensively* (in an all-inclusive manner) as this proposal do!"

Meanwhile, the assembled crowd has noticed on a 'Jumbotron', an aerodynamically built hi-speed train silently running along, with a strange-looking rail-system along the lake Vembanad connecting familiar towns, and places which people have never witnessed in real life. They are thinking that they have been missing out on some important things all along. It builds up great curiosity. It's actually a computer-generated '*graphic animation*' of the project being discussed here! "Quite impressive!" mused many.

"The Hydrogen commuter rail project is quite interesting to us from the Institute of mass transportation Technologies, Trombay. Would you care to elaborate on it, possibly the significant technical aspects?"

"Well, Mazumdhar Banerji is the expert on 'Green Hydrogen' technology. Hey, there, Mr. Banerji, could you take up this question please!"

"Of, course, gladly,... certainly!". It seemed that he has been waiting for his first opportunity all day long. He started, quite enthusiastically:

"Oh,...indeed.... I'm so excited to talk about it; for it is poised to bring about a radical *paradigm shift* (an important change that happens when the usual way of thinking about or doing something is replaced by entirely a new, and different way) of enormous multi-frontal consequences and benefits to humanity. Hydrogen is the most abundant element and cleanest fuel in the universe, in volume.

On this day, at this venue, we are unveiling this project for this region and other similarly sprawling landscapes on the banks of any broad waterways around our country. Interestingly, there are many such places in this great state. We like to call it HPSRS "Hydrogen-Powered Stilt Rail System". Hydrogen packs enormous power. Many don't know yet, mainly because most have not seen or felt it, except the few rocket launching scientists. It's actually invisible; being light, it travels straight upwards if leaked or released; of course, causing a streaking flame on its way skywards. You must have noticed it if you have watched rocket launching video broadcasts. Well, HPSRS doesn't take up any precious agricultural nor residential real estate. It makes use of the hitherto unused land; it goes over the bodies of water, and no one lives in the water except the fish. The rail system is built on raised 5.5 m above the waterline, T-shaped 'stilt' platforms erected at 5 m apart on the lake bed without disturbing any other means of transportation or any activities connected with daily life, business, or environment. Hydrogen fuel-based transportation is quite promising and exciting for its benefits such as zero-emission, extremely low noise, renewable energy usage & environmental friendliness with no pollutants. Its exhaust is pure water; means it's potable, yes, fit to drink! How so wonderful! Furthermore, it is uniquely suited for non-electrified and newly built lines."
"Oh, my God, What a great advantage, and cost saver!"

"These trains could run at a speed up to 140 K mph, hitherto unrivaled in terrestrial transport in this country. They are comparable with the equivalent sized Diesel trains in terms of distance range, reliability, and safety standards. The light train is powered by hydrogen fuel cell packs that generate electric power for traction, from a continuous supply of Hydrogen from an airtight, sealed tank. It runs completely emission-free, quiet, emitting only water as, vapor, and condensation. Its other prominent features are clean energy conversion, flexible energy storage in batteries, and intelligent management of power and available energy. Furthermore, it enables clean, sustainable traction while maintaining high performance. The other components and processes utilized in the chain are also efficient, renewable, and environmentally-friendly right from Hydrogen production, fuel cells to non-flammable Li-ion Vanadium phosphate battery packs, and traction motors. I shall focus on hydrogen production & Fuel cell, since others are already well known and have been widely in use.

In this context, let's remind ourselves of some processes taking place in nature around us. One of them is something we have studied back in the primary school and botany classes, and forgotten since; that the plants manufacture their food by the process of *photosynthesis*. It is the process in plants by which the chlorophyll on the green leaves synthesize and manufactures carbohydrates (food) from carbon dioxide captured from the air, and a source of hydrogen (usually water from the soil supplied through the roots), using only sunlight as the energy source for the synthesis. Another useful knowledge in this context, are that: a 'Molecule' is a group of atoms that are held together by chemical bond; whereas 'Atom' is the smallest unit of a matter that cannot be further broken down by chemical process; the 'Ion' can be atom or molecule that has gained or lost one or more of its valence electrons thereby altering its electrical charge.

In order to use H_2 as a fuel, we need to separate hydrogen from other sources, such as water. That is accomplished through a process called, electrolysis (Chemical decomposition, produced in an electrolyte by an electric current).

We can use the electricity produced from sunlight in electrolysis, precisely by splitting water molecules into 2 parts of hydrogen atoms, and 1 part of oxygen atom. A water (H_2O) molecule (a more or less stable grouping of two or more atoms held together by chemical bonds) has a slightly positively attractive (charged) oxygen end, and a slightly negative attractive (charged) hydrogen end, in the chemical bond, even though the molecule as a whole, is considered electrically neutral as the overall charges balance out. In the "artificial electrolysis", a hybrid module made up of light-absorbing '*triple junction Silicon photo-electro-chemical*' panel that is coated with 'oxygen catalyst' on one surface, and a 'hydrogen catalyst' material on the other is immersed in water. Here, unlike other electrolytes method, no costly electrolyte is needed! Then, current is applied. The 'slightly negatively charged oxygen' atoms are attracted towards the positive side, and the 'slightly positively charged hydrogen' atoms to the negative side of the modules. The hydrogen is collected in cylinders and piped to the specially designed and constructed storage tanks as fuel, while the pure oxygen is allowed to freely escape to join the atmosphere, that purifies the air in the immediate vicinity. In this process of *Artificial photosynthesis*, it is important to note that this chemical reaction occurs only between atoms, and therefore involves no radioactivity at all.

Thank God! [In contrast, the reaction *within* atoms is a nuclear reaction. It can occur naturally solely in stars and that involves harmful radioactivity]

Using this technique, H_2 can be produced at the stations wherever space is available for the solar panel array and tank installations. The required amount of water could be pumped in directly from the waterways along the rail system. Such installations need not be at every station, or even at close range, as the train can have a range of over 1000 km. By the way, the amount of H_2 is measured in its weight (Kg), while gasoline and most other fuels are measured by its volume in liters, for the simple reason that gases expand, and fill the container in which it's held.

Now, what must we do to make hydrogen useful for our needs?. We have already found out that in electrolysis, water is split into hydrogen and oxygen by sending an electric current through it. Thankfully, the reverse is also true, and practicable! A separate device, a fuel cell, accomplishes that function. Fuel cell is an *electrochemical, and energy conversion device*. It converts the two chemicals hydrogen produced by electrolysis, and oxygen from the air into water, in the process, producing electricity 'if a few other necessary criteria are met'. It's very much like a battery, which is also an electrochemical device. The major difference is that, a battery has all of its chemicals stored within itself; it converts those chemicals into electricity until the time its charge is depleted. Bear in mind that electricity is 'the flow of electrons' in an external circuit!

In a fuel cell, if a continuous supply of H_2 can be assured, an uninterrupted production of electricity can also be guaranteed. Most fuel cells in use today use hydrogen and oxygen, as chemicals. The principle is to capture the electrons in them and conduct it via an external circuit. So, we'll discuss the most promising polymer exchange membrane fuel cell (PEMFC) how it accomplishes the job. In the cell, a proton-conducting polymer membrane is used as the 'electrolyte' (described below). Hydrogen is typically used as the fuel. Remember also that proton is a positively charged elementary particle found in all atomic nuclei. The Fuel cell consists of at least four main elements: the negative anode (anode is the electrode that loses electrons in an oxidation reaction and therefore, becomes positively charged; cathode is just the opposite!), the positive cathode, the electrolyte (a solution or molten substance that ionizes and make it pass electricity) of proton exchange membrane, and catalysts. The membrane is usually made of nano-particles of precious metal platinum thinly coated onto carbon paper or cloth. It conducts only positively charged ions and blocks electrons. The catalyst, as it name implies, facilitates and speeds up the reaction of oxygen and hydrogen. When H_2 is forced through the catalyst under pressure from the anode side, it is split into 2 (H^+) ions and 2 electrons (e^-). The electrons are conducted through the negative element and the external circuit, returning via the positive side of the fuel cell, making an electric circuit and current flow. At the same time, oxygen gas (O_2) is being forced through the catalyst from the cathode end, where it is split into two oxygen atoms by a separate electrode of the precious metal. Each of these atoms carries a strong negative charge with it. This negatively charged atom attracts the two (H^+) ions passing through the membrane, where they combine with an oxygen atom and two of the electrons from the external circuit to form a pure water molecule (H_2O) which is emitted. This chemical reaction in a single fuel cell is quite feeble producing only very low voltage. Therefore, many fuel cells have to be combined to form '*stacks*' in order to produce the required amount of voltage and current. The electric charge thus produced is utilized to charge the Li-ion battery packs that drive the traction motors.

The interesting phenomenon in the entire process of H_2 production and using it in fuel cells is that it returns the lost heat, besides producing electricity. Pure water as the only byproduct, also making drinking water from waste water! It's all hard to believe; but true; just magical!; don't you see?"

"Superb, sounds great!,... likely, too good to be true! What are the hazards inherent in the Hydrogen fuel technology chain, and how are they managed and overcome for its regular use in human environment?" asks Jaison KudaMalloor.

"Thanks,... it's an extremely important question! Well, there are a number of problems and dangers with storing, handling and using Hydrogen. H_2 is the smallest atom, light, and very loosely spread, making it very difficult to contain. Besides, it is very combustible as it is very reactive. If even traces of it leak out, it will immediately come into contact with the oxygen in the air causing fire in a large release of energy, or explosion. It burns more readily than gasoline does. The prevailing methods of storing involve liquefying it at extremely low freezing temperature and under pressure which adds additional cost and logistical problems. Thankfully, the Hydrogen, being light and small, when leaks out, rises quite fast lifting its 'narrow-width flame' away from the ground without spreading it- unlike other fossil fuels such as gasoline, I might add! That means it is less likely to cause secondary fires, except may be in enclosed spaces such as garages and warehouses. For these reasons, we are hesitant to offer this technology to metro services which run through long tunnels, enclosed tubes or congested environment.

Quite importantly, in order to liquefy H_2, it will have to be cooled below its critical point of 22° K (Kelvin). A fully liquid state a*t atmospheric pressure* can occur at only extremely frigid temperature of at least [K=C+273; therefore at −253 °C]. Since Hydrogen's boiling point is exceptionally low at (-252 °C), in practical applications, both extreme pressure and freezing temperature are applied to maintain H_2 in liquid form for transport to the point of use. This necessity makes using liquid hydrogen very expensive at the moment.

There's also the potential for hydrogen to asphyxiate. Hydrogen isn't toxic *per se,* or poisonous, but if pure hydrogen is breathed in, *asphyxiation* could potentially cause fatality due to severe '*oxygen deprivation*'. H_2 has a strong affinity towards O_2. The reaction consumes all the available oxygen in its proximity, leaving us little to breathe in. That's the oxygen deprivation I'm talking about! Furthermore, much like oxygen, and Nitrogen, H_2 gas is also invisible, odorless, and flavorless. Normally, one will not know to be able to react before a catastrophe strikes. I will hasten to mention that in the event of a hydrogen leak in the open, it disperses itself rapidly vertically upwards, as it's a lighter than air. So the leaking H_2 drastically reduces the oxygen concentration in the air to cause serious breathing hazards in its immediate vicinity! Besides, as it is highly compressed under freezing conditions, liquid hydrogen is extremely cold. If it should escape from its container and come in contact with skin, it can cause severe '*frost bite*'.

In this regard, another element, Ammonia (NH_3) gas gains much interest. Ammonia can be stored under much lesser freezing temperature at (-33°C) compared with Hydrogen at (-253°C). To day, ammonia is distributed globally via pipelines, tankers & trucks, making it readily available, and cheaper. This gives a huge cost and logistics advantage in the production of Hydrogen from Ammonia. To top it all, Ammonia has an energy density 1.8 times that of Hydrogen, and holds 9 times the energy of Li-ion batteries. However, ammonia is more difficult to oxidize (the removal of electrons) than hydrogen, which causes ammonia fuel cells to produce less power than Hydrogen fuel cells.

Ammonia has many uses: as fertilizer (Ammonium nitrate fertilizer provides most essential nutrient 'nitrogen' to the soil; it is more stable than 'Urea', CH4N2O, and that it does not rapidly lose nitrogen to the atmosphere); also as a fuel for Ammonia powered ships; source for Ammonia fuel cell; driving gas turbines in power generation; also for producing Hydrogen by 'Ammonia cracking'. The released Hydrogen from ammonia can be utilized as fuel for Hydrogen fuel cells that produce electricity. There's a cyclic relationship in nature!

So far as India is concerned, I see the most interest in Railway transportation, shipping and power generation. Certainly, other applications will follow suite, as and when these gas fuel technology becomes affordable to the public use; the basic infrastructure for its use is built, and running.

By their natural properties, all fuels have varying degree of combustibility and danger associated with them. The safe use of any fuel focuses on preventing situations where the three combustion elements—*ignition source* (spark or heat), *oxidant* (air), and *fuel*—are simultaneously present at any point in time. Hydrogen uses require additional engineering controls to ensure its safe use. Adequate ventilation and leak detection are important elements in the design of safe hydrogen handling systems in its entire chain. Many of these hazards are largely mitigated by innovative design considerations and incorporating technical solutions; so much so, this technology could be implemented in most human environments, if stringent guidelines are strictly followed. Quick-release valves that are leak-proof, corrosion-proof; fortified high-pressure tanks such as carbon-graphene (a nanomaterial consisting of 'one-atom-thick' sheets of carbon atoms that are arranged in a honeycomb lattice structure) fiber reinforced tanks can protect against fires, and explosions. Any metals that can become brittle when exposed to hydrogen are totally excluded from all the hydrogen handling systems. Analyzing data from several controlled car crash tests, hydrogen safety when so used is considered potentially better than that of gasoline, or propane. It's also equal, if not better than natural gas. Besides, the substance is handled by only knowledgeable and trained professionals, and through well-controlled, fool-proof, and safe protocols & methods.

We have an ongoing research collaboration with Kalinga Institute of Industrial Technology in this specific matter".

.... At this time, some flurry of activities and commotion could be overheard over the PA (Public Address) system of the race venue, signaling that some official announcement of the decision is imminent. The crowd begins to hustle down towards the main stage to listen in, and join the excitement.

Chapter 38

The s. u. s. p. e. n. s. e!

The PA announcer comes on the air

Pin-drop silence!

"I can only tell you that it was a very close finish- a '*photo finish' indeed*! I dare to say that you may not find such a close one ever again in your lifetime".

The crowd murmurs;"....what?...ours?...theirs?.. Not my bitter nemesis yet again! No way, Gopala Krishna..."! It creates random guesswork as well as some tormenting anxiety in their minds. This is causing certain hype and tension, like the moments before the announcement of 'Mega lottery' winner.

A cameraman hurriedly positions his tripod; strenuously focuses his camera on the vacant podium. The Video control room has reviewed the '*Slo-Mo*', and frame by frame playback of the finishing footage repeatedly, over and over. The panel of judges discussed and argued over each scene presented to them. They have calculated times in hundredths and thousandths of 'split seconds and split frames,' and split their hair; teased their vision. "In a single second, there are 25 picture frames in a Television image," a smart guy explains. Yet, no one knows what the race official announcement is going to be. Only the chief judge knows it; or perhaps not. The video editors have quickly assembled footage and images of the finishing scenes with different perspectives from several vantage points, frontal, sideways, and overhead views: superimposed on measuring grids. They upsized and downsized by 'scaling' the image; zoomed in, and out to see the fine details; to locate the tip of the chundan Vallams crossing the finish line. Hopefully, they will all be shown during the course of announcing the results. However, it is the privilege and prerogative of the finish judge to make the final official announcement. But, everybody seems to be non-committal even at this late stage. Probably, in order to avoid widespread dissatisfaction and doubts, they are attempting to gauge the temperament and sensitivity of the fans before announcing a close and most likely a controversial decision.

The chief judge is now being introduced by the stage MC (Master of Ceremonies) Srimathi. Mahila Madhavan, reading from a litany of titles & qualifications the judge possess and has held since his elementary school days as the class monitor. She presents Sri. KuriaAchan KiliKkoottil, affectionately called Kuria-Chayan, a prominent personality among VallamKali enthusiasts; he is synonymous with the races. He was the first helmsman & captain for several consecutive years in his prime. He, as captain of 4 different chundan-vallams had won the champion's trophy, a record matched by none, yet! He was also an officer in the Indian Railway Services; knows a lot about speed, motion, punctuality, crowd management and event-management. Besides, his family was once royal physicians of the "Chembakassery" dynasty. The Chundans were among the most elite marine fleet of the same erstwhile Kingdom, bringing them a number of decisive and glorious victories.

In a mystical scheme of things, the association continues to date!

Sri. Kuriachayan takes up position behind the podium; greet the crowd; begins:

"I hope you all had a good time and have enjoyed the thrilling finish you have seen and experienced; and the extremely competitive races of all the other categories of boats.
The crowd roars in jubilant affirmation.

" I can tell you, in all sincerity and recollections, that I haven't experienced any race like this since my engagement with the races in the last 41 years!"
"It was a '*photo finish*',... Between four boats, Kavalam, St. George, Karu-vatta & Thayankari

Let's give four rounds of applause to each of the boat crews!" As the fans chant their respective anthem, smiles, and hope rise among the individual group. Their faces say it all. When one's anthem is sung and done with, their hope dissipates instantly, replacing with anxiety and suspense.

.........Now, it's my great honor and privilege to invite each one of you,... the esteemed and proud fans who are still patiently waiting here,...with great anxiety and expectations of course...,to review the final moments of the '*finish scenes*' on the TV screen,... along with me!"

"Come on, now!...just give us the result,... announce it!'"

"Why don't they declare the four as joint winners and done with?"
A section of the crowd shouts. The impatience is loudly evident!

It creates some jostling and commotion around the Television sets. Kuriachayan asks the Video controlroom to play back the few critical seconds of the finish. On the first go, people came up with conflicting results. Some argue and claim to have seen the boat in lane 2 as coming first; some disagree, totally! Quite a few say they saw the light flares in the 4th lane going off first-(remember the previously described technology in the trial). The judge asks to play the tape back in slo-motion a few more times until the crowd sees the finish sequence for their own belief and self-assurance. He further asks to freeze the frame at the exact moment of finish. Even for the slo-mo machines, it seems to be a challenging task. It is like catching a fish from the water with bare hands. It wriggles and wiggles away! The time difference surely shorter than the time to press 'pause' button! Smart people defensively think so. People complain of some jitter or shaking in the picture. The picture frame is not steady enough to determine its accuracy! A valid point of course, and needs to be addressed! This situation causes the picture to show a different boat crossing the line first at the very same instant. Besides, at one time, the tip of the chundan crosses the finish line first, at another time, the paddle of the first of the 'fabulous Fives' of another chundan breaching the line, and so on. The chief judge, Kuriachayan got the message. He stipulates "You should be focusing on the tip of the chundan that crosses the finish line first, ladies and gentlemen, nothing else matters!". That really helped to reduce the number of qualified parameters!

Even after a few slo-mo replays, it is still inconclusive and confusing, just as in the beginning! No progress is made so far. The video engineer routs the signal through a '*Frame Synchronizer* 'to eliminate the 'jitter'. The remedy works!; pacifies the crowd. Now the "*time code*" (or the 'Timestamps' in Television showing hours: minutes: seconds: frames/second in 2 digit formats in each domain. In the 'PAL' system of color adopted by Door-Darshan, there are 25 frames /second, giving the very last digit of the display a value of 4 hundredths of a second) digits are legible; displayed; can be read! He continues with the playback sequence a few more times, each time the judge interrogating the viewers. They call out and chant the names of Chundan they are supporting, disregarding which has come first, in competing loudness. Shri. KuriaAchayan reminds the crowd, he is not taking a '*voice vote*' asks them to be accurate and completely impartial like an umpire. A consensus '*of sort*' begins to emerge. It's an opportune time to go to the final step. The judge justifies the delay in announcing the results. He wanted to have a correct decision, a clear-cut, unanimous consensus among all judges and the video review experts. A few among them are surveillance, satellite, and drone photo analysts in their own profession. Independently, they scanned, and examined the frames carefully before submitting their opinion.

"The computer analyzed data. It chopped; cropped, color corrected the last 7 seconds of the race; measured speed and acceleration of the final seconds of the images, and resolved the most accurate timeline into another fractional unit of "fields". [In Interlaced scanning 2 fields, odd and even number of fields constitute a picture; as opposed, progressive scanning the lines are drawn in one at a time in sequential order. The entire single frame image is painted every 1/50th of a second, allowing for twice the detail to be seen in the same amount of time used in interlaced systems] He paused, creating another unwanted delay and suspense, as in the manner, and style of the announcement in "*Miss Universe*" pageant.

Paddlers on the Chundans are standing in rapt anticipation, paddles dropped aside, the other hand resting on head, staring at the closest Television monitor; squinting the eyes; wait for the 'moment of truth'; to hear straight out of the "cow's *mouth!*"

Few tense moments elapse; felt it much slower than what is on the usual time clock.
"The time difference between the first and second is just... 4 ...Hundredth... of a second;

The second and third are also as much.... 4 Hundredth... of a second;

Amazingly, again!.. Incredibly, between third and fourth......., 4 ... Hundredth... of a second; that's the finest resolution of the time code generator!

This, according to the video technology experts and in terms of the Television Frame analysis, the first 4 places are determined by a single 'field' (a field is one half of a full frame) difference on a TV screen!

I must say it is quite astonishing, incredible, never to be repeated,... never! Never, ever again! It won't be emulated in any competition, even in 100 m dash, period!"

The crowd says "hush!", "damn!"... A sigh; see contortions on faces; countenance tense up; ear lobes tighten as the gentle impalas sense danger of the approaching predators, the ferocious carnivores, the Gir lions, or the Bengal tiger.

A suppressed comment...a dream;....,or a curse;... a wish;.... ,and a prayer...! It's so...oh...agonizing!

The judge looks aside towards officials on the stage. As a matter of fact, he does not have any inside, privileged or prior knowledge. His job up to this point was simply interpreting the rules of the races and to explain each specific question. Mind you! he is the chief umpire. He was not privy to any back-room screening, judging, or deliberations. A judge shall be forthright, transparent, and impartial! Indeed, that's what he is! He's a rare breed! (Unlike the politically appointed, life-term supreme court judges!). In the solemn spirit and sentiment, he had recused himself; removed himself from the review process entirely; besides, his eyesight is somewhat impaired. He receives the final tabulation from Nair, Nambuthiri & Naik auditors in a sealed envelop with a reddish, embossed government seal affixed across its flap on a golden tray, once used in the royal court of the Maharajah of Travancore. All traditions and customs have to be followed with the 'pomp and circumstances' practiced in those days! That's a protocol without reproach!

It's an official document; a historical one, too! The winner gets the coveted trophy, a Million rupee prize-money from the race organization; besides, getting bragging rights for a whole year!

Kuriachayan asks for an envelope opener and waits casually. He is on '*live*' television; people are watching; must do it properly, stylishly as well. Another moment elapses. Suspense mounts, reminiscent of the "*arrival of queen Sheba*". The TV viewers all around the world are also witnessing the proceedings while tensely waiting for the announcement. It's nerve-racking, tormenting, to say the least!

An impatient person, also a hyper-sensitive hypochondriac (obsessive) in the crowd, cannot take the protracted delay, and the drawn-out suspenseful drama any longer. He has waged a huge stake in betting. He had a bold but extremely speculative '*perfecta*' bet on St. George, Kavalam & Maha Keralam Chundans in that order. His heartbeat has been rising, pushing blood pressure higher with it. It may be here promptly stated that he is an active high-stake '*options & futures*' trader in the stock market as well. He could be either a millionaire or hopelessly poor in an instant. Time is of the essence; gambling is his trait; his only career and passion in life! He is understandably nervous, and irritable on the "*Triple witching days*" which happen four times a year on the third Friday of March, June, September, and December on *Dalal Street*. One of them, falls tomorrow. He has some risky '*positions*' to settle. This one matters too, surely; indeed; everything, time, and money matters to him more than anything else! He climbs up to the stage and removes his "*Kaddari*" or dagger, a shiny sharp pocket weapon with a golden hilt as polished and shiny as the edge of the blade itself, concealed for personal protection against creditors, out of its ivory sheath, gilded with gold insignia of a fire-spitting dragon, his patron. It reflects his personality. People wonder what he's going to do! He shouts, "*here it is...take it, and use it*!"; he hands it over. Kuriachayan is shaken, understandably; promptly declines; gestures to him to go away, and get lost! The security personnel surround the man instantly. He is seen lead away in casual manner, sparing commotion and panic among the crowd. People say he may be drunk, crazy, or both. It feels like a potential tragedy is instantly averted, a time bomb diffused! These days, the security consideration is an essential part of staging any public event! Just recall the 1972 Munich Olympics incident!

A collective sigh of relief is heard!

Apparently, there was no envelop opener at the time of the King's rule! Hopefully, next time, the judge will indeed be given a golden 'envelop opener' along with the sealed envelop itself! Momentarily, the judge loses his poise and patience; tears it open by his nails and in some clumsy improvisations; reads aloud; tersely, in a reverberating tone: "**Karuvatta**!, Theirs,...first ever!" ...

Bravo! Congratulations, Karuvatta!

A collective roar of loud exultation rises from the paddlers of the Karuvatta chundan. As a customary ritual, they dropped their oars and jumped straight into the water. While doing so, some performed front dives in two, & one-half somersaults; others backward with reverse twists. The twins in the front did a synchronous dive; the first & the second helmsmen did the Pike routine with legs straight with the body bent at the waist. Not to be outdone, the rhythm makers did a group performance of Tuck, the body bent at the waist and knees, with thighs drawn to the chest and heels kept close to the buttocks; feet and knees kept together, and toes pointed in the true Olympic diving styles. The obese couldn't control their excessive weight, balance, and oblong figure; land in water rather awkwardly, belly first, turning it black and blue, their faces showing extreme pain. They had known the styles right from their unruly childhood, when they regularly took baths in the morning and evening in the river or canal with their 'bands of brat boys & buddies'. Disuse affects all the skills, the elegance, execution, and styles they mastered once, long ago! As a result, the boat yawed, tilted, and swayed threatening to overturn dangerously. But it didn't, thanks to its high arching, stable, stern; and sleek bow!

The team supporters, some totally drunk, on the land get excited. They simply threw themselves into the water, splash water on each other; some losing clothes in the melee. That freezes up their movements and adventure right on the spot! A few others swim towards the vacant boat; pick up the floating oars and paddles; clumsily row away in unbridled excitement, leaving the original paddlers in the water gasping, and bewildered. Their echoes, incoherent shouts, yell, screams, jeers, and commands no one wanted to heed. It faded in the revelrous pandemonium with rising din in the unexpected and historic upset-victory.

The other dejected groups sulk, balk; then begin to disperse in silence and visible disappointment, licking their wounds, like the Hyenas disperse on the appearance of the lion on the scene! Victories & triumphs are glorious and exhilarating. In contrast, the defeats are excruciatingly distressing and depressing; quite agonizing especially after coming so close to winning it all!
Luck certainly matters in all pursuits; it's a proven fact, folks!

Some spectator boats are seen racing in the vacated lanes and enjoy their moment of adventure. They ask the Television cameramen and friends to take their funny poses, and record their adventures. They want to be able to show the world and vouch that they were there at the very historical moment, and saw it all happening in front of them! We never know if they could become famous in an instant for this impetuous act! They know that stars are born every day on social media platforms unexpectedly for every silly, but uniquely queer reasons, and occasions. People who get more of it, suffer depression and participate no more; those who get a little, wants more of it. Similarly, they have also heard new stars are born every day in the sky, while some die away too; assume the shooting stars are the dying ones on their way to heaven, dispelling the notion that the paradise is somewhere up far beyond reach. Instead, now it seems to be somewhere closer to our planet earth, raising wide-spread hope, and optimism!

Talking about becoming a star, one could reap some riches besides some ephemeral fame. In the crowded fields of 'attention competition,' some could get noticed; others get ridiculed and sneered at. The noticed get some reward; others totally ignored. The point is, one has to be in the game and keep doing things without deliberating on uncertain outcomes. We are in a moment of fleeting fads, and transient cultural obsessions of strange sorts!
It's actually everybody's game; one must indulge and keep on playing; that's the new paradigm!

Chapter 39

Euphoria of winning and the agony of defeat

The long awaited results is eventually pronounced in great drama and suspense!

Exultations & Jubilation for the victorious; agony & despair for the vanquished, and the many other losers

Understandably, the losing sides look dejected, their valiant efforts fell a little too short, and in vain. It produced a lot of sweat and aches, but not the ultimate glory; revenge rages in their minds, vow atonement in the very next time; that's next year. The losers, generally, bemoan their cruel fate. People conjure up '*What if scenarios*' and suggestions:

"...if we had a little bit of luck!;...if the start were slimly better...!... Had we had picked any other lane ...! ... use a little more of Science!... should use automotive engine lubricating oil (10W-40) instead of the fish oil & egg yokes on the boat! Yeh, sir 'graphene' polymer would be better, it's a natural lubricant, offers little resistance to anything even to the flow of electrons!;... there was some fumbling at the sprint... find a new captain...how about raising the offerings to the patron deity...yeh?... Well, at the end of the day, they fought gallantly...... After all, it was just a paltry one-hundredth of a damning final second. That particular second will live in my mind in infamy,... with all said and done, it was a thrilling race... indeed, give some credit Yaar, where it's duly due; don't be too stingy, guys, let's go home, now!". So on and so forth!

After a short while, everyone seems reconciled and reluctantly inclined to accept the bitter results in the name, and spirit of good sport, or for lack of any other viable excuse.

The screen flashes the official results:

1. Karuvatta 04 : 29:59.53
2. Thayankari 04 : 29:59.57
3. St. George 04 : 30:00.01
4. Kavalam 04 : 30:00.05
5. MahaKeralam 04 : 30:00.17
6. Anari 04 : 30:00.23

It's now time to go home, perhaps there'll be a little bickering and squabbling all the way back. There was a lot of energy and enthusiasm coming in, emotions well spent; they were all used up; little left for the return trek; darkness has fallen. Friends are all gone. There will be a lot of people at the bus stand; must fight several crowd fights to get in, or hang on the footboard all the way; feel hungry too; didn't dare to go away and grab some 'grub' to eat; didn't want to miss the results. The food stalls are now closed and nothing is available; don't

know when I can reach home. The dogs in the neighborhoods are a fearsome nuisance. It is a '*no moon'* night, too. That's when wolves come out and start mourning for the missing moon, thinking it is dead, and will never resurrect. They howl, make the darkest night even more miserable. *Kayamkulam Kochunni* & '*Panichonni' Velu* (notorious robbers of the time) cleverly planned their biggest heists at these times, giving people sleepless night!
I don't want to be mistaken for any one of them!

Besides the 'Kavu' (the dense strip of woods) by the temple, where the serpent tends to come out in such horrible nights; the tall trees are haunted by harrowing nocturnes, owls & bats hissing and shrieking; yet a few other dreadful things to deal with. They cannot see, nor can I, in this pitched darkness! I am already weary and frightened beyond belief. It would have been better, if I could find a hotel room, but there is little money in the pocket; barely enough for bus fare. "Oh, well! One has to live with the means one has got, and that's what I have been, and should be doing for the rest of my life! So be it, maybe I'll be reborn as rich in the next life! Until then, I've to live my life, as it is!"

The crowd disperses itself with mixed emotions, but all seem to be gratified by having witnessed a thrilling race of a lifetime. I am witnessing a bigger crowd in the Toddy shops on the way than at the bus stop. It seems that the 11 pm curfew is relaxed today in the town because of the races. Some drink to celebrate victory, while others to forget the loss. Actually, the habitual drinkers never wait for any particular reason. Anything and everything is a good reason for them to indulge. I pity them! There could be some vandalism and trouble. The public will find them out from the morning newspapers in great detail. There will be more coverage and images about it than the exciting VallamKali itself!
Media rules our life and perceptions!

Chapter 40

Flotilla escorts the Chundan back to its original resting place

Meanwhile, the winning boat is tugged back by a motorboat. The frolicking fans join the flotilla tailing or trailing the procession back to the native hamlet, joking and '*laughing all the way*'. It is the equivalent of a '*ticker tape*' parade after winning the Football World cup and the cheering fans running after the carriage carrying the coveted glittering trophy. This one is on the vast, pristine backwaters on the shimmering waves under the blinking stars, and along winding canals. They stop by the Toddy shops on the banks of the Kayal for the best Toddy available on the planet. It's '*on the rich syndicate*' of the village. Individually, each one comes forward and in slurred musical drone, "*The toddy on my dul... ci...nea*!, my sweetheart...! Ha...ha...Haaa...!"; "*grilled lobster on the Poova-Paadam fa...mily...!*"...; "hot coffee on me, on myself alone...plenty of it!"; "*Karimeen ...a toast to my wedding next week... None of you're invited...aha ...ahaa...'cause you're all mean...and ugly...!*"; "*rejoice and make merry... there won't be any other day like today folks!*" "As another reward, you may end your 'vow' tonight!" the captain said in a sly remark; and on & on..., making a merry situation much merrier, each time drawing repartee, or shouts of '*Arpo... Hurray*', rhythmic claps, and stomping of feet. All of them feast on tender toddy, the choicest sweet water clams; well-cooked tapioca/coconut mix; sardine curry in red-hot chilly sauce on top of it; the "*Pollicha Karimeen*"- an exotic fish cuisine in a richly flavored spicy gravy of coconut milk, onions, garlic, ginger in the uniquely 'KuttaNaden' delicacy, and flavor.

It is a sumptuous meal, befitting the occasion. Lastly, hot brewed black coffee is served bland, without sugar or whitening cream to make them feel sober, and look good when they finally arrive home triumphantly to the waiting families. Surely, wives will be awake and awaiting; also the teenage sons, being curious to find out how to act and react when they achieve glory, and triumphs in the games they play. There is a lot of singing, shouting, and buffoonery act on the way, waking up people living along the remote banks. Usually, the simple, hard-working families go to bed early to get ready for their farm work at the dawn of the day, before the sun strikes it hot, and unbearable. Those folks, half awake, hurriedly light up their hurricane lamps, not knowing what had happened; check their children one by one, if they were all alright. All of them come out and look on; parade along the banks and accompany the flotilla in procession as far as the nearest "*Mata*"- a levee breach is. They don't get many sights to see or chants to hear, as they live on isolated strips of dikes. This is quite rare and not something they want to miss! Gladly, they follow in entourage. They are encouraged to sing along the *Vanchipatt* tunes with the frolickers. In the houses far across the lake, they see the lights suddenly coming on, one by one. Obviously, they are awakened too! Everybody senses that something strange and unusual is happening. Some gifts and souvenir articles are tossed out to the children on the banks for their 'sportsman spirit'. They haven't had any 'goodies' in their lifetime. They turn them over and upside down; shake and twiddle with it in great curiosity, as to find out what they are for, and what it can possibly do.

There are enough reasons to wake up early next morning to investigate them further!

Finally, in the wee hours of the dawn, they bring the boat to its usual resting place, after paying due respects, and thanksgivings at the local temple and church. A large crowd has already assembled at the jetty in anticipation and to take a fresh look at their newly crowned chundan. The members of the family are also eagerly waiting to receive their heroes home in the most fitting way possible. They are the ones who took care of the children and managed home affairs when the paddlers went out for the scheduled practices, and the races. Their contributions and services cannot be underestimated. They have participated actively in different ways too. The earned pride and joy are for everyone to enjoy. All the community events have been social, secular endeavors, since the advent of the race's history, and even prior to it!

By no means, this is not the end of the all the celebrations; few other events still to follow! This point in time, is a unique moment, perhaps never to be repeated, a historical one. The memories must be etched and preserved in the minds of this generation. They should grow up as 'winners'. A culture of success and winning must be their goals, and guiding lights. It will usher in many positive and lasting benefits. Without a doubt, it will rejuvenate the community; increase social participation; strengthen inter-cast relationships, closer interaction between people, a new collective sense of accomplishment, pride, recognition, unity, peace, and a tremendous boost in self-confidence. Villagers will become more amenable to "*Srama Danam*", or community volunteer work schemes to build footpaths, beautify the village, remove water hyacinth from the canals, help the needy; so on and so forth. It will definitely improve attitudes and courteous behavior towards one another. All bode well, and hope it will prolong, permeate and sustain for a very long time. Following the recent harvest, people were morose, gloomy, and dejected; but for now, they look happy; this gives them great cheer, and hope.

The fact alone should matter a great deal!

The folks at home must be busy, they have been told what to do 'when we finally arrive home triumphant. We wanted to be received in certain way, assembling every one in the village community, and in enthusiastic reception, and great fanfare'. They may be working overnight to fulfill the assignment. In the previous years, we would rather forget the outcomes. But, this year, our mindset is totally different; it's buoyant and sparkling! We want to get home as soon as possible, and celebrate with our own people. While we competed in the forefront, they toiled tirelessly in the background, managing many other necessary things, as well as making sacrifices. This victory is there's too; indeed in equal and full measure!

We cannot wait to see how we are going to be welcomed!

The veneration & coronation staging of the triumphant Chundan

In the morning, villagers flock to see the boat parked at the jetty near the modest local market square. Even people who were usually seen in a hurry, pose, cast a look, ponder, enable a coy smile, straighten up their posture, swagger (walk proudly) along the way, humming to themselves in a sense of pride, and joy. The winning boat has immediately become *the 'cynosure of all eyes'*. It will remain floating at the spot in ceremonious status in full view for the next couple of days, showing off its prestigious trophy, adornments, and ceremonious status.

In the evening, when the farmers return from the fields and the office workers have come back from work, there will be "Chenda-Melam & Pancha-Vadhyam" music mela in honor of the paddlers and the boat, curtsy of the race committee. The same musical group who performed at the launching will be invited back as a reward for their participation in the initial and auspicious launching. The richest family in the village has graciously offered to sponsor them. The 'Pancha-yath' (village governing body) President will preside. He happens to be the richest too, as is the case in most places. The school has suspended all homework exercises for the day. Women and children will make stage decorations and will bring fresh flowers picked from their home-gardens. Some have invited their 'kith and kin' for an overnight stay. Hurriedly, they iron their clothes and dresses; prepare themselves for the event; plans to finish up their chores early; fastidiously match their gold ornaments with the Sarees; ask around for opinions; shortlists a few, pending final decision later when the husband comes home. Standing in front of a large double-view mirror, spread the golden, Kanchipuram silk, and printed Sarees on the shoulders, one by one; do bare minimum tucks here and there; turn this way, and that; and around; strike a few poses; practices different styles and smiles; decide on a new one which no one has yet seen! Everything is ready! When family members arrive home, all will have a light supper, dress up and then go. The dog will guard the house. He is a good one, fearless; barks quite loud. He is feared in the neighborhood; no one can challenge him! This evening, the ducks, chickens, and the cattle in the village will get an unusually early feeding. That's the plan; all set!

The staging of the triumphant ChundanVallam, the great pride of the village community

Some lemonade and coffee stalls are already set up and serving. Volunteers are feverishly building a stage. No rain is forecast; therefore no precaution is taken. Everyone finds the other in good spirits. It becomes contagious. People mingle and crack jokes. Passengers in the passing buses cast inquisitive looks; shout questions as to what is happening there. The bus departs before an answer is given. One passenger gets off at the next stop and walks back, determined to find out for himself. He lives in a neighboring village; likes crowds, festivals, and events; doesn't like to miss any. He is nicknamed the

"Darsan, the Reuters" as his eyewitness account of the local events is the latest, truthful, and descriptive even to the most inquisitive minds.

In the meantime, a gang of pre-adolescent boys roam around the area without any supervision; finally assembling at the river bank. They poke fun; try to push each other into the water; soon become rowdy. Naturally, they see and sense the excitement in the air; discount the chances of getting punishments to nil. One suggests that they all jump into the water; race, and have some fun. But they don't have any swimming attire, nor even a bath towel. One goes to his father; waits around; soon, sneaks away with his "*Thorthu*", a loosely woven towel of cambric cloth, just good enough to wipe away sweat. It covers nor hides much, has tiny see-thru '*air holes*' in it; not adequate to cover modesty. He wraps it around his waist, removes his knickers. Other children stand in a circle, providing a semblance of cover. He poses like a long jumper at the top of the run-up to the pit while the rest of them line up one behind the other. The boy makes a short run to the edge of the bank and jumps off high and into the water, releasing his loin cloth in midair that floats in the air a little while; the next one in line acrobatically grabs it, quickly wraps it around, and throws it back to the next one right behind. In rapid succession, all the dozen or so boys repeat the same act in a skillful "release & catch" maneuver as they practiced in the relay race at their school. Each makes his own unique style of aerial somersault, half turn, full turn, reverse flip, twist, and combinations- a scene out of the Olympic platform diving competition. Some totally fail to enter the water properly and make belly flaps, raising big splash, turning the belly painful and noticeably bluish-black. One MIG fighter pilot trainee from the IAF (Indian Air Force) who is on a short vacation commented, "*it's like midair fueling*" we practice! He gives some quick unsolicited advice on moves as to how to do different group formations. The children cheer, clap, shout, and laugh; raise the din, noise, and spectacle. Not everyone enjoys it. Some are alarmed. The elders scold; tell them to get off. It falls on deaf ears! In the process of having fun, the children might have earned some form of punishment, scolding, or spanking from parents for indecent, unruly behavior, and violating the norms of forbidden public nudity when they finally return home, depending upon the temperament and sensitivity of each parent. Then again, on a rare cheerful day like this, they may get a reprieve, or pardoned altogether; For God's sake, "Let children be children!"

In a grand gracious gesture, the committee has extended invitations to other rivaling regional boat clubs of Anari, Cheruthana, Edathua, Aayaparampu,

Karichal, Vellamkulangara, Payippad, Thayankari, Sri Ganesh, Pandy, Thakazhi, NatuBhagam, and a few more. They were all once the subjects of the erstwhile Chembakassery dynasty. Their ancestors had fought several wars together, probably had paddled the same boats of the maritime fleet. It reflects the great camaraderie, unity, and spirit of the VallamKali, as well as the cordial neighborly relationship between nearby villages. Their immediate relatives are mostly spread around these areas. A special invitation was accorded to the winning "Women's team' with the most number of foreign nationalities, as a gratitude for their interest, contribution, passion, and participation. In all, a huge crowd is expected. For the locals, no formal invitation is needed. Those who are interested can join in and enjoy the celebratory events. The sale of Toddy is banned for tonight; this is considered a family gathering! The regular patrons have to make some adjustments or sacrifices; abstain and fast for just one night! The authorities don't want hooligans to create any awkward scenes or incidents, spoiling the fun and tarnish the new image the village has just gloriously amassed!

Well, so far as the Toddy is concerned, it will ferment and lose its taste and flavor within 24 hours. The shop owner will have to discover ways to save the liquor in some way; offering some promotional gimmicks like "Buy one, Get one free", a second glass at half price in the early afternoon, or possibly making vinegar out of it to cut his losses. He is a small business owner and entrepreneur, has a big family to support. He may stealthily pour the unsold inventory out into the river, making the fish drowsy, and promptly catching a few. Remember! The fried green chromide fish is a popular delicacy at the Toddy shop. The '*Panchayath*' is also a loser, forfeiting part of the revenue by way of its "*excise*' duty on alcohol. All being said, the patrons should reasonably expect to pay a little extra for the same glass in the coming days, but most likely a permanent price hike. He has ample justifications for such a big move, "everything is going up!" is everyone's refrain! In the business world, the consumer price that goes up never comes down! It's for eternity! The patrons might comply after some heated arguments, bickering, shouts, and filthy curses. But the seller won't mind them. Sometimes, one should sacrifice a little pride to make the money he is after. Business people would rather do it; the addicts have no choice!

Meanwhile, a faint sound of music emanates from a distance; it seems approaching at a regular speed. The assembled villagers train their ears and cast a longing look into the beyond. A decorated motorboat, escorted by a flotilla of riverboats, flanked by two PWCs (Personal Watercraft / 'special boat') leading the way, come in to view, as it emerges from behind the thick bamboo trees at the *hair-pin curve* of the meandering river. On them, there are embroidered silk parasols of different colors, red, green, blue, purple, and yellow, as commonly seen in religious processions. These parasols are craftily woven out of Burmese silk, and have pearls from farmed oysters, or freshwater mussels dangling from the perimeter fringes. In Hindu mythology, gods were carried around in chariots during heavenly processions where the parasols were used to provide sunshades. "What are good for the gods in heaven are also good for the kings on the earth!". The royal sycophants sang in those days. Readily, the emperors and kings began to use them elevating their self-status. It slowly spread to the folks-festivals, and weddings. Among the flotilla, there was a "Chengadam" of two ' Kettu-Vallam' tied together as in the iconic *catamaran* sail boats. It has a *bower*-canopy made out of *nature-friendly* materials of palm, plantain leaves and an assortment of renowned Kerala spice products. Its pavilion showcases items from the local environment, and farm produce, intended to provide shade from the rising sun and soothing cool for the people resting inside. A pair of ensembles of brass/wood musical instruments and the *Chend*a/drum percussion groups play romantic melodies, and lovely wedding songs. The sound of music reflects on the gentle waters; sweetly propagates over the moist air, reaching far beyond.

Soon, it becomes clear that it is indeed a wedding party returning home after the ceremony at the village temple. *"The marriage is made in heaven!"* according to universal belief. It has to be respected as such; must be celebrated with all the fanfare, reverence, and traditions. The crowd is amused and excited; they wave and shout congratulatory greetings at them; gesture to the party to stop by, and pay homage to the victorious Chundan. The flotilla slows down. Soon, they moor at the sprawling pier that extends far into the river, and one behind the other yielding to the main 'Celebrity-boat' to allow facile docking. It is a wooden jetty with beautifully carved and decorated handrails with flower bouquets, and arches. The members of the wedding party jump out quickly and line up forming a wide lane to the specially built dock for the triumphant Chundan. People are enthused to find members of the rowing crew of the epic race among them, and immediately recognize each one of them.

"Here comes the bride and the groom!" The couple alight from the boat and makes their elegant entrance. The crowd starts cheering, clapping, and dancing. They are Ravi & Ramani.
That's quite a surprise!
The bystanders are dismayed; can't fathom what they are witnessing. Everything seems incredible these days; makes no sense, at all!

In their adolescence, Teenager's dream the couple had unfortunately earned some notoriety in the community by sneaking away together to watch a boat race. On their way, to avoid detection, they took a secret path that they had to cross a flimsy 'one-log' bamboo bridge clumsily raised by school children on two feeble 'V- slotted' posts. As they reached the middle, it swung widely, broke into two equal pieces. Both fell in to the canal and caught in the tide. They were almost drowned. Although people quickly assembled on both banks, they were dissuaded by the local 'vigilante' group (self-styled 'morality enforcers') from offering them any help. Their two families blamed each other for the infamy, and the life-threatening escapade. They stopped talking to one another; remained incommunicado since then. It stirred up an ugly family feud which quickly escalated in to a full-blown factional conflict involving even their respective extended circles of relatives. The two families account for half of the population in the village. People know every story and never forget any of them; they only embellish in own figments of reckless imagination helping to drive the division only deeper. The boat race is the sole event they could agree to cooperate!

Peacock Ramani & BSF aspirant cadet Ravi

Ravi is the son of the chief helmsman of the winning Chundan. Ramani is the daughter of the head of the rhythm-makers of the "VallaPatt", the Kerala oarsman's song. Ravi is adept in ancient Kerala martial arts, the "Kalaripayatt". His exceptional acrobatic moves and skills earned him the glorious nickname "His Airiness". He represented 'Kerala Martial arts Troupe' in the Republic Parade at the magnificent and sprawling Janapath (People's promenade) in New Delhi, known among the Francophone diplomatic circles as "Avenue des Connaught circle" a design in town planning consisting of 8 radial streets resembling the "Avenue des Champs-Élysées". Ravi is fascinated by the 'high-kicking' Wagha military ceremony and the great spectacle at the daily opening and closings of India- Pakistan border where many amused citizens from both sides, and world tourists gather to watch the spectacle. Eventually, he wants to join the BSF (Border Security Forces) and hopes to be able to perform the ceremony some day in the same fashion, and tradition.

Ramani is a college graduate and an accomplished singer-dancer. Her gait, posture, searching looks and demeanor resemble those of the Peacock, the "National bird" of India. Her high-pitched voice resonates with the Peacock's call; her dainty dance resembles the *walk* of the bird, earning her a sweet nickname "Peacock Ramani". On her part, she was chosen as a member of the dance group to represent Kerala in the Independence Day parade at the same venue in New Delhi. She earned fame; give Television and stage performances; became a celebrity; received great acclaim, and accolades ever since.

As is the custom throughout the Asian culture, the groom leads the bride! Ravi is a well-built, well groomed young man. He has the physique of a gymnast. He is dressed in the traditional dress of "Kasavu Juba, and Mundu" with golden lining, graded, and folded pleats, showing off all the golden lines in equally arranged stripes. His gold medals hang from the chest as ostentatiously displayed by the military generals' across their broad chests, on their uniforms. His thick dark hair neatly combed back with a straight lining just off the center to the left, yet look symmetrical, as if he wears a 'split' black top-hat. He has a self-confident and beaming demeanor.

Ramani, is dressed in brightly colored peacock blue *Kanchipuram* silk Sari; embroidered with vivid blue & green shades; patterns of the "coverts" (tail feathers) with color coordinated fabrics of laces, and glittering sparkles stitched on to its borders. The end of the Sari hanging behind is in a flavescent yellow mango color; pleated with many folds; spread and pinned like an accordion bellows fully extended. She wears a gold ornament on her head; falling on forehead; the end protrudes like the head crest of the Peacock. She is adorned in ornaments, literally, from head to toe with fashionable necklaces, studded with 'aquamarine' jewels; the bangles crowding the slender arms with speckles of equidistantly embedded 'turquoise' gems. She wears 'bell-anklets' on both feet which jingle at every step she takes. In addition, she wears traditional and natural makeups; forehead with a heart-shaped 'dot spot' of sandal wood paste, etched in the center of her broad forehead; the face with "Kumkumum" 'shy blushes'; her dark-pupiled eyes sleekly extend to the sides in the shape of 'Lotus' flower petal; the darker than dark eyelashes pulled out and spread, as the peacock does to attract 'peahen'; the thick long black hair bundled and propped up like an arching mount, in the shape of the bow of the "Vaippu" race boat. She looked stunningly beautiful challenging the Bollywood film stars.

The couple is ceremoniously lead towards the Chundan that is rocking and floating on the gentle waves looking majestic. It has been garnering undivided attention until this moment. Two boys, one carrying a shield and the other carrying a sword, one of each item from the Kalari-ppayatt ,walk in front of Ravi. Two pretty little girls dressed in skirt and blouse in peacock color scheme, one of which carries a bouquet; the other a flower garland- the two indispensable iconic symbols of Indian wedding ceremonies lead the bride, Ramani.

The people look in awe and admiration; offer the couple pleasantries and greetings. In turn, they bow to them in polite, congenial, affable gestures, and traditional expressions to thank, and acknowledge the well-wishers.

The pair strolls down a short distance with measured graceful long and short steps. She straddles a bit to the left and then to the right with a momentary pause in each step as the beautiful Peacock does. After all, she's a peacock! The anklets oblige with a resounding jingle, making a pleasant marching rhythm; 1,2-3,4. It is evocative of the ritualistic elephant procession in "Thrissur Pooram" (a temple festival at Thrissur Vadakkunathan Siva temple, staging 101 specially trained elephants) where the '*GajaVeeran'* tuskers are decorated with "Thalappavu", a massive head worn ornament, (The biggest in the world in any ornament category!) as its bombastic status symbol. The "elephant bells" are tied to both of their front feet. As they are lead around the processional route, the 'pitched' bells make a rhythmic 'bell symphony' of 202 blending sounds, bettering all the other 'glockenspiel' symphonies composed. They spare the "ChendaMelam" group a number of rows for its cymbal ensemble. Austrian tourists once commented that it reminded them of the *March Militaire* (Military March) by their Viennese composers *Franz Schubert & Hayden's Military symphony*. The elephants are intelligent, as well as musical; seems to have memorized the hymns and chants remarkably well. The sound blends seamlessly with the ChendaMelam & Nadaswaram *repertoire* by timing the tempo; plays crescendo by stamping the feet; tremolo by shaking the bells, and reverberations by dragging the feet in certain way. No wonder why Keralites love elephants, and processions! Seeing them together will instantly corroborate the fact. Elephants have almost three times more neurons than humans, [no wonder, they have the biggest head of all animals] which must have given them additional faculties, and information processing capabilities to learn skills quickly! In a fitting analogy, "A computer with a hi-speed microprocessor and more bits of RAM memory helps to do tasks done faster!"

The couple arrives at the Chundan. The attentive escorts hastily make a broad '*bridge-platform'* for their safe entry; a few others hold the boat steady from rocking; tell them they can tread boldly onto it. They enter the chundan vallam with great reverence and solemnity. Ravi offers an additional helping hand to Ramani. They turn to the stern; together, they bow their heads three times in deep respect, in perfectly timed sequence. They murmur a personal

thanksgiving prayer; cast a longing look up and towards it. At the very moment, the both become intensely emotional; followed by a few suppressed sobs, tear drops of joy and thankfulness trickle down the cheeks and fall on the boat. It is rather dramatic, telling the poignant story and blessings of their young lives! Ravi looked at Ramani and asked something. She nodded in consent with a pleasant smile. Promptly, Ravi summoned his dad; handed his wedding garland over to him asking him to put it up on the towering stern of the great chundan. As a helmsman, he is no stranger to the pinnacle platform on the chundan; promptly climbed over, and obliged to his son's wishes. The crowd cheered and praised them for their respectful gesture.
People are surprised; felt emotional. Some women shed tears with them to join in their happiness. Others ask loudly "why do they cry in the happiest moments of their conjugal life?". All are baffled, spiking the mystery to yet another level!

Most of the invited members of the wedding party were the crew of the winning Chundan. They could hold the secret no longer! Actually, it does no longer qualify as one! The *pledge* of '*confidentiality*' is no longer relevant! One by one, each begins to reveal to the crowd their own version of what had transpired in the previous night on their triumphant return from the race. They huddle together in several groups; listen with absorbing interest and rapt attention, as children listen to ancient bedtime stories narrated in mellow voice by their grandparents.

Chakkappan Chandanakkari, a local Television reporter began his explanation of the puzzling mystery: "Well! On that beautiful night, the sky was magnificent; just a few white drifting clouds; stars shining through them brightly; moon play hide 'n seek behind them casting shining halo around it. The scattered light bounced off the gentle waves and were shining on our faces. We were singing and chatting under the nightly sky, enjoying its glamor. Soon, the flotilla arrived at the exquisite Toddy shop in R-Block Resort, the one with the 'Finest Toddy on the Planet'. We were celebrating, making ourselves merry; enjoying few drinks with the meals; self-congratulating, enthusiastically complimenting each other for their grit and guts, sharing each other's drinks, and telling jokes. Conspicuously, the two most important men of the crew, did not respond to each other when their turns came up. They looked sullen, and indifferent to one another in strict silence. One stared at the floor, the other at the ceiling. It was quite awkward, ugly, incongruous, and embarrassing, given the exuberant mood prevailing around them. The teammates pleaded with them individually to relent; forget the pride and ego; make peace to make this moment truly special, and joyous for everyone in the community. Without exception, everyone uttered identical sentiment, and the familiar cliché: "Let the bygones be bygones! Whatever happened, had happened! It's all our human

frailties; that should not ruin happiness forever, and so on. It happened a very long time ago; nobody remembers it, unless reminded of it.
Our own behavior alone can do that!"

Suddenly, the signs of hostilities of the feuding men melted away like butter in a hot frying pan. The faces turned to grins of mutual respect and recognition. Without any further ado, or persuasion, they walked towards each other, smiled heartily, hugged, and embraced. Apparently, they wanted to be considered as noble men of the lot, and felt chosen to set good example for the entire community. It was remarkably a beautiful moment! The Toddy shop erupted in several rounds of loud applause, praising them in chants of "Arpo Hurray", "Ki Jai! Karuvatta Chundan", Brave men in the village of Karuvatta!, Glory to God, in heaven!
"Oh Manuja. Madhava...Mahadeva!...; ...Blessed, St. George of Edathua!... Ente, Patachone!.."; and so on. *Some wanted to "drink to the occasion" by ordering another round of half-glass as the snobbish wealthy do on TV shows*. A totally intoxicated crew member (name with held!) came staggering forward. Two others had to rush to help him from falling down. In a slurred, broken voice, and *undulating* pitch, he suggested pointing to the lead rhythm-maker: "Now, on this memorably happy occasion, and in the solemn witness of the braves, and winning teammates of our Chundan, why won't you propose your daughter in marriage to the helmsman's son?... yeah,... Mister?". The man lost his balance; instantly collapsed to the floor from severe intoxication. Everybody was frightened. Frantically, we helped him with first aid; fanned; sprinkled water / Toddy on his face trying to revive him; waited in great anxiety. Finally, when the fainted man opened his eyes and slouched up, the captain commented "Well, yours is a fantastically bright idea, pal!" The president of the boat club hastened to say "I will bear the cost of the wedding dress". Each pledged his own able contribution, along with some funny remarks going from one end to the other. Everyone committed their support, and cordial wishes. All of them shouted in the same voice "What do you say to it, your M*aestro*?" with inquiring looks. They continued with rounds of persuasive clapping like in the Television re-play of '*lbw*' call in cricket matches, and the line calls in Tennis tournaments.

They asked the two men to please come to the front, and face the audience. The two meekly obliged; looked and stared at each other, tilted heads probing; soon, sensed each other's readily agreeable expressions. They smiled at one another, broadly and cordially. "I'll say Yes, to it!". The other retorted "a fat shouting **Yes** from me!". "We want our children to be happy. They are the most precious souls in our lives. We want to reward them for their exemplary self-restraint, patience, and equally commendable talents. They will make a wonderful couple. We should be proud of them!" Everyone nodded, approvingly! "We want to perpetuate the memory of this day and victory. Vallam Kali has

been the greatest passion for both of us. Therefore, it has to be tomorrow itself! And has to be part of the whole celebration! How about that, mates?.
Oh, the most merciful Bhagavane! Ayyappa,.. Sa-ra-na-me!". "Listen guys, everyone must keep this secret ultra confidential overnight; don't even tell your wife; the greatest secret of all times. It should be our village-shaking surprise!" said with a sly wink, and crackling laughter.

Among the crew, was the highly reverenced *Jolsian* (the astrologer) who was the "poojari", and spiritual leader who conducted the rituals at the boat launching ceremony. He is otherwise, also the village "Kaniyan", dabbling in indigenous medicine. He is a 'Jack of all trades' and aptly nicknamed "Kaniyan Nostradamus". It's common knowledge that Nostradamus was a French apothecary who had the reputation for clairvoyance. A number of his predictions from his "'Quatrains' have indeed come true. It's a distinct style of all fortune-tellers that they use contemplative mysticism; cryptic, and ambivalent symbolism; use plausible conjectures. It's not much different here either. The Astrologer was asked to determine the 'Sakunum' for the hastily arranged wedding plan. He will not enter any 'place of sin' such as a Toddy shop. So, he located a dark spot on the fully vacated chundan; turned towards the 'bow'; took out his shoulder scarf, folded neatly; spread it on its shapely hull; positioned few markers here and there on it; stood still, continuously chanting scriptures. Sternly, he scolded the drunkards who interfered with his holy chants. No one wanted to antagonize the vengeful magician. He foretells the casualties before it befall, within a certain period of time, that no one likes to hear and brood about. People trust his words better than the weather forecast. On the other hand, the skeptics call him Kaniyan "Duritham" (Dr. doom!). He practices *black magic or witchcraft;* can cause people real harm. He ordered complete quietness. The Toddy shop came to a *pin drop silence* for the first time ever during the business hours. He then proceeded; chanted some more unintelligible verses that no one else understood; dragged and repositioned and shuffled the markers a few times; referenced his 'Book of Wisdom' which he always carries in his long pocket of his loose Juba. The Kaniyan does not like to miss an opportunity to earn extra cash 'any time such a moment should pop up'. He can determine the position of 28 Nakshatras (stars) at any given time; knows *Synchrony & Ascendance* of the 7 *classical* planets that are named after the 7 Greek gods for easier understanding, and their specific characteristics of vibrant *energy radiation, and influence*. In Indian mythology, gods are sometimes identified based on their influences. He scribbled a few things using some Babylonian number figures and Greek astrological symbols, presumably of Mars, Mercury, Saturn, and the Sun; closed his eyes in deep contemplation; made some calculations on the fingers of his wrinkled hand, almost snapping them off. As a common knowledge, Mars, Mercury & Saturn are mostly evil; Venus is the most auspicious; liked by all. He announced in solemn voice commanding respect and *adherence:* "Venus has entered into its ascendancy in full force, it bodes well; the proposed day is the most auspicious one in the whole calendar year, as evidenced by this historic win we achieved today. However,... the period of "Rahu" (the inauspicious period) enters precisely at Nazhika 13:06 & 33 Vinazhika (fraction of seconds)" cautioning with an ominous 'Mantra': "All the ceremonies must be completed; the wedded couple should reach home before then. Everything must be carried out with utmost precision and punctuality. This Rahu is wicked, vicious, and unforgiving; will not tolerate any breach, whatsoever! Understood? He paused for a moment; then said: "It's up to you to go forth with it, or not"!

The two men looked at each other; made cryptic gestures and then unanimously affirmed with resounding "**yes**". The two young people, nor their mothers know nothing about it! They may be peacefully asleep at this time of night. The assembled assured their full support; volunteered to work through the rest of the night, provided they are invited. All conditions were readily met. They went back to the boat and resumed the homeward journey!

It's a custom and tradition in the society that the parents prudently prepare very early for the wedding of their daughters, by buying ornaments and wedding dresses whenever they find extra income, monetary gifts on birthdays, and special occasions. No one really knows when the good luck and the blessed calling come knocking. "Time and tide waits for no one!" One has to be ready at the instant! They have heard this wisdom repeatedly told by their ancestors.

In the life of the village, in spite of the fiasco and the acrimony, the two young hearts have been growing in love and affection for one another. They showed great control and restraint in keeping it within their hearts. Consciously, they pretended that they are purposely avoiding each other, making the gullible villagers think that they have become rivals in seeking fame, name recognition, glory, and reward. But, intense feelings of love were churning and burning within, similar to what movie-goers used to see dramatically depicted on large screens in the perpetually romantic Indian movies.

Now, in real time, here!...

As the bride stepped out of the Chundan, the ladies and young women, who have been marveling at her beauty, dress, and collection of Jewelry surrounded her. They liked to touch and feel. They held each ornament in their palm one after the other, tossed and juggled a few times; estimated the weight and judged the *carat* value. One woman held her arm in hers and swiftly strung through the bangles as if playing an "arpeggio" on a harp. She said "23". Another lady who is a cashier at the town's jewelry shop did the same and confirmed the accuracy of the count. “She has 23 bangles on each of her arm” they shouted to the inquiries from other curious women looking on from a calling distance away.

It caught the attention of a fashionably dressed middle-aged lady, who was passing by, and hurrying towards the bus stop, showing no interest in the crowd, and the unusual excitement at the jetty. She is Mrs. Mariaamma Mathullah Mappila, a professor at the convent college; has two grown-up daughters at the same college. The professor is a 'liberated woman and a champion of women's causes in all its forms, and rights. She never likes to miss an opportunity to voice her opinion and altruistic principles, or build awareness on the themes she cherishes the most. Suddenly, she stopped; turned around and murmured to herself “did I hear 23?”. It’s very clever indeed! A very thoughtful piece of imagination! I'm impressed! I never expected anyone in this village would know the science of “Genetics” except my husband to whom I explained the fact repeatedly”. Furthermore, she was surprised to see many of her students huddled around the bride and excitedly chatting with her, forgetting the time. As if in a classroom setting, the Madam Professor asked “what is the significance of 23?; anyone?”. The students shouted back in unison “She is 23 years old; Ha ha ha!” showing off their smartness and quick wit. Politely, but abruptly, Ramani corrected in an innocent voice “I just turned 25 today! The Kaniyan had written in my “Jathakam” (a customized and elaborate horoscope written at the birth of a child) that I will be married at the age of 25. She continued in the same tone: “only yesterday, my oldest uncle came for a visit to inquire about me. (it is a local custom that uncles and aunts play a dutiful role in finding suitable matrimonial matches for their prospective nieces & nephews. A favorable selection reflects well upon them and boost their standing in the family!). He started blaming my mother for having selected this 'hopeless' village Kaniyan to write my horoscope”. Uncle also scolded and berated harshly, while Amma (mum) stood there helplessly, crying. “He knows nothing, nothing about stars, or astrology; I’m not sure if he can read or write; he’s simply a con artist, a swindler; a fake, a quack, a consummate scoundrel; his chants are not in Sanskrit to the best of my knowledge; what he carries in his bulging pocket is a *book of lies*; ...phooh...he should be thrown out of this place; then it will prosper; just waisting people's time, and money. I can't stand silly and gullible people; Mind you, things won't fall from the sky; You have to do the right things... always. I’m leaving right away!” Obviously, he is an easily irritable, 'hot, ill, and short-tempered' individual; so are also most of his peers, to be fair. He rushed out of the home; got into the waiting AutoRiksha and left, discharging thick trailing puff of diesel smoke towards the beautifully planted flower garden in the front yard.

At the moment, understandably, Ramani is relishing the adoration of the villagers; the new-found recognition, immediate, and warm acceptance in the community. For a long time, she has been missing these wonderful experiences of inclusiveness, and respect. It helps her to overcome self-doubt and mental reservations. She began to chatter and babble freely. “Shortly after my uncle left, this good word came!..and...” she said in a flurry of bashful blushes!

The madame Professor has been waiting for Ramani to finish her story, but grew increasingly impatient. She was visibly disappointed in the answer from her students and concerned that they are not applying heuristic, or *creative thinking*. Only in the very last session, she talked about this specific subject in the Science class. She interjects picking up on the same question posed earlier. Mrs. Mathulla said “No my dears, it's the count of the chromosomes one inherits from each of the parent. “Read the chapter a few more times!” she said in plain exasperation, and continued: “Humans have 23 pairs of chromosomes. Don’t you remember the two interwoven ‘double helix’ I showed in my slide? (cats have 19, dogs 29, horses 32, elephants 28 pairs etc.. Etc). One set of chromosomes of each pair comes from a person’s mother, and the other set comes from the father. The number of bangles on one arm signifies the set of 23 chromosomes from her mother, and the ones on the other indicate the other complementary set of 23 chromosomes from her father. She is the daughter of both of them in equal measure! Ramani wants to show that her parents love her equally. All of you, like Ramani have inherited chromosomes in the same way and in the same count. Sometimes, a father may blame mother that the children's faults, and shortcomings are all because of 'her'; all the admired qualities are from the ‘proud’ father himself!” The professor seem to be holding a ‘field class’ on her way. “You understand the logic of the world? It is the rule of men, and tyranny of their thoughts that we are regularly subjected to. They make rules, traditions, and customs to keep their authority, and superiority. Women have to always follow! Indeed, we do for the sake of peace! Well, that's beside the point, although I just expressed it so plainly. This beautiful bride teaches you a lesson you shall never forget in life and I hope you will take it to your heart. You will treat all people accordingly, most importantly your own children, irrespective of gender, whenever that time comes around for you!”

In fact, most of the newly discovered facts are well-known among the people in ancient cultures before it was corroborated in the laboratories, and under the electron microscope. The massive mystery remains as to exactly how our ancestors amassed the knowledge, and how they built lasting social institutions to preserve their legacies. They were actually far advanced than we give them credit for. In those times, since fellow citizens were mostly illiterates, they made up stories to develop fascination and imagination which later evolved as Mythology- not as Science. Over centuries passed, these pieces of knowledge were accepted as ‘simply stated by someone’. Gradually, it descended into mere cliché. The scholars, inventors, and discoverers of ancient times had never claimed credit or patents for them, then and there. All were considered as their noble and meaningful contributions for the betterment of humanity of generations.

Mrs. Mathulla is now a 'professor emeritus' of Genetics at the convent college. She has actually done lab-research in “Human genome Sequencing & Mapping” at the “Human Genome” facility in Rockville, Maryland, USA. She returned to her native place to get married; later, decided to stay back, rediscover her native culture, and contribute to the society dear to her. Now, she makes presentations and offer consultation services for a fee to various medical, research & pharmaceutical companies.

Punctually, the “Ladies Only” college bus has come on time as usual. It has been waiting at the bus stop for some time now. The *college Kumaris* (the young college girls) are lost in time, and still animatedly talking to the bride. The irate bus driver honks the high-pitched horn repeatedly in indignation. It's annoyingly loud and shrill! He loses his patience, and begin to drive off. The people at the stop shout at him and berate him for his inconsiderate behavior. He stops at some distance away from the stop. He is not expected to drive an empty bus to college! There is still at least one more thing the girls wanted to do before leaving the scene. One of Ramani's statement has been resonating in their minds. It is something vitally important. They took out their cell phones; entered password, accessed address book; sweetly asked: “by the way Ramani, you're a lucky girl; what's this Kaniyan's name?”. Ms. Geethakumari asked in haste; promptly entered it carefully spelling the long name. “Come on, girls let's go, the bus...”. “Got his address?” Ms. Rejini asked. Geetha answered in a ‘thumbs up’!.

Being rest assured that Geetha will generously share this valuable info, the girls scrambled to catch the bus before it leaves without them. In the melee, some stationary items got dropped, and spilled on the road. Some decided it's better to leave them there, hoping a good Samaritan will pick them up and entrust with the shopkeeper at the bus stop for safe keeping until they return wiser in the evening. They know very well that the Rev. Principal won't stand for any tardiness, nor lame excuses whatsoever. She is very earnest; enforces discipline, diligence, and punctuality with *brutal efficiency,* like the unforgiving Amazon warehouse supervisors. Well, it clearly shows up glamorously in the results at the college. It boasts 100% graduation rate in each of the past years since the Rev. Sr. Scholastica has assumed the Principal's role. The driver, on his part, does not want to get any derogatory reprimand or stern *citation.*

Her peers have left, now Ramani is in the company of mostly home-makers and mothers. Quietly, they begin to approach her. “I have three daughters, marriageable ages; would like to know this Kaniyan who wrote your Jathakam; what' his name; where does he live; how far?". Quickly, a surge of mostly women and mothers came forward and closed ranks around Ramani; some fathers showed interest too. They listened to her story with rapt attention, frequently interrupting her, asking details, including the fees he normally charges. They are quite impressed and wanted to meet with the Kaniyan ASAP. Instantly, the Kaniyan gets a celebrity status, though, he is totally oblivious of what is currently taking place in the cyberspace on his name and behalf!

"In the age of social media, the news will spread like wildfire. People will search Kaniyan on the Internet on the WWW (Wide World Web) from far and beyond. Soon, he will be the richest man in the village in the shortest time, faster than the founders of tech companies". A '*computer geek*' made his qualified assessment. An elderly lady, KalyaniKuttiamma, well known in the village for her outspoken wit and wisdom said "Kaniyan seems to know many things; does people good deeds; he should be elected as the chief minister. At this moment everyone will vote for him. The current one is useless; the entire band of politicians is ignorant- totally; they only talk and insult each other incessantly the whole day while people like us work hard for a living. Citizens live in constant anxiety as to how to get through a day!" It encouraged other people to speak up their minds, not fearing any retaliation from political goondas (thugs). "Good people bring good luck to everyone; we need to be surrounded by only good people, you agree?".

"Patachone! (yeh, Allah!) what is happening here?; everything changed, in just one day. My brother KoyaKutty was on this Vallam: he was in the front; Masha Allah!". Fathima Beevi from the neighboring *Vazhiambalam* junction on NH 47 exclaimed. "My only grandson Jose-Kuttan also rowed this boat. Did anyone see him? We own 2 boats, like everybody else in our neighborhood. He has been a VallamKali fan from childhood. I don't know where he sat, but I spotted him on TV, paddling hard. I got goosebumps watching him. I prayed throughout the race and pledged a *Novena* to '*Mother of perpetual help*' of Karuvatta church to reward my little darling boy. I'll be busy for the next 9 consecutive Saturdays" said Achaamma, Aattina-kkara from the village of Kara-Muttu.
"My nephew too!" another lady claimed in some pride.

"This Chundan (a boat) is taking over our lives. Are we ready for this? Asked Vaidhyan Vattaparambil. I pledged a *Vazhipadu* at Haripad Murugan temple! It's strange that the attention, has shifted from the *Chundan* to the B*ride & groom*, then to the godforsaken *Kaniyan,* and abruptly back on to the same Chundan that is the central figure of all the excitements felt around here today!"

Oh, my Lord! Karutha, Ponne! Karumadi Kutta!" exclaimed Gaythri Devi, a businesswoman from Purakkad. Karumadi Kuttan is a dark granite, sculptured statue of the idol of the young meditating Buddha. It is consecrated at the bank of the waterway at Karumadi, near Ambalapuzha, from where it was found after having been missing for several centuries. It's a universal fact that the colonialists covet ancient idols and artifacts for their intrinsic, and historic value. They possess facilities and expertise to conduct 'carbon dating' that can accurately determine the age of such objects. The older, the better in its value! People assumed that they must have smuggled it into their country avoiding detection. The religious icon of Budha is revered as the patron and protector of all boats & boatmen passing through. He may have something to do with this victory too. Taking advantage of all the history and facts, Srimathi Devi is today offering a special discount for her Karumady Kuttan brand of Banana products for the upcoming Onam festival season. Promptly, she dishes out discount vouchers from 'Coupon book' she carries in a basket along with sample packets. The signs of commercialism and its gimmicks have slowly creeped into remote villages.

People have become lively, buoyant, and talkative unlike any other time before. They enjoy a rare feeling of communal harmony, company, and unity among themselves as one people. All feel a high sense of belonging to the same native village with mutual respect, and common cause. They relish their new status and fame together; feel proud of the legacy that is passed on.

At this moment, the lovely couple is escorted back to their boat to arrive home before the 'Rahu'. As soon, it departs. The crowd shout-outs "kurava," a yodeling- *voice cheer* reserved for celebration and exultations of joy. Cordially, they keep on waving until the boat turns to a narrow canal and disappears behind the thick coconut groves and the rows of beautiful river-side bungalows.

Gradually, the crowd comprehend the full meaning and context of the mystery. The winning of the race by the Chundan was equally a serendipitous event in the personal lives of Ravi & Ramani, leading up to their blessed marriage. It is hard to imagine that it could have happened without it. The tears the couples shed were actually their spontaneous and genuine reaction to the gratefulness towards the Chundan. The ensuing peace and harmony brought to the community is solely due to its victory! People begin to wonder what individual good tiding is there in store for each of them!

"The intricate link and connection among humans, the creatures, plants, environment, the machines, tools, and objects we use are apparent; often not fully understood, as it is amazingly mysterious. The thorny, nagging issue our bright people could not have solved, was 'automagically' and amicably settled by an inanimate, mute object, our Chundan. It strived hard during the races, responded well to the rowing, rhythm, and prompts; endured many taunts. Eventually, made the crew and our village happy, and proud. It played a significant role as a 'Peacemaker'; converted foes into friends; healed deep divisions; ended bitter rivalries for good. There are certainly many other handsome contributions made by this Chundan, that has become a darling, and the 'apple of our eyes'! Indeed, it has!

Our own magnificent Chundan, the Karuvatta Chundan!"
With this reverberating message, the Panchayath President ended his speech. In the background of joyous applause and cheers, he announced:

"Now, I declare the celebrations of the evening may begin!

My fellow citizens of the village, let's celebrate, and be happy from this time forward in great humility and goodwill to each other!"

The evening of celebrations started with the introduction of the entire crew members on the stage. They brought the glittering trophy lifted high above the shoulders of the tallest man with euphoric jumping chants of "Arpo hurray!" from the villagers crowding the stage. The ecstatic scene was accentuated with rousing fanfare; the drummers embellished their drumbeats with a 'flourish of fast drumsticks' followed by the wind instrumentalists blowing their high-pitched horns in a lively celebratory repertoire in exultant flourish; the bell metal cymbal ending the musical serenade with a stylish, loud, reverberating clang of the two golden, round & large metal plates colliding with each other fast and furious in a boisterous bang! In the nightly calm, its tonal rumbles resonate longer than its accustomed persistence; seems trapped in the moist air; the echoes become slower to disappear; reverberating much longer than usual, defying physics and acoustics.

The program went well without a hitch. There was a lot of musical interludes periodically waking up infants in mothers' lap, and lying on fathers' shoulders. The old and tired who were past their usual bedtimes wonder where they are after all, after a momentary spell of nodding off. There were fanciful performances of group dances by school children; the screening of the taped race finals in its entirety; rich variety of food, spectacular fireworks, and a lot of cheer all around. The villagers and the guests thoroughly enjoyed the occasion.

Senior citizens discussed their illness; boasted how well their children are standing on the corporate ladder; sparing them the toil in the muddy fields, as they had to in their whole lives; the care they are getting from the adorable new daughter-in-law. The young people rekindled their acquaintances, flame, and friendship. Some fresh faces, mainly of in-laws and newborns were introduced. Elders gave expert advices and useful tips as to how to quickly integrate themselves into the unique village life; a couple of marriage proposals came forth; a few, sound promising; ornaments exchanged; so also good deeds, word, and wishes; sorrowful were consoled; some disputes were mediated; promises were made for the repayment of long over-due debt; the long borrowed items shall be promptly returned; return invitations accepted; job opportunities discussed. The bankers solicited generous deposits. Local business offered discounts. All went home in good spirits and a fine state of mind!

The Chundan strikes a pose of regal elegance and grace in its triumph, and veneration, without pretending any air of the adventure or conquest displayed only hours earlier. It is quite humbling to see the boat resting there serenely on the gentle waters; the string of bells around its stern jingle with the waves in the familiar refrain of the Vanchippatt: '*thi thai tho, thi* ***thai*** *thom; thi* ***thi*** *thara thi* ***thai*** *thom*'; the long dangling garlands at the bow reach down to the water level. They still look fresh; appears to be playing '*hide-and-seek*' by diving in and out of the water among the waves, renewing its freshness; the Chundan-head nodding at times with the bigger waves, as if acknowledging the praise and attention it is currently drawing from the crowd assembled at the bank. A team of "Animation & Computer Graphics students from St. Joseph College of Communication (SJCC), Chy made and fitted an embossed 3D 'hologram' (A three-dimensional image of an object made by holography technic) on the totempole figure. It shows various visual illusions of mischievous grins (like the 'winking woman' on the back of the playing card), when looked at different angles. The stern of the chundan also got a new adornment of a crown that scatters light when the waves rock the boat, and as stars blink in the nightly skies.

Many come to see the boat, to thank, to talk to, take '*selfies*' and group photos in its background. They know they are not allowed to climb on it. Suddenly, it has attained a sacred status! People offer their comments freely, narrate as to how they saw the regatta live, exactly what has transpired during the race, for the benefit of those who were unable to see it 'live'; the distinguished personality and dominance of this Chundan during the race compared to its peers, the valor of the paddlers; the nerve-racking suspense they endured in waiting for the results, all with characteristic embellishments and anecdotes in such great eloquence, and ebullience. The school children come early in the morning, needing none of the usual persuasions, or repeated wake-up calls; assemble at the decorated jetty. They experience goosebumps; silently recall actions and images in their minds exactly as how they saw them in the extended TV coverage. "You must strive to win anything!" was the refrain they heard from everyone. That axiom is still reverberating in their minds.

They know they have now a relic to brag about, cherish, worship, and pray for.

They wonder if they could ever emulate and accomplish something similar in their lifetime!

Meanwhile. The Chundan. the pride & joy of the villagers

earns a well-deserved rest. until the time for the race next year!

At the end of the ceremonious and stately-staging.

The Chundan shall be moved back to its modest hanger.

It will remain there out of view. and unattended in fast forgotten glory!

Glossary (contextual meaning)

A

adjunct	something added/jointed to another, but not a main or essential part
alacrity	willingly ready
alms giving	making voluntary contributions, charitable donations, giving alms
alter ego	another aspect of the personality of oneself
ambrosial	exceptionally delicious or fragrant
anathema	a thing detested, curse, a damned thing
arable	that can be cultivated
artifacts	errors that occur during signal transmission or compression of a signal
astringent	tending to tighten
Aurora Borealis	Sunset reflections and light waves seen at the North Pole regions
Aurora Australis	Sunset reflections and light waves seen at the South Pole regions
avaricious	insatiably greedy

B

Beam	usually the widest point of the boat ; here it means cross platforms, Chundan has many cross-beams for stability and seating the paddlers
beholden	owing gratitude, favor ,etc. to another; indebted.
biosphere	entire ecosystem comprising earth's crust, waters & the atmosphere that supports life
biota	ecological collection of plants & species
bow	the fore-end of a vessel
bract	the brownish covering fold of the banana blossom protecting the buds that carry honey
buses	a bank of switches in rows, on a video production switching console

C

cadence	rhythmic style of speaking with a long and short or varying accent
cackleberry	(slang) egg
chronicle	a narrative detailed record
contextual	depending on the context, not necessarily literary
countenance	facial expression

D

dissemination	spreading widely
dulcinea	a ladylove; sweetheart

draft — the distance between the waterline and the deepest point of the hull (keel);the hull (the body) of the boat that is immersed in the water

E

ecstasy — exalted delight or joy

ectotherms — cold-blooded animals, whose body temperature adjusts to the environmental temp.
endotherms — hot-blooded animals, who must keep a constant body temperature

elucidate — explain clearly

ersatz — (Ger.) substitute, temporary replacement

exo/meso/ endo/pericarp — the various layers of fruit or seed

expatriate — temporarily leaving outside native land

euphemism — a mild and agreeable expression for an otherwise offensive condition

F

fag-end (Br. Inf.) the last and usually least appealing or interesting part of something

fealty — faithfulness to the landlord in a feudal system

fuzzy logic — an instinctive logic, using truth and probabilities

Freeboard — the area of the hull that is above water

G

Gunnel / gunwale — the rim, or 'the ridge edge the upper edge of a boat's sides that runs all along the top edge of the boat

H

Hangar — a shed to do maintenance work

heteronym — words spelled the same but have different meanings & pronunciations

heuristic — helping to learn; guiding in discovery or investigation

holt — a hidden hole carved out by rodent or pest

homographs — words spelled the same, but have different meanings

homage — respect & obedience of tenant to landlord

hull — main body of a boat

I

incessant — without interruption

ingenious — cleverly inventive or resourceful

irrepressible — difficult to be contained, restrained, uncontrollable

intuitively — knowing by intuition or natural instinct

immersive — a virtual feeling of being there, the ambiance created by the application of multimedia images, narration, etc

intonation — Accent, varying of pitch or style in musical or speech presentations

isthmus — a narrow piece of land that joins two larger areas and has water on both sides

J

Jumbotron	A large display screen consisting of alternate vertical arrays of RGB (Red,Green,Blue) LED bulbs

K

keel	the structural 'backbone' of a boat, extending from the bow to the stern, giving the boat greater stability and control when moving forward;
Kinematics	a branch of Mechanical engineering, dealing with motion, joints, fulcrum, levers, linear, rotary and any other kind of motion
Kinetics	a branch of Mechanical engineering, dealing with motion, forces, torques, effects, etc.

L

Laser	Light Amplification by Stimulated Emission of Radiation, extremely narrow beam

M

meta-	about itself
morsel	a mouthful bite of food

N

nectar	'honey' from the flowers, trees

O

omentum	(anatomy), used here as "pot-belly'
ornithology	study about birds

P

pedagogue	a school teacher
pedantic	ostentatious and meticulous in words, knowledge and its details
Peduncle	stalk of plant bearing fruit, as on a banana tree
Phyto-genesis	the origin and development of plants (phyto means plants)
pileated	(ornithology) feathery head crest of a bird, typical of the rooster
pitch	(marine) to dip and raise its bow and stern alternately
port side	(marine) Left side of the boat
postulate	put forth as 'possibly true'
presenter	one who presents
precentor	lead singer, chanter
prevail	win out, succeed
primal	first & original
proselytize	preach and convert to another religion
punting	also known as 'poling', to propel a small boat or barge with a light bamboo pole

Q, R

Race progress data	Data of current location, speed, leading positions of the boats etc.
Rachis	the lower serrated banana bunch bearing stalk, extending to the bud
regatta	Boat race
renaissance	a historical period for such revival of interest & activity in music, Literature & art forms
repertoire	list of performance of singer, actor or any skilled performer
respite	relief
root	cheer on and supporters
rotator cuff	anatomy: shoulder joint & associated muscles, tendon, and ligaments
S	
serendipitous	unexpected but favorable result; fortunate and timely discovery
starboard	(marine) Right- side of the boat
stem	the forefront, bow of vessel
stern	the aft end or the rear end of the vessel, opposite end to bow
T	
TelePrompter	Television monitor/reading system, attached to the camera meant for reporters to read from, while looking into the lens
tweedle	Modulated high-pitched sound bytes of birds or musical instruments
U, V, X, Y	
Vassal	tenant in the feudal system
wanderlust (Ger.)	natural instinct to rove around, also domiciling in other places and countries
yaw	To turn about the vertical axis (aircraft, spacecraft, or projectile)
yoke	the cross-beam in the center of the that connects the two sides;

Lexicon of Vernacular words

Aanjili	a hard wood tree, preferred for making race boats in Kerala
Annum	cooked rice
Ayurveda	One of the 7 Vedas, ancient book on Health sciences based on natural, herbal medicine
bund	a raised stretch of land around a rice field built to prevent water from entering
Devi & Devika	female deities
Choru	rice
houseboat	refurbished cargo boat for sightseeing, casual, excursion, stay, offering luxury, comfort & relaxation
jibe	triangular sail of a yacht
Juba	a collar-less shirt made from fine, airy, transparent fabric

Jaathakam Custom horoscope made in the accuracy of time of birth to the second

Kaitha Water cactus

KalariPayatt an ancient, acrobatic, aerial & martial art- form using sword & shield

Kavani a long shoulder scarf made of cotton fabric worn as a formal dress

Kara hamlet, reclaimed from a water body

Kayal Lake

Katta Mud cake 'cut-out' from the floor of a water body

KettuVallam Huge wooden 'joined and stitched together' cargo boat, custom-built

Kochu-Vallam a single seat paddle boat; small wooden canoe, sleeker and narrower

koomb apical buds of a plant-like (palm Cabbage) of coconut flowers extracted
from the tip of young palm trees or banana blossoms with honey petals

Kuttanaad Low lying area in the central coast of Kerala, lying at or below MSL

Kuttanadiens people living in the region of Kuttanad lake basin

Maharshi A venerable sage, who is bestowed with mystic wisdom & knowledge;

Malar flower

Mela festival, jamboree (boisterous & ostentatious)

Mundu a tropical dress, loin cloth worn around waist, common to Kerala,
Sri Lanka, Myanmar, Indonesia, Malaysia, Yemen

Munshi a laureate in Malayalam literature

Neram point in time, usually related to telling an auspicious moment

Oduva Vallam A KettuVallam with a single rectangular shaped sail with a diagonal
bamboo cross-beam made of natural fiber mat, used for cargo transport

Pallathi small fish (green /yellow chromide)

Panchayath village Governing body

PataSekharam agricultural Landmass, reclaimed from the Vembanad lake

Payal Water Hyacinth, floating moss

Pooja Hindu religious rituals, equivalent to Christian 'Mass'

Puncha fields reclaimed fields from the lake

R-Block A 1400 acre raised block in the middle of the lage, a tourist attraction

Sakunam a devotional ritual in determining the auspicious moment based on
Astrological positions, calculations and religious considerations

Sadhya Banquet, a lavish multi-course 'squat down' dinner served on plantain leaves,
eaten with bare hands, and sitting on handwoven floor mats

Sloakam Art song. poetic prose or poem in the style of recitals of Sanskrit & Veda

Thalappav a massive head adornment pu ton the processional Elephant

Thoni punt, an open flat-bottom boat with also flat ends, used in crossing a canal by pulling on a cord tied between the Thoni and the two banks across

Vedic period (c. 1500-500 Before Common Era)when Vedic/Sanskrit Vedas were compiled

Vanchipatt a genre of boatman's song composed for and sung in Kerala VallamKali based on simple themes on rustic scenes & life

Vembanad name of the largest Lake in Keralam, also the longest in India

Ulbhavam Genesis, creation, advent, the origin of the world

Ulsavam festival, generally temple procession

Urula a handful morsel

VallamBoat boat propelled by a straight, long bamboo pole, punting boat

Vidwan a laureate, an accomplished scholar

Viruppu fields flat land fields

The Pallithanam Heritage Bungalow, Homestay

An impressive and venerable floating relic

On the shimmering waves of the

Greater Kuttanadan waters

Book information

Published!	**Available** globally for sale on IngramSpark Content Group's 40,000 retail and library partners, both in person and online!
Book	Text-Pages 400 Picture collages (86)
File size	155 MB; 420 MB 600dpi PDF; Word count: 133,525
Publisher info	Book: Hard copy, E-book; global distribution Kindle, Nook, IPAD, & Digital Publishing & "*Download from website*"
Author	Aurora Bobbyalis
Order Info	**ISBN: 978-0-578-25679-5** [Book Trim size: **8.5" *11" Print**] **ISBN: 978-108-799-0613** Format:-**EPUB**

Distribution & Availability: **Global** distribution by IngramSpark retail network

Digital download in compliance with "DMCA" guidelines

Copyright information: protected under the copyright laws

Preface/Synopsis	online, free reading
Chapters	described under "Synopsis"
Glossary	English Native words
Index	provided
ISBN	**978-0-578-25679-5**
Barcode	

Disclaimer	listed on website
Terms & conditions	Listed on website
Privacy Policy	listed on website
Web site	https://www.desi-paradesi-diaspora.net/

Notes :

1. *The author reserve the rights to edit, add, alter, modify content as appropriate without notice!*
2. *The online readers are advised to ensure that their devices can access and download the specific file formats!*

Other Books listed on: https://www.desi-paradesi-diaspora.net/

- ***The first Concert***

About a magnificent concert taking place in an acoustically, aesthetically

Appointed and splendidly equipped grand amphitheater

Culminating in dramatic and mystical happenings

- ***Singing simplified!***

Tips to improve singing abilities with the actual resources each of us possess

"Each sound is unique and is music to one's own ears!"

- ***Namasthe, India! Let's play the Amazing Accordion!***

Book offering text and video instructions for playing the instrument

Coming soon!

Vincent, the magnificent

A story of a boy from the fishermen's community on the Arabian Sea cost, who has grown more practical and accomplished than his peers, but became largely isolated by lack of recognition. *Please check back for new additions!*

The multimedia website include literary works of:
books, anthologies, essays, articles, stories, poems
music works of lyrics, songs, parody music of internationally well-known songs, music score sheets, music videos, music translations,
***Video tutorials,**
presentations, mp3 audio files, and graphics

Contributing YouTube channels:

https://www.youtube.com/results?search_query=bobbyalis

https://www.youtube.com/results?search_query=bodoll+mm+llc

https://www.youtube.com/results?search_query=bobbyalis

https://www.youtube.com/channel/UCHaTFGLY184cLN9y4U2d57w

Acknowledgements, Attributions & citations

Photos

https://www.nasa.gov/

https://www.thefreedictionary.com/

https://commons.wikimedia.org/wiki/Category:

https://images.search.yahoo.com/

https://www.google.com

https://www.pinterest.com/pinterest/

documentaries

http://www.pbs.org/wgbh/nova/

NOAA

National Geographic

Free documentaries

BBC documentaries

wildlife documentaries

Various other free Internet Resources

https://www.britannica.com/

The golden coconuts, as good as Gold!

AUTHOR BIO

I'm now a retiree; feel continuously blessed to be able to live active. Prior to that, I was an engineer in Electronics & Telecommunications, and had worked in Television, Medical Cable, ITV, Community TV, Satellite TV, Audiovisual & Multimedia, "A jack of all trades, master of none". I worked in 4 countries in 3 continents; had to learn to speak 4 languages. The highlight of my career and life was when I worked in the ME for a stretch of 17 consecutive years among an International, close-knit residential community of scientific, medical, research professionals and their families. I had the coveted opportunity to work in advanced state-of-the art technologies in professional work environments receiving recognition. My work & study in Germany among great many numbers of multinational 'Gast Arbiter' workforce and Germans perhaps paved the way for my selection. Both have been stimulating, rewarding and beneficial experiences. I decided to share the wealth of knowledge amassed, and the experience gained during these years, before the memories fade away and the time runs out. Some may benefit from them!

There is no better way than telling them in stories written in eyewitness authenticity, feeling global vibes, pulse, perspective, sensitivity, and sensibility. An ancient oriental saying is: "Sharing knowledge and experiences, are aggrandizing and enriching; benefits others. It's a convenient path to attain immortality!". Currently, I'm preoccupied with writing books; altogether, wrote 4. Sensing some dormant literary talents, I extended my foray into essays, short stories, poems, parody lyrics for internationally well recognized songs; scoring & recording music videos of songs and genre of different countries for the enjoyment of the global community. They are posted on my website. For these activities, I chose "Aurora Bobbyalis" as my pseudonym, while my immediate relatives and dear former friends can still recognize me by the original, uncorrupted, baptismal name!

Tennis, and track running are my favorite hobbies. Sometimes, I get a chance to accompany my daughter on accordion while she plays violin, until it becomes too annoying to her; also join my son for piano and Tennis, if he ever finishes his work. Regularly, I perform chores &work at home; toiling in the garden; wife doing the planting and weeding alongside, chattering profusely, but falling on deaf ears. I try to help her in the kitchen claimed to be her sovereign domain, often causing unwanted troubles, once again proving my penchant for creating problems and then struggling to find solutions; that life is a lifelong experiment of trial & error; hit, or miss with unpredictable outcomes!

www.ingramcontent.com/pod-product-compliance
Lightning Source LLC
LaVergne TN
LVHW082111200926
848484LV00008B/1940